Studies in Computational Intelligence

Volume 1233

The series "Studies in Computational Intelligence" (SCI) publishes new developments and advances in the various areas of computational intelligence—quickly and with a high quality. The intent is to cover the theory, applications, and design methods of computational intelligence, as embedded in the fields of engineering, computer science, physics and life sciences, as well as the methodologies behind them. The series contains monographs, lecture notes and edited volumes in computational intelligence spanning the areas of neural networks, connectionist systems, genetic algorithms, evolutionary computation, artificial intelligence, cellular automata, self-organizing systems, soft computing, fuzzy systems, and hybrid intelligent systems. Of particular value to both the contributors and the readership are the short publication timeframe and the world-wide distribution, which enable both wide and rapid dissemination of research output.

Indexed by SCOPUS, DBLP, WTI AG (Switzerland), zbMATH, SCImago.

All books published in the series are submitted for consideration in Web of Science.

Alma Yunuen Raya-Tapia ·
Francisco Javier López-Flores ·
César Ramírez-Márquez · José María Ponce-Ortega

Machine Learning and Clustering for a Sustainable Future

Applications in Engineering and Environmental Science

Springer

Alma Yunuen Raya-Tapia
Universidad Michoacana de San Nicolás de Hidalgo
Morelia, Mexico

Francisco Javier López-Flores
Universidad Michoacana de San Nicolás de Hidalgo
Morelia, Mexico

César Ramírez-Márquez
Universidad Michoacana de San Nicolás de Hidalgo
Morelia, Mexico

José María Ponce-Ortega
Universidad Michoacana de San Nicolás de Hidalgo
Morelia, Mexico

ISSN 1860-949X ISSN 1860-9503 (electronic)
Studies in Computational Intelligence
ISBN 978-3-032-03875-3 ISBN 978-3-032-03876-0 (eBook)
https://doi.org/10.1007/978-3-032-03876-0

This Springer imprint is published by the registered company Springer Nature Switzerland AG
The registered company address is: Gewerbestrasse 11, 6330 Cham, Switzerland

Preface

The urgent need to address global sustainability challenges such as climate change, resource scarcity, and urban transformation has brought new relevance to data-driven approaches in engineering and environmental sciences. In recent years, machine learning has emerged as a powerful tool to manage complexity, optimize resource use, and inform decision-making. Within this field, clustering methods have proven particularly useful for identifying patterns, revealing system behavior, and supporting strategic planning across multiple domains.

This book was conceived as a response to the growing need for applied, interdisciplinary, and accessible resources that connect unsupervised learning techniques with sustainability-driven research and practice. It reflects several years of collaborative work at the Universidad Michoacana de San Nicolás de Hidalgo, and is grounded in both academic inquiry and practical problem-solving. The result is a comprehensive volume that combines theoretical foundations with hands-on tools, enabling readers to apply clustering in real-world scenarios across energy systems, environmental science, and urban analysis.

Organized to support both learning and application, the book begins with a thorough introduction to clustering principles, including distance metrics, evaluation criteria, and key algorithmic approaches such as k-means, hierarchical, and density-based clustering. It then moves into a series of applied chapters that demonstrate how these methods can be used to address concrete sustainability challenges. These include the clustering of greenhouse gas emission profiles, the optimization of energy systems through representative period selection, the spatial analysis of gentrification in urban zones, and the use of clustering to explore interdependencies in the energy—food nexus during geopolitical and health crises.

To support practical engagement, the book integrates programming exercises in Python, R, and MATLAB. These examples are designed to enhance the reader's ability to implement and adapt clustering techniques using widely available tools and libraries. Whether for academic research, industrial application, or policy support, the material provides both depth and adaptability.

What distinguishes this work is its commitment to bridging the gap between data science and sustainability science. While much of the technical content is rooted in chemical and process systems engineering, the broader framing makes it relevant to professionals and scholars working in environmental management, urban planning, data analytics, and public policy. Our goal has been to create a resource that is technically sound, pedagogically useful, and globally relevant.

The development of this book was possible thanks to the academic environment and institutional support of the Universidad Michoacana de San Nicolás de Hidalgo. We are especially grateful to Prof. Fabricio Nápoles-Rivera and Prof. Luis Fernando Lira Barragán for their leadership, encouragement, and critical input throughout this process. Their guidance has helped us shape a project that aspires not only to advance academic knowledge but also to generate meaningful solutions in the context of sustainability.

We also extend our deepest thanks to Prof. Mahmoud M. El-Halwagi of Texas A&M University, whose mentorship and scholarly influence have had a significant impact on our work. His commitment to interdisciplinary research and sustainable systems design has inspired several of the frameworks and ideas presented in this volume.

We wish to acknowledge the many colleagues, students, and research collaborators who contributed directly and indirectly to this work. Their insights, questions, and participation in workshops and seminars were invaluable in refining the scope, content, and clarity of the book.

We hope this book will serve as a valuable resource for a wide range of readers interested in the intersection of machine learning and sustainability. It is our intention that the ideas and methods contained here will inspire further innovation and foster a deeper integration of data science into sustainable engineering and environmental practice.

Morelia, Mexico

Alma Yunuen Raya-Tapia
Francisco Javier López-Flores
César Ramírez-Márquez
José María Ponce-Ortega

Competing Interests The authors have no competing interests to declare that are relevant to the content of this manuscript.

Contents

Chapter 1
Artificial Intelligence, Machine Learning, and Clustering in Sustainability

1.1 Introduction

Artificial Intelligence (AI) and Machine Learning (ML) are becoming increasingly relevant in efforts to understand and respond to complex environmental and social dynamics (Shandilya et al., 2024). Their growing use reflects a shift in how knowledge is produced and applied in sustainability contexts, where traditional methods often struggle to keep pace with rapid changes in ecosystems, resource use, and human behavior. AI offers not only new tools for data analysis but also novel ways of interpreting patterns, detecting anomalies, and supporting adaptive responses across different scales (Himeur et al., 2023).

As environmental pressures intensify (through phenomena such as climate instability, biodiversity decline, and land degradation) there is a growing need to understand large volumes of diverse and often unstructured data. These may include satellite images, sensor outputs, public records, or social media content, all of which can be difficult to process using conventional approaches. Machine learning, especially unsupervised techniques like clustering, provides opportunities to explore such datasets without predefined assumptions, uncovering structures or trends that might otherwise remain hidden (Tirthyani et al., 2024).

In urban waste management, for example, clustering techniques have revealed how factors like age, education level, and environmental awareness influence household participation in recycling programs (Taweesan et al., 2025). Similarly, in natural disaster monitoring, AI systems have been used to model the behavior of wildfires by combining weather forecasts, vegetation indices, and land use data (Albahri et al., 2024). These examples demonstrate that AI does not simply improve efficiency; it introduces qualitatively different ways of engaging with environmental and social phenomena.

Despite this potential, challenges remain. Technical limitations, such as access to high-quality data or computational resources, can constrain the performance and generalizability of AI models. Ethical concerns, including data privacy, algorithmic

A. Y. Raya-Tapia et al., *Machine Learning and Clustering for a Sustainable Future*,
Studies in Computational Intelligence 1233,
https://doi.org/10.1007/978-3-032-03876-0_1

bias, and the transparency of decision-making processes, are also increasingly part of the discussion. Moreover, the environmental cost of running large-scale AI systems (particularly the energy demands of training complex models) has raised important questions about the sustainability of the technologies themselves.

Understanding the implications of AI in sustainability requires moving beyond a purely technical perspective. It involves examining how these technologies interact with institutional frameworks, local knowledge systems, and policy environments. The effectiveness and legitimacy of AI-driven tools depend not only on their accuracy but also on their capacity to support equitable, inclusive, and context-sensitive decision-making.

As the field evolves, the integration of AI, ML, and clustering into sustainability efforts is less about replacing human judgment and more about augmenting it, offering deeper analytical perspectives, enabling proactive interventions, and optimizing decision-making under uncertainty. This chapter is motivated by the growing need to consolidate and critically examine how these technologies are being applied across different domains of sustainability. While numerous studies have explored AI's potential in isolated contexts, there remains a lack of integrative understanding that connects technical advances with social, ethical, and environmental considerations. The novelty of this chapter lies in its interdisciplinary perspective, highlighting not only methodological progress but also the challenges of equitable implementation. By synthesizing current evidence and practical insights, the aim is to offer a clearer roadmap for harnessing AI-based approaches to support sustainability transitions in meaningful and responsible ways.

1.2 Conceptual Framework

Understanding the intersection between AI, ML, and sustainability requires a clear conceptual foundation that links technological capabilities with the multidimensional goals of sustainable development. This framework outlines how AI and ML methods, particularly clustering algorithms, contribute to sustainability-driven strategies through data-driven decision-making, system optimization, and environmental monitoring.

At its core, AI refers to systems designed to perform tasks that traditionally require human intelligence, such as reasoning, learning, and adapting to new information (Sajja & Sajja, 2021). ML, a subfield of AI, focuses on developing models that learn from data to identify patterns, make predictions, or support decision-making processes (Adeyeye & Akanbi, 2024). These models range from supervised techniques (e.g., neural networks, support vector machines) to unsupervised methods like clustering, which plays a critical role in understanding complex, unlabeled data in environmental and societal contexts.

Clustering, as an unsupervised learning technique, groups data points based on similarity without prior knowledge of categories (Chander & Gopalakrishnan, 2023). This is particularly useful in sustainability applications where labeled data may be scarce or difficult to obtain. For example, density-based clustering methods like Density-Based Spatial Clustering of Applications with Noise (DBSCAN) have been effectively used to uncover behavioral patterns in community waste management (Fauzan et al., 2024), while hierarchical and k-means clustering techniques assist in segmenting energy usage profiles, predicting resource demand, and identifying at-risk ecological zones (García-López et al., 2024).

The framework employed in this chapter recognizes sustainability as a multifaceted concept encompassing environmental integrity, social equity, and economic viability (Anitha, 2024). AI and ML techniques do not operate in a vacuum but interact with human systems, institutional structures, and physical infrastructure. Therefore, the framework positions these technologies as socio-technical tools, capable of enabling transformation, yet shaped by the values, access, and governance structures within which they are deployed.

Three operational domains are considered in this framework: (1) environmental management, (2) societal impact, and (3) systemic governance (see Fig. 1.1). In environmental management, AI and ML support the monitoring of ecological parameters such as deforestation, biodiversity loss, or carbon sequestration, often through remote sensing and computer vision technologies. In societal contexts, clustering and predictive models enable more effective urban planning, healthcare delivery, and public service optimization by identifying behavioral trends and community-level vulnerabilities. Within governance systems, AI tools offer decision-support mechanisms to simulate policy outcomes, improve risk communication, and inform strategic adaptation plans, particularly under climate-related uncertainty.

Another key component of this framework is the recognition of feedback loops (Sjödin et al., 2021). AI systems, particularly those involving reinforcement learning or continuous data streams, are embedded in dynamic environments where actions influence future states. As such, ethical considerations, model interpretability, and fairness are not peripheral concerns but central to ensuring sustainable and equitable

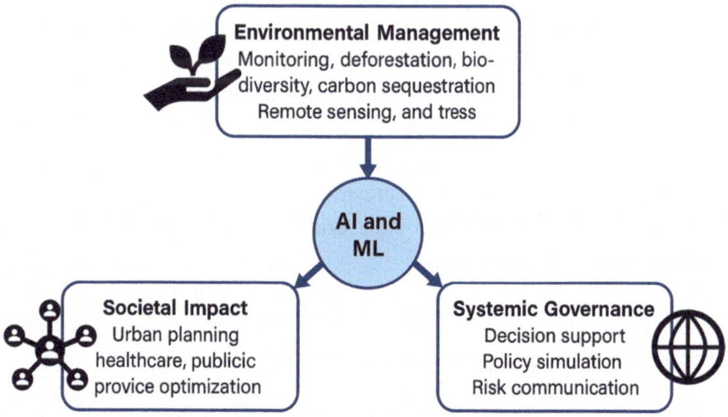

Fig. 1.1 Artificial intelligence applications across environmental, societal, and governance domains

AI deployment. Bias in training data, opaque decision-making, or limited access to digital infrastructure can exacerbate existing inequalities, making inclusive design and transparency essential components of sustainable AI (Sargiotis, 2024).

This conceptual structure also emphasizes the value of interdisciplinary integration. Addressing sustainability challenges through AI and ML requires the convergence of expertise from environmental science, computer science, social sciences, and policy studies. The effectiveness of technological solutions hinges not only on their computational accuracy but also on their contextual relevance, cultural acceptance, and alignment with long-term sustainability objectives.

This framework provides a lens through which to understand the synergistic role of AI, ML, and clustering in promoting sustainable development. It lays the groundwork for the subsequent sections by connecting algorithmic functions with sustainability outcomes and by highlighting both the transformative potential and the systemic limitations of these technologies. This integrative approach is essential for developing scalable, equitable, and resilient solutions in the face of increasingly complex environmental and social challenges.

1.3 Applications in Sustainability Domains

The integration of AI, ML, and clustering techniques into sustainability domains has gained traction due to their ability to manage large datasets and derive meaningful patterns that guide informed decisions. These technologies have shown considerable potential in enhancing environmental management, promoting social equity, and enabling adaptive strategies in the face of climate change.

One of the most prominent applications is in waste management, where AI has been applied to improve the efficiency of municipal services. For instance, in the city of Wollongong, Australia, survey data related to food and garden waste (FOGO) was analyzed using unsupervised ML techniques, particularly DBSCAN (Taghikhah et al., 2022). This clustering approach helped uncover behavioral patterns among households regarding their participation in waste sorting and composting programs. The study revealed that key determinants of program success were awareness, environmental beliefs, social media influence, and education levels. Residents in one cluster showed low awareness and engagement, while another group displayed high levels of environmental concern and were more active in using FOGO bins. These findings support the use of AI-based behavioral insights to shape more targeted and effective waste reduction strategies.

In the area of ecosystem conservation and biodiversity monitoring, ML techniques have been used to analyze satellite and ecological data to identify regions at risk due to deforestation, habitat fragmentation, and biodiversity loss (Shivaprakash et al., 2022). For example, machine learning models have been trained on remotely sensed imagery to detect illegal logging activities or monitor reforestation progress. These tools provide conservationists and policymakers with near real-time insights, allowing for quicker interventions. Such capabilities are vital in regions like the Amazon or

Southeast Asia, where rapid land-use changes threaten biodiversity and ecosystem services (Velastegui-Montoya et al., 2022).

AI and ML also contribute to climate change prediction and mitigation (Amiri et al., 2024). Predictive models built using these tools can simulate complex environmental systems by learning from historical data and detecting non-linear interactions among climate variables. Applications include forecasting extreme weather events, sea-level rise, and shifts in precipitation. These models are critical for national and regional planning efforts, as they support risk assessments and help allocate resources for climate adaptation (Hamdan et al., 2024). For instance, ML algorithms have been used to evaluate bushfire risks in Australia, integrating variables such as topography, vegetation type, and historical fire data to map vulnerable areas (Sharma et al., 2023). This allows for better land management and fire-prevention strategies.

In the energy sector, AI is playing an increasingly important role in optimizing both production and consumption (Ukoba et al., 2024). Applications include predictive maintenance of renewable energy infrastructure, demand forecasting, and real-time grid optimization. Wind turbines, for example, are now equipped with AI-driven systems that can adjust performance based on predictive algorithms for wind behavior, thereby increasing energy output and reducing operational downtime. Similarly, ML is used in battery management systems to enhance the lifespan, efficiency, and safety of lithium-ion batteries, a crucial component in the global energy transition (Krishna et al., 2024).

Water resource management is another domain where AI and clustering algorithms provide value (Ghobadi & Kang, 2023). By analyzing data on water quality, consumption patterns, and availability, these technologies help municipalities optimize water distribution, detect leakage, and reduce waste. AI-powered platforms also support environmental monitoring initiatives, such as identifying contamination sources or predicting drought events, which are critical in water-scarce regions (Jana, 2024).

Definitively, AI applications in urban planning and societal services include tools for traffic management, energy-efficient building design, and smart infrastructure. In public health and education, AI systems enable more equitable service delivery through demand modeling, behavioral analytics, and resource allocation. For example, AI models are used to map healthcare needs in underserved regions and inform interventions based on real-time data.

Across these domains, the adoption of AI, ML, and clustering is not only enhancing technical performance but also helping to align sustainability goals with actionable strategies. These technologies allow stakeholders to anticipate changes, prioritize interventions, and optimize the use of limited resources. However, realizing their full potential requires addressing challenges such as data availability, model transparency, and equitable access, issues that remain critical as sustainability agendas continue to evolve.

1.4 Clustering Techniques in Sustainability Science

Clustering is a key unsupervised learning method used in AI and ML to uncover structure in unlabeled data. In the context of sustainability science, clustering has become an essential analytical tool to identify patterns, segment populations, classify environmental zones, and support decision-making under uncertainty. These techniques allow researchers and policymakers to work with heterogeneous data, including behavioral, geospatial, temporal, and ecological information, to inform interventions that are both adaptive and context-specific.

Unlike supervised learning methods, clustering does not require predefined labels or outcomes (Nasteski, 2017). Instead, it detects similarities among observations based on distance metrics or density thresholds, grouping them into clusters that represent latent structures or typologies within the data. These groupings can reveal operational modes of energy systems, citizen behavior profiles in waste management, zones of environmental risk, or patterns of biodiversity loss. Clustering is particularly useful when sustainability challenges involve complex, high-dimensional, and multivariate data.

Several clustering algorithms have been successfully applied in sustainability-related domains. The most common include K-means, Hierarchical Clustering, DBSCAN, and more advanced approaches like Gaussian Mixture Models (GMMs) and Fuzzy C-means (Zangana et al., 2023). Each technique has different assumptions, strengths, and limitations, and their applicability depends on the structure and scale of the dataset, as well as the goals of the analysis.

For instance, K-means has been widely used for classifying household energy usage profiles, land-use categories, or climate zones due to its computational simplicity and effectiveness in detecting spherical clusters. However, it struggles with noise, non-convex clusters, and varying densities. In contrast, DBSCAN is robust to outliers and capable of identifying arbitrary-shaped clusters, making it suitable for environmental spatial data, such as deforestation zones or flood risk areas. Table 1.1 presents a comparative summary of clustering techniques commonly used in sustainability science, highlighting their core features and application domains.

Each technique supports different aspects of sustainability research. For instance, fuzzy clustering has been instrumental in modeling nexus systems (e.g., water–energy–food) where sectoral boundaries are fluid and interactions are dynamic. Similarly, GMMs are valuable for modeling mixed or overlapping environmental regimes, such as combined land-use zones with transitional ecotones.

The use of clustering in sustainability science also supports stakeholder engagement and decision transparency. By visualizing clusters (whether of socio-economic vulnerability, consumption profiles, or ecosystem health) decision-makers can better prioritize actions and allocate resources equitably.

Clustering techniques offer a flexible and powerful toolkit for sustainability science. As environmental and social datasets grow in complexity and size, the demand for robust, interpretable, and scalable clustering solutions will continue to increase. Future work should focus on hybrid approaches (e.g., combining clustering

Table 1.1 Comparative Summary of Clustering Techniques in Sustainability Applications (adapted from Omari et al., 2025)

Clustering method	Key characteristics	Strengths	Limitations	Typical applications
K-means	Partitions data into *k* clusters based on centroids	Fast, easy to implement, good for large datasets	Sensitive to initial seeds, assumes spherical clusters	Energy consumption profiling, land-use classification, urban planning
Hierarchical clustering	Builds a dendrogram based on proximity (agglomerative/ divisive)	No need to predefine number of clusters, good interpretability	Computationally expensive for large datasets	Biodiversity analysis, policy segmentation, climate typology
DBSCAN	Groups data based on density; identifies noise points	Detects arbitrary-shaped clusters, handles noise well	Performance declines in high-dimensional data	Spatial risk mapping (e.g., floods, fires), citizen behavior in waste sorting
Fuzzy C-means	Assigns membership probabilities rather than hard labels	Captures overlapping group membership	Requires predefined number of clusters, sensitive to noise	Water-energy-food nexus modeling, participatory governance categorization
Gaussian mixture models (GMMS)	Probabilistic model assuming data are generated from a mixture of Gaussians	Captures complex, non-linear boundaries, uncertainty modeling	Sensitive to initial conditions, can overfit	Emissions pattern detection, ecological habitat classification, transportation emissions

with network analysis or deep learning), adaptive clustering methods that incorporate feedback over time, and the integration of domain knowledge to improve the relevance and actionability of the results.

1.5 Challenges and Ethical Considerations

As AI, ML, and clustering techniques are increasingly integrated into sustainability-related efforts, several technical, social, and ethical challenges arise. While these tools offer significant promise, they also bring complex questions about reliability, fairness, transparency, and accountability. Understanding these challenges is essential to ensure that the deployment of AI technologies in sustainability aligns with societal values and long-term environmental goals.

One of the primary technical obstacles involves data availability and quality. Sustainability datasets often come from fragmented sources such as satellite sensors, field observations, citizen science, and institutional reports. These sources can vary in resolution, accuracy, and update frequency. Many developing regions lack continuous monitoring systems, which results in limited or outdated information for training AI and ML models. Moreover, the absence of standardized data formats complicates data sharing and model reproducibility across contexts and regions.

Model interpretability is another pressing issue. Many AI and ML algorithms, especially deep learning models, operate as "black boxes" where the internal decision-making logic is not easily understandable. In domains like climate policy, biodiversity protection, or disaster response, where decisions affect ecosystems and communities, explainability becomes critical. Without transparent reasoning, model outputs may be distrusted or misused by decision-makers.

In the context of clustering, ethical concerns emerge when algorithms are used to group populations or territories without sufficient understanding of local dynamics. For example, applying clustering to identify "high-risk" communities for waste management interventions or energy access planning could unintentionally reinforce stereotypes, exclude marginalized groups, or lead to biased policies if the input data reflect existing social inequalities.

Another key consideration is bias in algorithmic design and training data. AI models learn from historical records and current observations, which may encode systemic biases. For instance, if deforestation monitoring tools are trained predominantly on data from tropical rainforests, they might underperform in dryland ecosystems. Similarly, models for energy forecasting or food security might be skewed by underrepresentation of indigenous knowledge systems or rural lifestyles. Such biases can lead to unfair decisions, resource misallocation, or missed opportunities for inclusive development.

Digital inequality also poses a barrier to the equitable use of AI in sustainability. Advanced AI applications often require high computational power, internet access, and specialized technical knowledge, resources that are unequally distributed across and within countries. This disparity can widen the gap between regions that benefit from AI and those left behind, especially in the Global South.

Ethical data governance is central to responsible AI deployment. Many sustainability challenges involve sensitive or personal data, such as water consumption, health records, or geolocated environmental behavior. Ensuring privacy, consent, and secure storage is vital. However, existing legal and institutional frameworks often lag behind technological advancements, creating regulatory vacuums. The potential for surveillance or misuse of environmental monitoring data must be carefully addressed.

Finally, there is a growing concern regarding the environmental footprint of AI itself. Training large-scale ML models demands significant energy, particularly in data centers with high computational loads. If not sourced from clean energy, these processes may contradict sustainability goals. Research on energy-efficient algorithms and low-resource AI is gaining traction but remains underdeveloped in many application areas.

While AI, ML, and clustering provide valuable tools for advancing sustainability science, they are not neutral or infallible. Their development and application must be guided by ethical frameworks that emphasize inclusivity, transparency, and justice. Multidisciplinary collaboration (with input from environmental scientists, ethicists, social scientists, and affected communities) is essential to ensure that technological innovation supports equitable and resilient futures.

1.6 Future Research Directions

Bridging the gap between data-driven insights and actionable sustainability outcomes remains a central research frontier. While recent advancements in AI, ML, and clustering have demonstrated significant promise, there are still critical questions to explore and barriers to overcome in order to fully harness their potential in sustainability science.

One priority is to develop methodologies that move beyond predictive accuracy and support explanatory power. Models that can uncover causal mechanisms (rather than simply identifying statistical associations) are essential for informing effective policy and intervention strategies. This shift from correlation to causation is particularly relevant in complex systems such as climate adaptation, food security, and public health, where interventions have long-term and often non-linear effects.

Another important research avenue lies in adapting algorithms to low-data or noisy environments. Many regions most affected by environmental degradation or climate risks lack consistent, high-quality datasets. Research must focus on building robust, transferable models that can operate effectively under uncertainty, data sparsity, or infrastructural limitations. Approaches such as transfer learning, federated learning, and synthetic data generation may prove particularly useful in these contexts.

There is also a growing need for dynamic, context-aware clustering methods that can adjust to shifting patterns in environmental and social data. Unlike static clustering, these techniques would account for evolving user behavior, seasonal variations, or changes in policy or land use, offering more relevant groupings over time. This is especially useful in monitoring energy consumption trends, urban mobility patterns, and ecosystem resilience.

As sustainability challenges often span across multiple sectors and disciplines, research must also emphasize the integration of multi-source and multi-scale data. Combining satellite imagery, sensor data, social media, and administrative records requires models capable of managing heterogeneous inputs while maintaining transparency and interpretability. Future work should explore how to fuse these diverse data types into coherent frameworks that support local and global decision-making alike.

Improving algorithmic transparency and trust remains an urgent task. As models are increasingly used to guide critical environmental and social policies, the ability to communicate their logic, limitations, and confidence levels becomes essential. Research on explainable AI (XAI), interactive visualizations, and participatory design processes can help ensure that these tools are not only technically sound but also socially acceptable and inclusive.

Collaboration between technical experts and practitioners is also a major opportunity. Future research should explore co-creation frameworks, where scientists, engineers, policymakers, and communities collaboratively define problems, develop models, and interpret results. This approach strengthens the relevance and legitimacy of AI applications and supports ethical and place-sensitive innovation.

Another emerging direction is the study of environmental externalities of AI systems themselves. The energy consumption, carbon footprint, and material requirements of training large models must be critically assessed, particularly in applications that aim to contribute to sustainability. Research into green AI (models optimized for efficiency and low resource use) should be encouraged and evaluated alongside conventional performance metrics.

Lastly, regulatory frameworks and governance models for AI in sustainability contexts need further development. Establishing mechanisms to ensure accountability, equity, and adaptability of AI systems will be crucial as their deployment scales. This includes defining standards for environmental data use, auditability of algorithms, and safeguards against reinforcing existing inequalities.

By addressing these research gaps, the scientific community can support the responsible evolution of AI and ML for sustainability. The path forward depends not only on computational progress but also on interdisciplinary dialogue, social sensitivity, and a long-term commitment to ethical and impactful innovation.

1.7 Conclusions

Taking into account the transformative capabilities and inherent limitations discussed throughout this chapter, it becomes evident that AI, ML, and clustering techniques offer both promise and complexity when applied to sustainability challenges. Their capacity to process vast, multidimensional datasets and extract meaningful insights has already enabled advances in areas such as climate adaptation, resource management, renewable energy integration, and biodiversity monitoring.

Rather than serving as standalone solutions, these technologies must be seen as enablers, tools that can augment human decision-making, improve foresight, and guide evidence-based policies. Their utility is shaped not only by algorithmic sophistication but by how they are embedded within social, environmental, and ethical contexts. Issues related to data bias, accessibility, explainability, and governance remain central to their responsible deployment.

Sustainability science increasingly demands interdisciplinary collaboration, where technical innovation is aligned with ecological resilience, social inclusion, and long-term systemic thinking. In this evolving landscape, AI, ML, and clustering should be further explored not merely for optimization, but for enabling adaptive, just, and transparent pathways toward sustainability transitions.

Ultimately, the next phase of research and practice must focus on aligning these intelligent technologies with equitable and regenerative models of development, where innovation is driven not only by efficiency, but by purpose.

References

Adeyeye, O. J., & Akanbi, I. (2024). Artificial intelligence for systems engineering complexity: A review on the use of AI and machine learning algorithms. *Computer Science & IT Research Journal, 5*(4), 787–808.

Albahri, A. S., Khaleel, Y. L., Habeeb, M. A., Ismael, R. D., Hameed, Q. A., Deveci, M., ... & Alzubaidi, L. (2024). A systematic review of trustworthy artificial intelligence applications in natural disasters. *Computers and Electrical Engineering, 118*, 109409. https://doi.org/10.1016/j.compeleceng.2024.109409

Amiri, Z., Heidari, A., & Navimipour, N. J. (2024). Comprehensive survey of artificial intelligence techniques and strategies for climate change mitigation. *Energy*, 132827.https://doi.org/10.1016/j.energy.2024.132827

Anitha, K. (2024). Emerging trends in sustainability: A conceptual exploration. *Global Sustainability: Trends, Challenges & Case Studies*, 19–35. https://doi.org/10.1007/978-3-031-57456-6_2

Chander, B., & Gopalakrishnan, K. (2023). Data clustering using unsupervised machine learning. In *Statistical Modeling in Machine Learning*, 179–204. https://doi.org/10.1016/B978-0-323-91776-6.00015-4

Fauzan, A., Fadillah, G., Fitria, A., Adriana, H., & Bariklana, M. (2024). Cluster mapping of waste exposure using DBSCAN approach: Study of spatial patterns and potential distribution in Bantul regency. *JOIV: International Journal on Informatics Visualization, 8*(2), 751–759. https://doi.org/10.62527/joiv.8.2.2475

García-López, J., Domínguez-Amarillo, S., & Sendra, J. J. (2024). Clustering open data for predictive modeling of residential energy consumption across variable scales: A case Study in Andalusia. *Spain Buildings, 14*(8), 2335. https://doi.org/10.3390/buildings14082335

Ghobadi, F., & Kang, D. (2023). Application of machine learning in water resources management: A systematic literature review. *Water, 15*(4), 620. https://doi.org/10.3390/w15040620

Hamdan, A., Ibekwe, K. I., Etukudoh, E. A., Umoh, A. A., & Ilojianya, V. I. (2024). AI and machine learning in climate change research: A review of predictive models and environmental impact. *World Journal of Advanced Research and Reviews, 21*(1), 1999–2008. https://doi.org/10.30574/wjarr.2024.21.1.0257

Himeur, Y., Elnour, M., Fadli, F., Meskin, N., Petri, I., Rezgui, Y., ... & Amira, A. (2023). AI-big data analytics for building automation and management systems: a survey, actual challenges and future perspectives. *Artificial Intelligence Review, 56*(6), 4929–5021.https://doi.org/10.1007/s10462-022-10286-2

Jana, P. (2024). AI-powered IoT solutions for sustainable water management in cities. *Uncertainty Discourse and Applications, 1*(2), 158–169. https://doi.org/10.48313/uda.v1i2.37

Krishna, G., Singh, R., Gehlot, A., Almogren, A., Altameem, A., Ur Rehman, A., & Hussen, S. (2024). Advanced battery management system enhancement using IoT and ML for predicting remaining useful life in Li-ion batteries. *Scientific Reports, 14*(1), 1–18. https://doi.org/10.1038/s41598-024-80719-1

Nasteski, V. (2017). An overview of the supervised machine learning methods. *Horizons. b, 4*(51–62), 56.

Omari, M., Kaddi, M., Salameh, K., & Alnoman, A. (2025). Advancing image compression through clustering techniques: A comprehensive analysis. *Technologies, 13*(3), 123. https://doi.org/10.3390/technologies13030123

Sajja, P. S., & Sajja, P. S. (2021). Introduction to artificial intelligence. *Illustrated Computational Intelligence: Examples and Applications*, 1–25. https://doi.org/10.1007/978-981-15-9589-9_1

Sargiotis, D. (2024). Fostering Ethical and Inclusive AI: A Human-Centric Paradigm for Social Impact. *Available at SSRN 4879372.* SSRN: https://ssrn.com/abstract=4879372

Shandilya, S. K., Datta, A., Kartik, Y., & Nagar, A. (2024). Role of artificial intelligence and machine learning. In *Digital Resilience: Navigating Disruption and Safeguarding Data Privacy* (pp. 313–399). Cham: Springer Nature Switzerland. https://doi.org/10.1007/978-3-031-53290-0_6

Sharma, S. K., Aryal, J., Shao, Q., & Rajabifard, A. (2023). Characterizing topographic influences of bushfire severity using machine learning models: A case study in a Hilly Terrain of Victoria, Australia. *IEEE Journal of Selected Topics in Applied Earth Observations and Remote Sensing, 16*, 2791–2807. https://doi.org/10.1109/JSTARS.2023.3249643

Shivaprakash, K. N., Swami, N., Mysorekar, S., Arora, R., Gangadharan, A., Vohra, K., ... & Kiesecker, J. M. (2022). Potential for artificial intelligence (AI) and machine learning (ML) applications in biodiversity conservation, managing forests, and related services in India. *Sustainability, 14*(12), 7154. https://doi.org/10.3390/su14127154

Sjödin, D., Parida, V., Palmié, M., & Wincent, J. (2021). How AI capabilities enable business model innovation: Scaling AI through co-evolutionary processes and feedback loops. *Journal of Business Research, 134*, 574–587. https://doi.org/10.1016/j.jbusres.2021.05.009

Taghikhah, F., Erfani, E., Bakhshayeshi, I., Tayari, S., Karatopouzis, A., & Hanna, B. (2022). Artificial intelligence and sustainability: solutions to social and environmental challenges. In *Artificial Intelligence and Data Science in Environmental Sensing* (pp. 93–108). Academic Press. https://doi.org/10.1016/B978-0-323-90508-4.00006-X

Taweesan, A., Kanabkaew, T., Surinkul, N., & Polprasert, C. (2025). Integrating clustering algorithms and machine learning to optimize regional snapshot municipal solid waste management for achieving sustainable development goals. *Environmental Advances, 19*, Article 100607. https://doi.org/10.1016/j.envadv.2024.100607

Tirthyani, D., Kumar, S., & Vats, S. (2024). Clustering and unsupervised learning. In *Recent Trends and Future Direction for Data Analytics*, 119–139. https://doi.org/10.4018/979-8-3693-3609-0.ch005

Ukoba, K., Olatunji, K. O., Adeoye, E., Jen, T. C., & Madyira, D. M. (2024). Optimizing renewable energy systems through artificial intelligence: Review and future prospects. *Energy & Environment, 35*(7), 3833–3879. https://doi.org/10.1177/0958305X241256293

Velastegui-Montoya, A., Montalván-Burbano, N., Peña-Villacreses, G., de Lima, A., & Herrera-Franco, G. (2022). Land use and land cover in tropical forest: Global research. *Forests, 13*(10), 1709. https://doi.org/10.3390/f13101709

Zangana, H. M., & Abdulazeez, A. M. (2023). Developed clustering algorithms for engineering applications: A review. *International Journal of Informatics, Information System and Computer Engineering (INJIISCOM), 4*(2), 160–182. https://doi.org/10.34010/injiiscom.v4i2.11636

Chapter 2
Fundamentals of Clustering: Methods, Metrics, and Optimization

2.1 Introduction

Clustering is a fundamental technique that aims to automatically organize a dataset into groups based on the similarity between its elements. Each cluster groups objects that share common characteristics, while those that differ significantly are assigned to distinct groups, even allowing the identification and separation of noise. Unlike supervised methods, clustering operates without prior labels, discovering intrinsic structures in the data in an unsupervised manner. The usefulness of clustering transcends multiple fields, such as medicine, biology, economics, sociology, or engineering, the latter being a key technique for analyzing complex data, optimizing processes and improving decision making.

In the era of big data, the challenge is amplified by the high dimensionality of modern data sets, where each object can be described by hundreds or thousands of variables or features (numeric, categorical or mixed). This phenomenon, driven by massive storage and proliferation of data generated by sensors, bioinformatics or users on the web, demands robust methods to handle complexity. However, there is no consensus on what constitutes "high dimensionality": while some studies define it from 10 variables, others work with thousands. In this scenario, clustering serves not only to explore hidden patterns, but also as a preprocessing step for supervised machine learning techniques, such as classification or regression, by reducing dimensionality through the assignment of cluster IDs, or improving predictive capability through additional features or generating labels that enrich the database when there are few labels for training.

There are multiple clustering algorithms, such as those based on distance (e.g. K-means), density (e.g. DBSCAN) or hierarchical approaches, reflecting the diversity of problems and contexts in which it is applied. However, one of the biggest challenges of clustering is the lack of universal criteria for assessing efficiency. As an unsupervised method, there is no "absolute truth" against which to compare results,

A. Y. Raya-Tapia et al., *Machine Learning and Clustering for a Sustainable Future*,
Studies in Computational Intelligence 1233,
https://doi.org/10.1007/978-3-032-03876-0_2

and the choice of algorithm and number of clusters depends on the researcher's objectives.

Clustering faces limitations, such as sensitivity to noise, subjective choice of similarity metrics, and difficulty in determining the optimal number of clusters. Validation techniques such as the Jambu elbow or the silhouette index attempt to address this, but effectiveness varies depending on the context. Ultimately, clustering is a versatile and powerful tool, but successful application requires a balance of domain knowledge, appropriate algorithm selection, and critical interpretation of the results to extract meaningful insights.

2.2 Stages of Clustering Analysis

Data preparation is a fundamental stage in clustering analysis, since it determines the effectiveness of the algorithms and the quality of the results obtained. The clustering analysis process starts with stage 1 (Fig. 2.1), which involves the definition of the objective (what will it be used for?). Stage 2 is data preparation and preprocessing. This stage begins with the selection of the set of features that describe the objects of study, since these are the ones that allow differentiating between groups and defining patterns of similarity. As the dimensionality of the data increases, many similarity measures lose effectiveness, which is a phenomenon known as the curse of dimensionality. Moreover, an appropriate choice of features not only improves the discriminant ability of the algorithms, but also influences the interpretability of the results. Therefore, it is crucial to opt for feature selection techniques (which choose a relevant subset of original variables) or feature extraction techniques (which create new artificial variables, as in Principal Component Analysis (PCA) or t-Distributed Stochastic Neighbor Embedding (t-SNE)), depending on the nature of the problem.

Stage 2 also includes data cleaning, where missing values are handled either by eliminating features with too many null values or by imputing values, taking into account not distorting the distribution of the data. This stage also includes the normalization of the data to avoid biases due to different scales in the features. In summary, meticulous preparation combining intelligent variable selection, appropriate transformations and rigorous cleaning is the basis for obtaining meaningful clusters and efficient algorithms, thus facilitating the extraction of valuable insights in practical applications.

In stage 3, the clustering algorithm and the similarity measure are chosen. In the selection of the algorithm, the advantages and limitations of each one must be taken into consideration, for example, DBSCAN detects irregularly shaped clusters ignoring noise however it is sensitive to parameters (ε and min_samples), agglomerative hierarchical clustering creates a dendrogram of nested clusters, but is computationally expensive; while K-means divides the data into K spherical groups quickly, but is sensitive to outliers. On the other hand, the selection of the similarity measure can be based on the type of data, numerical, mixed or categorical.

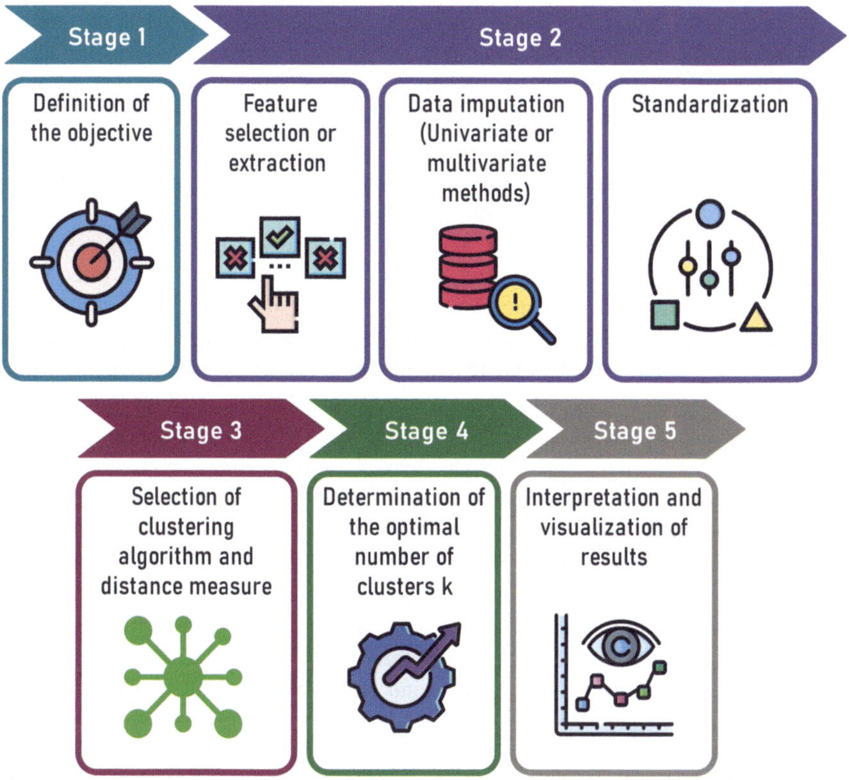

Fig. 2.1 Stages of clustering

In step 4, the optimal number of clusters is determined, using different methods such as elbow, silhouette, Gap, among others. This stage is applied only in algorithms where the number of clusters must be initially defined. In stage 5, the clustering algorithm is applied, and the results are interpreted and visualized (using 2D or 3D graphs, among others).

2.3 Missing Data Imputation

The imputation of missing values is an essential process in data analysis, where artificial values are generated to replace missing observations. This procedure is classified into two main methodological categories according to the strategy: univariate and multivariate. Univariate methods (such as SimpleImputer) treat each variable independently, using basic statistics (mean, median or mode) or constant values, being fast but limited as it may distort the original distribution of the data, underestimate the variance and ignore correlations between variables. On the other hand, multivariate methods (such as IterativeImputer) consider correlations between variables, employing predictive models (e.g., linear or logistic regression, decision trees or KNN) to impute values more accurately by exploiting patterns in the observed data to predict missing values. This second category is ideal for complex data sets, although it requires more computational power. The choice between the two strategies depends on the balance between speed, accuracy, and data structure.

The choice of imputation method must fundamentally consider the mechanism generating the missing data, classified by Little and Rubin (2019) into three categories: MCAR ("missing completely at random", where the absence is not related to any variable), MAR ("missing at random", where the probability of absence depends on observed variables), and MNAR ("missing not at random", where the absence is linked to unobserved values). Most techniques assume MCAR or MAR, this distinction being crucial as methods such as k-nearest neighbor (kNN) imputation or regression can produce biased results in MNAR scenarios. Imputation is a powerful tool, but with challenges. So, success depends on understanding the missing data mechanism, selecting techniques according to the nature of the data, and rigorously validating the results to avoid biased conclusions. As Rubin (1976) points out, the objective is not to predict missing values accurately, but to generate plausible imputations that allow valid analyses.

In handling missing data, SimpleImputer, KNNImputer and IterativeImputer (from the scikit-learn library in Python) stand out as fundamental approaches, each with specific advantages depending on the complexity of the problem. SimpleImputer is a univariate method that replaces missing values using simple statistics (mean, median, "most frequent" mode) or a constant value (such as zero or other) along each feature (column), being efficient but limited by ignoring relationships between variables. For example, if mean is selected, missing values are replaced using the mean of each column. It should be noted that the mean and median are used only with numerical data, while the mode and constant value can be used with numerical and categorical data. KNNImputer is an efficient univariate method for filling in missing values using the k-Nearest Neighbors (k-NN) algorithm.

This technique imputes missing values by taking the average of the nearest neighbors (n_neighbors) in the feature space, using by default a Euclidean distance. For each sample with missing values, the method identifies the nearest neighbors and averages their values, allowing different weighting schemes: uniform (all neighbors contribute equally), distance (higher weight to nearest neighbors) or customized (according to a user-defined function). Thus, the imputation is based on the similarity between observations, preserving the structure of the data. It is ideal for data where similar samples share close values. Advantages include simplicity and efficiency; however, it is sensitive to the scale of the data and has a high computational cost with large data sets.

IterativeImputer is a multivariate method that iteratively models each variable with missing values as a function of the other variables, using predictive algorithms such as LinearRegression (for linear relationships), DecisionTreeRegressor (for nonlinear and categorical patterns), or even advanced methods such as RandomForest, among others. The flexibility of IterativeImputer makes it ideal for datasets with intricate multivariate structures, although it requires higher computational cost. Some predictors (estimators) available for use in IterativeImputer are discussed below.

2.3.1 LinearRegression

The integration of LinearRegression as an estimator in IterativeImputer allows accurate multivariate imputations by taking advantage of linear relationships between variables. LinearRegression uses Ordinary Least Squares (OLS), fitting coefficients by a closed analytical solution, minimizing the sum of squared residuals between predictions and actual values. Key attributes include: simplicity and speed for small/medium data sets; interpretability as the coefficients indicate the magnitude and direction of the relationship between variables; although it assumes linearity, homoscedasticity and independence of errors. However, it is sensitive to outliers and multicollinearity. The combination with IterativeImputer iterates sequentially over missing variables offers a robust approach for numerical data.

2.3.2 BayesianRidge

The combination of BayesianRidge and IterativeImputer provides an advanced method for imputing missing values, incorporating uncertainty through Bayesian inference. BayesianRidge is a Bayesian variant of linear regression that estimates model parameters as probabilistic distributions, providing automatic regularization by avoiding overfitting through coefficient restrictions; uncertainty estimation by calculating credible intervals for predictions; and robustness by being less sensitive to outliers than OLS (ordinary least squares). So, the combination of BayesianRidge and IterativeImputer considers not only the linear relationships between variables, but also the inherent uncertainty in the estimated values. This approach is ideal for data with noise or small samples and/or cases where uncertainty quantification is critical. However, one of the most important limitations is the high computational cost.

2.3.3 DecisionTreeRegressor

The combination of DecisionTreeRegressor as an estimator of IterativeImputer provides an advanced method for imputing missing values. DecisionTreeRegressor, a decision tree-based model that can model complex interactions and nonlinear relationships between variables, improves the accuracy of imputations on data with intricate patterns. This approach is ideal for data sets where relationships cannot be adequately modeled with linear methods, such as LinearRegression or BayesianRidge. Important aspects include nonlinearity and robustness as it is less sensitive to outliers and multicollinearity compared to linear methods. While it faces limitations by having a high risk of overfitting (if the tree depth "max_depth" is not controlled) and high computational cost (especially with deep trees or large datasets).

2.3.4 *RandomForestRegressor*

The combination of RandomForestRegressor with IterativeImputer offers an advanced method for handling missing values, particularly effective when the relationships between variables are complex and nonlinear. This multivariate approach combines multiple decision trees (ensemble) as the base estimator in the iterative imputation process, providing several key advantages such as Robustness to outliers and noise thanks to the averaging mechanism between multiple trees, which also impacts on the reduction of overfitting compared to a single decision tree; and Automatic handling of interactions between variables without the need to specify them manually. The implementation is simple, although it requires more computational power than linear alternatives.

2.3.5 *ExtraTreesRegressor*

The combination of ExtraTreesRegressor (Extremely Randomized Trees) (also known as extra trees) and IterativeImputer offers a robust method for missing value imputation, highlighting the computational efficiency and ability to handle nonlinear relationships. Unlike traditional RandomForest, ExtraTreesRegressor introduces more randomization when selecting splitting thresholds, which speeds up the process while maintaining accuracy. Among the advantages are robustness and improved speed since the random selection of thresholds reduces the computational cost compared to RandomForest. However, a disadvantage is that there is a higher variance in small data sets and two important parameters must be adjusted to optimize the results, which are the number of decision trees (n_estimators) and the maximum number of variables considered to split a node in each tree (max_feratures).

2.3.6 *GradientBoostingRegressor*

The integration of GradientBoostingRegressor with IterativeImputer allows for sophisticated missing value imputation, taking advantage of gradient descent boosted tree-based models that sequentially optimize predictions. In order to capture nonlinear relationships, complex interactions between features and non-random patterns

(MNAR/MAR). Among the advantages is the flexibility to adjust key parameters such as n_estimators, learning_rate and max_depth to achieve a balance between accuracy and overfitting. This is ideal when the quality of the imputation is critical for subsequent analysis or predictive performance. However, the computational cost is higher than alternatives such as RandomForest or ExtraTrees.

The tree-based algorithms GradientBoostingRegressor (GBR), RandomForestRegressor (RFR) and ExtraTreesRegressor (ETR) differ fundamentally in the assembly approach: GradientBoostingRegressor uses boosting, a sequential technique that builds trees one after another, where each new tree corrects the residual errors of the previous one, optimizing a loss function by gradient descent; this approach achieves high accuracy, but requires further tuning to avoid overfitting. On the other hand, RandomForestRegressor and ExtraTreesRegressor employ bagging, a parallel method that trains multiple independent trees on random subsets of the data (bootstrap sampling) and combines the predictions by averaging them; while RandomForestRegressor selects the optimal splitting points at each node, ExtraTreesRegressor chooses them randomly, which makes it faster but slightly less accurate. So, GradientBoostingRegressor excels in accuracy with structured data, RandomForestRegressor offers robustness and balance, and ExtraTreesRegressor prioritizes speed and efficiency, especially useful on noisy or high-dimensional data sets.

2.3.7 KNeighborsRegressor

The combination of KNeighborsRegressor and IterativeImputer provides an efficient method to impute missing values using proximity patterns in multivariate space. This non-parametric approach uses the distance (Euclidean, Manhattan or custom) between complete observations to estimate missing values using the average or median of the nearest neighbors. It has key advantages such as adaptability to complex patterns since it captures non-linear relationships without assuming a specific functional structure. To optimize the results, it is recommended to calibrate the number of nearest neighbors (n_ neighbors).

This approach can be confused with KNNImputer and although both methods use the concept of k-nearest neighbors (k-NN), they are not the same due to differences in implementation and operating logic. Firstly, KNNImputer is a univariate method (it does not consider the relationship between variables) since it calculates the (Euclidean) distance between rows (samples) using only one feature (column), imputing each missing value by the average of the nearest neighbors. While the combination of IterativeImputer with the KNeighborsRegressor estimator is a multivariate method, since it predicts the missing values of one variable as a function of all other variables. In addition, KNeighborsRegressor is more flexible as it allows customization of the model (weights, distance measures, etc.).

The choice between these missing data imputation methods depends on the balance between accuracy, interpretability and available resources, with SimpleImputer and KNNImputer being suitable for fast imputations or IterativeImputer being preferable when the relationships between variables are critical to the quality of the imputed data. Below is a code that generates a dataset with missing errors type MCAR, MAR, MNAR which is then used to compare the different algorithms for data imputation (addressed previously) in Python.

```
************************************************************

## 1. Import libraries
import numpy as np
import pandas as pd
from sklearn.impute import SimpleImputer, KNNImputer
from sklearn.experimental import enable_iterative_imputer
from sklearn.impute import IterativeImputer
from sklearn.linear_model import LinearRegression, BayesianRidge
from sklearn.ensemble import RandomForestRegressor,
ExtraTreesRegressor
from sklearn.tree import DecisionTreeRegressor
from sklearn.neighbors import KNeighborsRegressor
from sklearn.metrics import mean_squared_error
from sklearn.ensemble import GradientBoostingRegressor
import time
import matplotlib.pyplot as plt
import seaborn as sns

## 2. Synthetic data generation with MCAR, MAR and MNAR patterns
np.random.seed(42)
n_samples = 500
n_features = 6

# Create base data with complex correlations
data = np.random.randn(n_samples, n_features)

# Establish relationships between variables
data[:, 1] = 0.7 * data[:, 0] + 0.3 * np.random.randn(n_samples)  #
Strong linear relationship
data[:, 2] = 0.5 * data[:, 1]**2 + 0.5 * np.random.randn(n_samples)
# Non-linear relationship
data[:, 3] = 0.3 * data[:, 0] + 0.7 * data[:, 2] #Mixed relationship
data[:, 4] = np.log(np.abs(data[:, 1]) + 1e-10) + 0.5 * data[:, 3]
# Another non-linear relationship
data[:, 5] = 0.4 * data[:, 2] * data[:, 4]  # Interaction between
variables

df = pd.DataFrame(data, columns=[f'X{i}' for i in
range(n_features)])
df_completo = df.copy()

## 3. Enter missing values with different patterns

### MCAR (Missing Completely At Random)
mask_mcar_x0 = np.random.random(n_samples) < 0.25
mask_mcar_x1 = np.random.random(n_samples) < 0.25
df.loc[mask_mcar_x0, 'X0'] = np.nan
df.loc[mask_mcar_x1, 'X1'] = np.nan

### MAR (Missing At Random)
mar_cond_x2 = df['X1'] > np.nanpercentile(df['X1'], 75)
mask_mar_x2 = mar_cond_x2 & (np.random.random(n_samples) < 0.6)
df.loc[mask_mar_x2, 'X2'] = np.nan
```

```
mar_cond_x3 = df['X0'] < np.nanpercentile(df['X0'], 25)
mask_mar_x3 = mar_cond_x3 & (np.random.random(n_samples) < 0.5)
df.loc[mask_mar_x3, 'X3'] = np.nan

### MNAR (Missing Not At Random)
mnar_threshold_x4 = np.percentile(df_completo['X4'], 30)
mnar_mask_x4 = (df_completo['X4'] < mnar_threshold_x4) &
(np.random.random(n_samples) < 0.7)
df.loc[mnar_mask_x4, 'X4'] = np.nan

mnar_threshold_x5 = np.percentile(df_completo['X5'], 70)
mnar_mask_x5 = (df_completo['X5'] > mnar_threshold_x5) &
(np.random.random(n_samples) < 0.6)
df.loc[mnar_mask_x5, 'X5'] = np.nan

## 4. Verification of missing patterns
print("Percentage of missing values per column:")
print(df.isna().mean().round(4) * 100)


## 5. Comparison of imputation methods

# Missing values mask for evaluation
missing_mask = df.isna()
results = []

# -----------------------------------------------------------------
# Imputation of missing values
# -----------------------------------------------------------------

# a) SimpleImputer (mean)
start_time = time.time()
imputer_mean = SimpleImputer(strategy='mean')
df_mean = imputer_mean.fit_transform(df)
mse_mean =
mean_squared_error(df_completo.values[missing_mask.values],
                             df_mean[missing_mask.values])
time_mean = time.time() - start_time
results.append(['SimpleImputer (mean)', mse_mean, time_mean])

# b) SimpleImputer (median)
start_time = time.time()
imputer_median = SimpleImputer(strategy='median')
df_median = imputer_median.fit_transform(df)
mse_median =
mean_squared_error(df_completo.values[missing_mask.values],
                               df_median[missing_mask.values])
time_median = time.time() - start_time
results.append(['SimpleImputer (median)', mse_median, time_median])

# c) KNNImputer
start_time = time.time()
imputer_knn = KNNImputer(n_neighbors=5)
df_knn = imputer_knn.fit_transform(df)
mse_knn =
mean_squared_error(df_completo.values[missing_mask.values],
```

```
                              df_knn[missing_mask.values])
time_knn = time.time() - start_time
results.append(['KNNImputer (k=5)', mse_knn, time_knn])


# d) IterativeImputer "LinearRegression"
start_time = time.time()
imputer_lr = IterativeImputer(estimator=LinearRegression(),
max_iter=15, random_state=42)
df_lr = imputer_lr.fit_transform(df)
mse_lr = mean_squared_error(df_completo.values[missing_mask.values],
                              df_lr[missing_mask.values])
time_lr = time.time() - start_time
results.append(['LinearRegression', mse_lr, time_lr])


# e) IterativeImputer "RandomForest"
start_time = time.time()
imputer_rf =
IterativeImputer(estimator=RandomForestRegressor(n_estimators=50,
random_state=42),max_iter=15, random_state=42 )
df_rf = imputer_rf.fit_transform(df)
mse_rf = mean_squared_error(df_completo.values[missing_mask.values],
                              df_rf[missing_mask.values])
time_rf = time.time() - start_time
results.append(['RandomForest', mse_rf, time_rf])

# f) IterativeImputer "DecisionTree"
start_time = time.time()
imputer_dt =
IterativeImputer(estimator=DecisionTreeRegressor(max_depth=5,
random_state=42),max_iter=15, random_state=42)
df_dt = imputer_dt.fit_transform(df)
mse_dt = mean_squared_error(df_completo.values[missing_mask.values],
                              df_dt[missing_mask.values])
time_dt = time.time() - start_time
results.append(['DecisionTree', mse_dt, time_dt])

# g) IterativeImputer "ExtraTrees"
start_time = time.time()
imputer_et =
IterativeImputer(estimator=ExtraTreesRegressor(n_estimators=50,
random_state=42),max_iter=15, random_state=42)
df_et = imputer_et.fit_transform(df)
mse_et = mean_squared_error(df_completo.values[missing_mask.values],
                              df_et[missing_mask.values])
time_et = time.time() - start_time
results.append(['ExtraTrees', mse_et, time_et])

# h) IterativeImputer "KNeighbors"
start_time = time.time()
imputer_kn =
IterativeImputer(estimator=KNeighborsRegressor(n_neighbors=5),max_it
er=15, random_state=42)
df_kn = imputer_kn.fit_transform(df)
mse_kn = mean_squared_error(df_completo.values[missing_mask.values],
```

```
                                df_kn[missing_mask.values])
time_kn = time.time() - start_time
results.append(['KNeighbors', mse_kn, time_kn])

# i) IterativeImputer "BayesianRidge"
start_time = time.time()
imputer_br = IterativeImputer(estimator=BayesianRidge(),
max_iter=15, random_state=42, sample_posterior=True)
df_br= imputer_br.fit_transform(df)
mse_br= mean_squared_error(df_completo.values[missing_mask.values],
                                df_br[missing_mask.values])
time_br = time.time() - start_time
results.append(['BayesianRidge', mse_br, time_br])

# j) IterativeImputer "GradientBoostingRegressor"
start_time = time.time()
imputer_gb = IterativeImputer(estimator=GradientBoostingRegressor(),
max_iter=20)
df_gb = imputer_gb.fit_transform(df)
mse_gb = mean_squared_error(df_completo.values[missing_mask.values],
                                df_gb[missing_mask.values])
time_gb = time.time() - start_time
results.append(['GradientBoostingRegressor', mse_gb, time_gb])


## 6. Presentation of results
df_results = pd.DataFrame(results,
                              columns=['Méthod', 'MSE', 'Time
(s)']).sort_values('MSE')

print("\n Comparison of Imputation Methods:")
print(df_results.round(4))


## 7. Viewing results
plt.figure(figsize=(14, 6))

# MSE Graphic
plt.subplot(1, 2, 1)
bars = plt.barh(df_results['Méthod'], df_results['MSE'],
color='skyblue')
plt.bar_label(bars, fmt='%.3f', padding=3)
plt.title('Imputation Error (MSE)\n Lower is better')
plt.xlabel('Mean Squared Error')
plt.xlim(0, max(df_results['MSE']) * 1.1)

# Time Graphic
plt.subplot(1, 2, 2)
bars = plt.barh(df_results['Méthod'], df_results['Time (s)'],
color='salmon')
plt.bar_label(bars, fmt='%.2f s', padding=3)
plt.title('Execution Time\n Less is better')
plt.xlabel(' Seconds')
plt.xlim(0, max(df_results['Time (s)']) * 1.1)

plt.tight_layout()
```

```
plt.show()

## 8. Visualizing missingness patterns
plt.figure(figsize=(10, 6))
sns.heatmap(df.isna(), cbar=False, cmap='viridis')
plt.title('Distribution of Missing Values\n(MCAR, MAR and MNAR
patterns)')
plt.show()
```

****

Figures 2.2 and 2.3 are generated with the above code.

2.4 Normalization

Data preprocessing is fundamental to guarantee the quality of the results in cluster analysis since it depends critically on the adequate preparation of the data. Therefore, it is essential to consider the scales of measurement. A common challenge in distance-based methods (such as hierarchical clustering or k-means) is sensitivity to units of measurement; attributes with wide ranges can unfairly dominate over others, distorting the true similarity. To mitigate this, normalization (such as Z-score or Min–Max scaling) homogenizes the contributions of each feature. The choice of normalization method should consider the nature of the data, the subsequent algorithm and the presence of outliers, ensuring that the clusters reflect genuine patterns and not scale artifacts. Some methodologies for data normalization available in the sklearn library are discussed below.

2.4.1 *MinMaxScaler*

MinMax normalization is a scaling method that linearly transforms the original data to a defined interval, typically [0, 1]. It uses the minimum *min(X)* and maximum *max(X)* values of each attribute to rescale the data using Eq. 2.1 that preserves the original relationships between them.

$$X_{norm} = \frac{X - \min(X)}{\max(X) - \min(X)} \tag{2.1}$$

However, this method is sensitive to outliers, since their presence distorts the values of min*(X)* and max*(X)*, artificially compressing the rest of the data into a narrow subrange (e.g. [0, 0.006]). This means that MinMaxScaler does not eliminate the impact of extreme values but incorporates them into the range limits. It

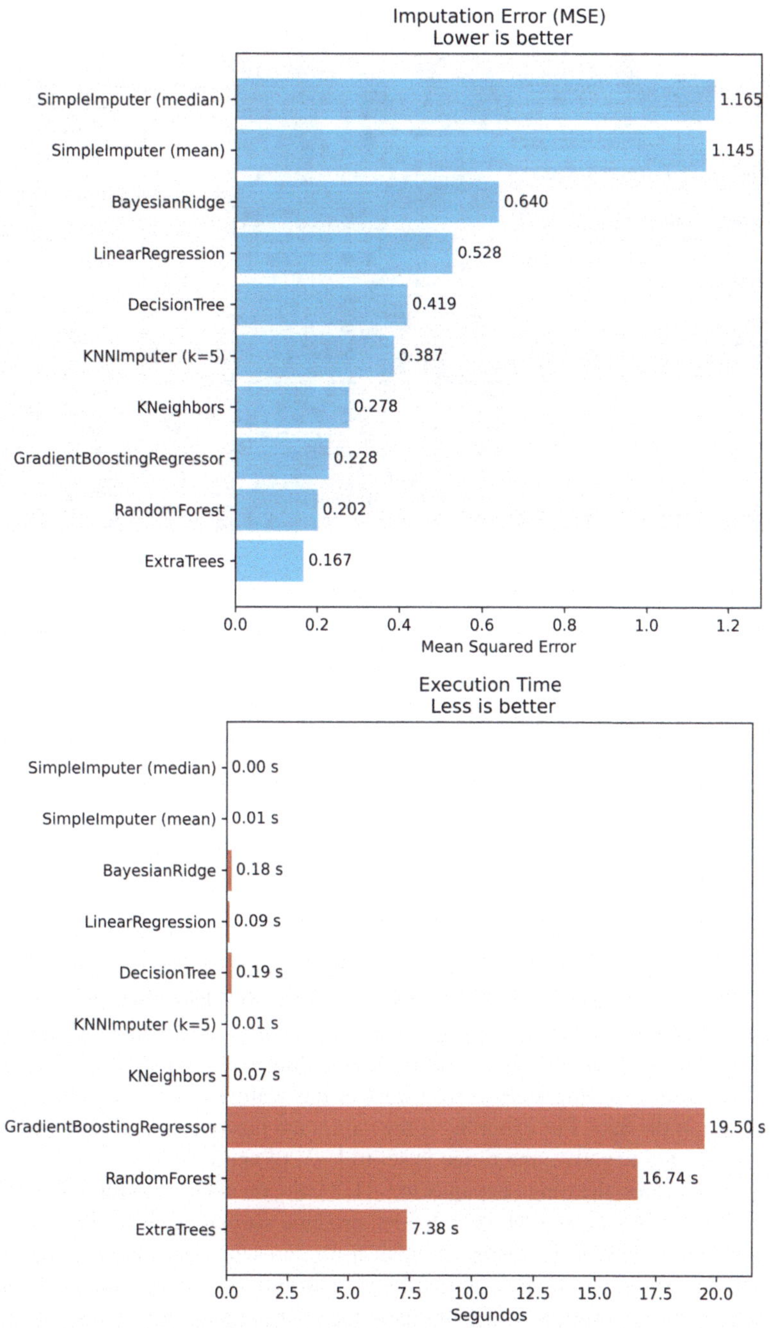

Fig. 2.2 Comparison of missing value imputation methods

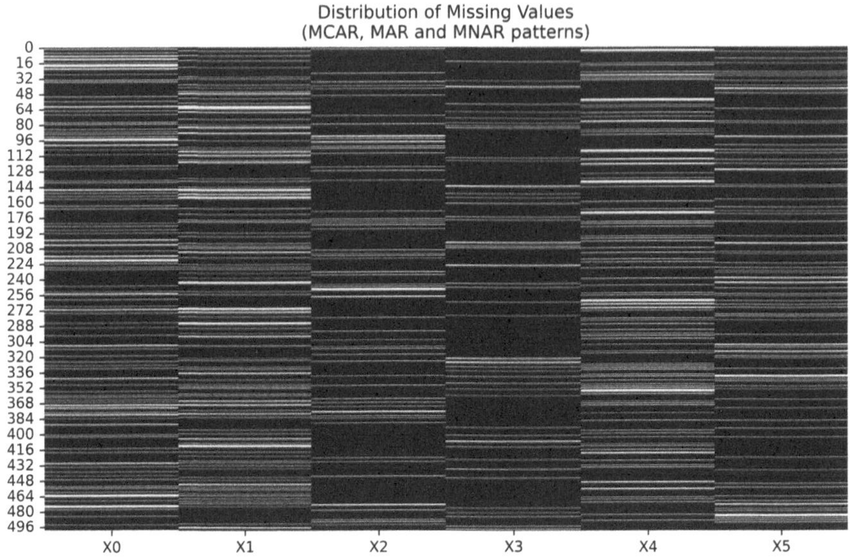

Fig. 2.3 Distribution of missing values in the variables

is therefore recommended for data without significant outliers or when a straightforward interpretation of the original scales is required, but less suitable when the distribution has heavy tails or pronounced anomalies.

2.4.2 MaxAbsScaler

MaxAbsScaler is a scaling method that normalizes each feature individually to its maximum absolute value, fitting the data so that this maximum value is 1. Unlike other scaling techniques, it does not center or shift the data, thus preserving its original dispersion. However, this approach does not mitigate the impact of outliers, as it only reduces them linearly without eliminating them. MaxAbsScaler shares similarities with MinMaxScaler, but differs in the scaling range depending on the distribution of the data. For example, if the values are positive, the resulting range is [0, 1], if they are negative, the range is set to $[-1, 0]$ and if there are both positive and negative values, the range expands to $[-1, 1]$. So, on data sets with only positive values, MaxAbsScaler and MinMaxScaler produce similar results. However, like MinMaxScaler, MaxAbsScaler is sensitive to the presence of extreme outliers, as these can distort the scaling by dominating the absolute maximum value. Therefore, the use is more appropriate when the data does not contain significant outliers or when the original dispersion structure needs to be preserved.

2.4.3 StandarScaler

Data standardization (Z-score) transforms features to have zero mean ($\mu = 0$) and unit variance ($\sigma = 1$). This approach, implemented in scikit-learn's StandardScaler, calculates the standard score for each value using Eq. 2.2.

$$X_{norm} = \frac{X - \mu}{\sigma} \tag{2.2}$$

where μ is the mean and σ the standard deviation of the attribute. By centering the data at zero and scaling it according to its dispersion, all features are made to contribute equally to scale-sensitive algorithms. However, StandardScaler is sensitive to outliers, since in the presence of outliers, both the mean and standard deviation are distorted, causing each feature to scale unequally. As a result, balanced scaling between features is not guaranteed when outliers are present, compromising the effectiveness of scaling for algorithms using normalized data. Unlike MinMaxScaler, which rescales to a fixed range, StandardScaler assumes a near-normal distribution and is ideal when compatibility with statistical or machine learning techniques that assume centered data and homogeneous variance is required. For sets with extreme values, it is recommended to combine them with previous outlier treatment techniques or to use more robust alternatives such as RobustScaler.

2.4.4 PowerTransformer

PowerTransformer is a technique that applies power transformations to data to bring its distribution closer to a Gaussian (normal) distribution, which is especially useful in cases of heteroscedasticity (non-constant variance) or when normality is required to improve the performance of statistical and machine learning models. This method employs parametric and monotonic transforms, the most common being Box-Cox and Yeo-Johnson, which automatically estimate the optimal parameter by maximum likelihood to stabilize variance and reduce skewness. A key difference between the two transformations is that Box-Cox can only be applied to strictly positive data, while Yeo-Johnson supports both positive and negative values, making it more flexible. By default, PowerTransformer normalizes the transformed data to zero mean and unit variance, facilitating its use in scale-sensitive algorithms. However, it is important to consider the limitations: if the data contain non-positive values, Box-Cox is not applicable, so Yeo-Johnson becomes the preferred alternative. This technique is especially valuable when seeking to improve the linearity of relationships between variables or to meet normality assumptions in advanced statistical models.

2.4.5 RobustScaler

RobustScaler is a technique that scales features using statistics robust to outliers. Unlike other scaling methods, such as Z-score standardization, RobustScaler eliminates the median and scales data according to the interquartile range, which is the difference between the third quartile (75%) and the first quartile (25%). This approach ensures that centering and scaling are performed independently for each feature, using statistics calculated from the data set.

The main advantage of RobustScaler is that it is based on percentiles, and this means that it is not significantly affected by a small number of extreme outliers. So, the resulting range of transformed feature values is larger compared to other scalers. It should be noted that outliers are still present in the transformed data. If it is required to remove them completely, a nonlinear transformation is necessary. This technique is especially useful in machine learning models where the presence of outliers can distort the results.

2.4.6 QuantileTransformer

QuantileTransformer is a method that transforms features to follow a uniform or normal distribution by redistributing values according to their quantiles. This technique disperses the most frequent values and mitigates the impact of outliers, thus making it a robust option for data processing. The process is performed independently for each feature: first, the cumulative distribution function (CDF) is estimated to map the original values to a uniform distribution; then, these are transformed to the desired distribution (uniform or Gaussian) using the inverse quantile function. If the data contains values outside the fitted range, they are automatically mapped to the limits of the output distribution (0 and 1 for uniform, or the ends of the normal curve for Gaussian).

Unlike RobustScaler, which only reduces the influence of outliers by robust statistics (such as the interquartile range), QuantileTransformer forcibly compresses them within the predefined limits of the output range. However, this abrupt compression can create saturation artifacts, i.e., artificial clusters of data at the edges of the range. For example, if multiple very high outliers are mapped to 1, they will lose their original variability and appear as a single repeated value. This can distort subsequent analyses, especially in models sensitive to the distribution of the data. Another important consideration is that, being a non-linear transformation, it may disrupt linear correlations between variables of the same scale, although it facilitates direct comparison between features measured at different scales. Compared to RobustScaler, both methods are robust to outliers, but QuantileTransformer offers tighter control over the final distribution of the data.

2.5 Distances

In clustering techniques, the distance metric is an essential component, since it quantifies the similarity or dissimilarity between instances of a data set. These metrics define a metric space, among the most commonly used are the Euclidean distance, the Manhattan distance, the Chebyshev distance, and the Minkowski distance, each with different properties depending on the context.

Cluster analysis groups multivariate data using a proximity index, commonly based on distance measures. However, the semantic interpretation of these measures can vary even with identical numerical data. For example, in scaled features, the arithmetic mean can be used to represent sets of values, but the choice of metric influences the results. In algorithms such as k-means, the chosen metric (e.g., Euclidean distance) determines how points are assigned to clusters and how their centroids are updated. However, this method requires a priori knowledge of the number of clusters and is sensitive to initialization. Moreover, the selection of features, metrics, and algorithms can generate completely different results, as highlighted by Theodoridis et al. (2010). In the following, some commonly used distances in clustering are discussed.

2.5.1 *Minkowski Distance*

The Minkowski distance is a generalization of various distance metrics in f-dimensional spaces (Han et al., 2022). For two points $G\left(g_1, g_2, ..., g_f\right)$ y $H\left(h_1, h_2, ..., h_f\right)$ (e.g., a data point and the centroid), it is defined by Eq. 2.3.

$$d_p(G, H) = \left(\sqrt{\sum_{i=1}^{f} |g_i - h_i|^p}\right)^{1/p} \tag{2.3}$$

where p is a parameter that determines the type of distance.

2.5.2 *Manhattan Distance*

Manhattan distance ($p = 1$, Norm *L1*) is the sum of absolute differences (Eq. 2.4). It is used in robust analysis to outliers.

$$d_1(G, H) = \sum_{i=1}^{f} |g_i - h_i| \tag{2.4}$$

2.5.3 Euclidean Distance

Euclidean distance ($p = 2$, Norm $L2$) is the root of the sum of squares (Eq. 2.5). The Euclidean distance is widely used in clustering methods as it is ideal for continuous spaces (e.g., k-means clustering).

$$d_2(G, H) = \sqrt{\sum_{i=1}^{f} (g_i - h_i)^2} = \sqrt{(g_1 - h_1)^2 + (g_2 - h_2)^2 + \ldots + \left(g_f - h_f\right)^2} \tag{2.5}$$

2.5.4 Chebyshev Distance

Chebyshev distance ($p \to \infty$, Norm $L\infty$) is the maximum difference between coordinates (Eq. 2.6). It is useful in optimization problems and grid motions.

$$d_\infty(G, H) = \max_i |g_i - h_i| \tag{2.6}$$

The Euclidean distance is more sensitive to outliers than the Manhattan and Chebyshev distance. These distances are fundamental in clustering, classification and pattern search algorithms. The choice depends on the nature of the data and the objective of the analysis. However, it is recommended to normalize the data before applying these metrics to balance the weight of the variables and avoid biases due to different scales.

In addition to traditional metrics such as Euclidean or Manhattan distance, there are other important measures to quantify the similarity between data, including some that are not strictly metric. For example, Pearson's correlation coefficient, which measures the linear relationship between variables, and cosine similarity, which assesses the angular orientation between vectors, regardless of magnitude. These measures are particularly useful in high-dimensional spaces and for normalized data. In cases where features are correlated, the Mahalanobis distance is more appropriate, as it incorporates the covariance structure of the data to adjust the weights of the dimensions.

2.5.5 Mahalanobis Distance

The Mahalanobis distance measures the distance between a point and a cluster (or centroid μ), considering the covariance structure of the data (Eq. 2.7). It is ideal for non-spherical clusters or clusters with correlations between dimensions.

$$d_M(G,\mu) = \sqrt{(G-\mu)^T \Sigma^{-1}(G-\mu)} \tag{2.7}$$

where Σ^{-1} is the inverse of the cluster covariance matrix.

2.5.6 *Cosine Similarity and Cosine Distance*

These metrics measure the similarity between two vectors based on the angle they form, ignoring their magnitude. They are useful in clustering text, images or sparse data (Wierzchoń & Kłopotek, 2018). Cosine similarity using the angle formed between two vectors G and H (Eq. 2.8).

$$sim_{\cos}(G,H) = \cos(\theta) = \frac{G \cdot H}{\|G\|\|H\|} = \frac{\sum_{i=1}^{f} g_i h_i}{\sqrt{\sum_{i=1}^{f} g_i^2}\sqrt{\sum_{i=1}^{f} h_i^2}} \tag{2.8}$$

where the range of values is $[-1, 1]$ where 1 represents identical vectors, 0 represents orthogonal (unrelated) vectors, and -1 represents opposite vectors. It is ideal for high-dimensional spaces. Meanwhile, the cosine distance is derived from cosine similarity to become a measure of dissimilarity (Eq. 2.9).

$$d_{\cos}(G,H) = 1 - sim_{\cos}(G,H) \tag{2.9}$$

where the range of values is [0, 1], with 0 representing identical vectors and 1 representing completely different vectors. The cosine distance is used in algorithms such as k-means, DBSCAN, and hierarchical clustering.

2.5.7 *Pearson Correlation Coefficient and Pearson Correlation Distance*

The Pearson correlation coefficient (r) measures the linear relationship between two f-dimensional vectors $G(g_1, g_2, \ldots, g_f)$ and $H(h_1, h_2, \ldots, h_f)$ (which can be data points or cluster centroids) (Eq. 2.10).

$$r(G,H) = \frac{\sum_{i=1}^{f} (g_i - \overline{g})\left(h_i - \overline{h}\right)}{\sqrt{\sum_{i=1}^{f} (g_i - \overline{g})^2}\sqrt{\sum_{i=1}^{f} \left(h_i - \overline{h}\right)^2}} \tag{2.10}$$

The range of values is $[-1, 1]$, with 1 representing a positive perfect linear correlation, 0 indicating no linear correlation, and -1 representing a negative perfect linear correlation. The Pearson correlation distance is derived from the Pearson correlation coefficient (r), which is used in clustering to measure dissimilarity between vectors based on their linear relationship (Eq. 2.11).

$$d_r(G, H) = 1 - r(G, H) \tag{2.11}$$

where the range is [0, 2], with 0 representing a perfect positive correlation ($r = 1$), 1 indicating no linear correlation ($r = 0$), and 2 representing a perfect negative correlation ($r = -1$). In clustering, it can be used to measure similarity between points (instead of distances), evaluate the cohesion of a cluster (correlation between its members), and compare clusters (e.g., correlation between centroids).

The selection of the distance metric is not trivial as it defines the notion of "proximity" in the clustering, directly influencing the quality of the clusters generated, and must be adapted to the type of data and objective of the analysis. For example, the Euclidean distance stands out for its intuitive geometric interpretation and its effectiveness in spaces with compact distributions, while the Manhattan distance is more robust to outliers, and the Mahalanobis distance considers correlations between variables. Therefore, understanding its impact is key to obtain meaningful clusters.

2.5.8 The Dynamic Time Warping (DTW)

DTW is an advanced time alignment technique that compares sequences of data using a dynamic programming approach (Keogh & Kasetty, 2002). The main advantage lies in the ability to align similar patterns that may have variations in speed or duration, which makes it particularly useful in applications such as speech recognition, motion analysis or industrial diagnostics. The algorithm works by calculating a cost matrix between all the points of both sequences and then determining the optimal deformation path that minimizes the accumulated distance, respecting three fundamental conditions: boundaries (alignment of extremes), monotonicity (preservation of temporal order) and step size (restriction of displacements).

Unlike rigid metrics such as Euclidean distance, DTW allows comparing series of different lengths and compensating for temporal shifts, mainly evaluating similarity in the shape of patterns (Fig. 2.4) (Keogh & Ratanamahatana, 2005). However, its quadratic computational complexity limits its application in very large data sets, where optimization techniques such as constraint windows or approximations are required. In time series clustering, DTW outperforms conventional methods by capturing structural similarities independent of local time variations, although its performance depends critically on an appropriate selection of the local cost function and alignment constraints.

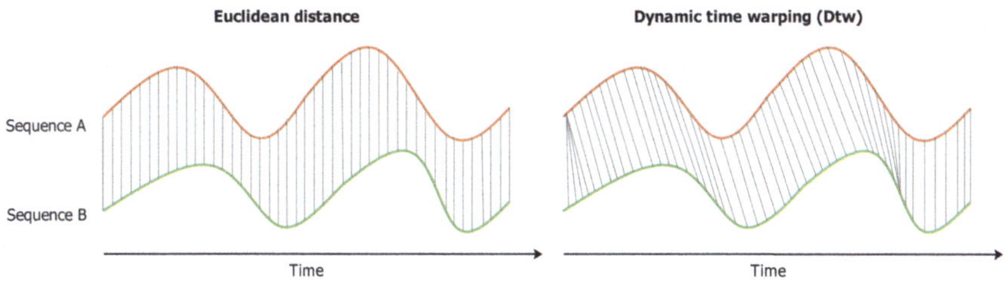

Fig. 2.4 Comparison between Euclidean distance and Dtw with two data sequences

2.6 Determination of the Optimal Number of k Clusters

The identification of the optimal number of clusters is a critical step in clustering analysis, since many algorithms, such as k-means, require this parameter as input. An incorrect choice can lead to misinterpretation of the data structure, such as underestimation of relevant clusters. To address this challenge, internal validation metrics, which assess the quality of clusters without external information, are employed. Among the most commonly used is the Sum of Squared Errors (SSE), which measures cluster compactness: a low value indicates greater cohesion within clusters. However, SSE alone is not always sufficient, so it is complemented by indices such as the Davies-Bouldin index, which considers both internal homogeneity and separation between clusters, favoring configurations with a lower Davies-Bouldin index. Other key indicators are the Calinski-Harabasz Index, where a high value suggests good separation between clusters, and the Silhouette Index, which ranges from -1 to 1, with a value close to 1 being preferable. To determine the optimal number, methods such as the elbow method (based on the marginal reduction of the SSE with increasing k), silhouette analysis (which maximizes the overall silhouette index), or gap statistics (which compare the observed dispersion with a reference distribution) are applied. These approaches allow selecting the optimal value of k that balances compactness and separation, ensuring a robust and meaningful clustering for the analyzed data set. In the following, different methods for determining the optimal number of clusters are discussed.

2.6.1 Elbow Method

The elbow method is a graphical technique widely used to determine the optimal number of clusters (k) in clustering algorithms such as k-means. This method is based on analyzing the Intra-Cluster Sum of Squares (WCSS) also called Sum of Squared Errors (SSE), which measures the dispersion of the points within each cluster (Ci) with respect to its centroid (μ_i). The central idea is to select the value of k at which the reduction of the WCSS starts to stabilize, forming a curvature similar to a "elbow" in the graph. Mathematically, the WCSS is calculated as the sum of the squared distances between each point and its corresponding centroid, expressed with Eq. 2.12.

$$WCSS = \sum_{i=1}^{k} \sum_{x \in C_i} \|x - \mu_i\|^2 \tag{2.12}$$

where x represents the cluster points C_i and μ_i its centroid. The procedure plots WCSS (y-axis) for different values of k (x-axis). Where optimal k is identified, the curve presents a pronounced break (the "elbow"), indicating that adding more clusters does not significantly improve data compactness. However, this method has limitations: the choice of the elbow can be subjective in cases where the curve does not have a clear inflection point, and it is not always applicable to data sets with complex structures. Despite this, the simplicity and visual approach make it a fundamental tool for internal validation in clustering, complementing it with other metrics such as the silhouette index or the Davies-Bouldin for a more robust decision.

2.6.2 *Silhouette Method*

The silhouette method is a widely used technique for evaluating the quality of clusters and determining the optimal number of groups k in clustering algorithms (Rousseeuw, 1987). This method is based on the silhouette index or coefficient, which measures how well each point fits its cluster compared to neighboring clusters. For each point j, the coefficient is calculated with Eq. 2.13.

$$SC(j) = \frac{b(j) - a(j)}{\max\{a(j), b(j)\}} \tag{2.13}$$

where $a(j)$ represents the average distance between point j and all other points within the same cluster (cohesion), and $b(j)$ is the average distance between j and the points of the nearest cluster (separation). The value of $SC(j)$ ranges from -1 to 1, where a value close to 1 indicates that the point is well assigned to its cluster, 0 suggests that it is close to the boundary between two clusters, and -1 signals a possible misassignment. The average silhouette coefficient, obtained by averaging the silhouette coefficients of all points, provides an overall measure of clustering quality, where a high value indicates well-defined and separated clusters. In addition, the graphical representation of the coefficients allows for visualizing the internal consistency of each cluster and the distance between them, facilitating the identification of the optimal k value. Unlike the elbow method, the silhouette method not only evaluates the compactness of the clusters, but also their separation, which makes it a more robust tool for validating clustering structures, especially in complex data sets. However, sensitivity to outliers and higher computational cost are limitations to consider.

2.6.3 Gap Statistics Method (GAP)

The gap statistics method efficiently estimates the optimal number of clusters in a data set. This versatile approach can be applied to any clustering technique and distance measure, being particularly useful with the K-means algorithm. The method compares the dispersion (sum of Euclidean distances between points and their centroids) of the real data set with multiple reference sets generated by Monte Carlo sampling from a null distribution (Tibshirani et al., 2001). The gap statistic for each number of clusters k is calculated as the logarithmic difference between the expected dispersion of the reference data and the dispersion observed in the original data, following Eq. 2.14.

$$Gap(k) = \mathrm{E}_n^*\left[\log\left(W_k^*\right)\right] - \log(W_k) \tag{2.14}$$

where W_k represents the sum of intra-cluster distances (also called sum of intra-cluster squares or intra-cluster inecia “WCSS”) of the original data with k clusters. $\mathrm{E}_n^*\left[\log\left(W_k^*\right)\right]$ is the expected (average) value of the logarithm of the intra-cluster dispersion (${W_k}^*$) in the randomly generated reference data. The symbol * indicates that it comes from reference data and not from the original data, while n is the sample size. The optimal number of clusters is selected as the smallest value of k where *Gap(k)* $\geq$ *Gap(k + 1)–SE(k + 1)*, *SE* being the standard error also called standard deviation. This criterion identifies the point where the actual clustering structure deviates significantly from what is expected in random data, thus maximizing the quality of the clustering. Unlike methods such as the elbow or silhouette method, the gap statistic provides an objective assessment by directly comparing the results to a reference distribution, making it particularly robust for complex data sets where other methods may fail to clearly identify the optimal number of clusters.

2.6.4 Calinski-Harabasz Index

The Calinski-Harabasz Index (CHI), also known as the variance ratio criterion (VRC), is an internal evaluation metric for assessing the quality of clustering algorithms. This index quantifies the ratio of inter-cluster separation (BCSS) to intra-cluster dispersion (WCSS), normalized by degrees of freedom (Caliński & Harabasz, 1974). It is calculated using Eq. 2.15.

$$CHI = \frac{BCSS/(k-1)}{WCSS/(n-k)} = \frac{\sum_{i=1}^{k} n_i \|\mu_i - \mu\|^2/(k-1)}{\sum_{i=1}^{k} \sum_{x \in C_i} \|x - \mu_i\|^2/(n-k)} \tag{2.15}$$

where k is the number of clusters, n is the total number of points, n_i is the number of points in cluster, μ_i is the centroid of cluster C_i, and μ is the global centroid of the data set (global mean). A high CHI indicates well-defined and separate clusters, as it reflects high variance between clusters and low dispersion within clusters. In Scikit-learn, it is implemented with sklearn.metrics.calinski_harabasz_score, allowing to compare models or to select the optimal number of clusters. It is especially useful for centroid-based algorithms (such as k-means), where minimizing WCSS is a key objective as it is ideal for validating partitions where the data geometry is spherical and separable by centroids, providing an intuitive measure of clustering quality. However, it tends to favor a larger number of clusters, so it should be combined with other metrics (such as silhouette or Davies-Bouldin) for robust validation.

2.6.5 *Davies-Bouldin Method*

The Davies-Bouldin Index (DBI) is an internal validation metric used to evaluate the quality of clustering algorithms. This index measures the average similarity between each cluster and its most similar cluster, where similarity is defined as the ratio of the intra-cluster distance (dispersion of points within a cluster) to the inter-cluster distance (separation between clusters) (Davies & Bouldin, 2009). Mathematically, the DBI is calculated with Eq. 2.16.

$$DBI = \frac{1}{k}\sum_{i=1}^{k}\max_{j\neq i}\left(\frac{s_i - s_j}{d\left(c_i, c_j\right)}\right) \tag{2.16}$$

where k is the number of clusters, s_i is the average dispersion of the points in cluster i (distance to the centroid), and $d\ (c_i,\ cj)$ is the distance between the centroids of clusters i and j. When the value of DBI is close to 0 it indicates a better partitioning, with compact and well-separated clusters. Since it reflects greater internal homogeneity (little dispersion within clusters) and greater separation between clusters. In the Scikit-learn library, the Davies-Bouldin index is implemented by the sklearn.metrics.davies_bouldin_score function, being an effective tool for comparing clustering models and selecting the optimal number of clusters, especially useful when seeking to minimize the overlap between clusters. Its simplicity and focus on the cohesion-separation relationship make it a popular choice for internal validation in algorithms such as k-means or DBSCAN.

2.7 Clustering Algorithms

Clustering is a process that seeks to organize data sets into homogeneous groups, where points within a cluster are more similar to each other than to those in other groups (Witten & Frank, 2005). Similarity is commonly measured by distances in multidimensional spaces. Figure 2.5 shows a classification of clustering algorithms,

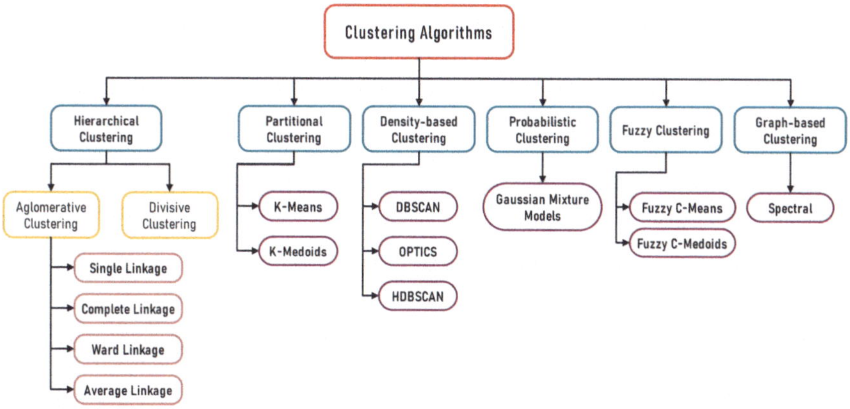

Fig. 2.5 Taxonomy in clustering

organized by fundamental criteria and based on their mathematical principles and practical applications.

2.7.1 *Partitional Algorithms*

Partitional clustering algorithms divide a set of objects into k groups, so that the elements within each group are as similar as possible to each other, and at the same time, different from the elements belonging to other groups. This type of clustering is posed as an optimization problem, since it seeks to organize the objects into k clusters by minimizing the intra-cluster variance (WCSS or SSE) using centroid (k-means) and medoid (k-medoids) prototypes, which is a previously defined criterion function. The clustering process can be terminated when a local optimum is reached or after a given number of iterations. They are efficient but require specifying the number of k clusters a priori. In the following, partitional algorithms will be explored, starting with the most representative of this approach: k-means.

2.7.1.1 K-Means

The K-means algorithm is a prototype-based partitional clustering method that divides a data set into k homogeneous groups. As one of the simplest and most widely used clustering algorithms, however, it requires the user to pre-specify the number of k clusters for which prior knowledge or auxiliary methods (elbow method, silhouette, among others) are necessary. Its main objective is to find k centroids, which are prototype points representing each cluster (i.e. the average of the group) and assign each data to the nearest centroid, thus forming compact partitions. The stages of the algorithm are shown below.

1. Initialization: k points are randomly selected as initial centroids.
2. Assignment: Each data point is associated with the nearest centroid using Euclidean distance (the commonly used metric) (Kotu & Deshpande, 2019).
3. Update: The centroids are recalculated as the average of all points assigned to the cluster.
4. Convergence: Steps 2–3 are repeated until the centroids stabilize, i.e., they do not change significantly.

Mathematically, K-means minimizes the Sum of Squared Errors (SSE) or also known as intra-cluster sum of squares (Eq. 2.17).

$$SSE = \sum_{i=1}^{k} \sum_{x_j \in C_i} \left\| x_j - \mu_i \right\|^2 \tag{2.17}$$

where μ_i is the centroid of the cluster C_i. While x_j is a data point. The K-Means algorithm is implemented in Python using the sklearn.cluster.KMeans module of the scikit-learn library. Among the advantages of k-means are its simplicity, efficiency, Interpretability as the centroids provide clear representation of each cluster and Versatility as it is applicable to various domains. K-means has a sensitivity in initialization because it can converge to local optima so multiple random restarts or smart initialization (k-means++) is necessary. Where k-means++ is an initialization scheme designed to improve the performance of k-means by selecting initial centroids distant from each other, thus reducing the risk of convergence to suboptimal local optima (Arthur & Vassilvitskii, 2006). Unlike random initialization, this method uses a probabilistic approach that favors points far from the already selected centroids, which generally produces more consistent and efficient results. However, its sequential nature implies a higher computational cost, as it requires multiple passes over the data set to select the k initial centroids. In addition, k-means++ is not robust to outliers, as these can distort centroid selection. Although k-means always converges, its success depends critically on this initialization, so in practice multiple runs with different random seeds are often performed.

The shape of the clusters is spherical and isotropic, so it is limited for complex shapes. Moreover, it is sensitive to outliers because they distort the position of the centroids. If categorical or mixed (numerical and categorical) data are used, adaptations such as k-prototypes can be used. Thus K-means remains a fundamental tool in clustering due to its balance between simplicity and effectiveness, although its application must carefully consider its inherent limitations. For complex problems, its variants or alternative algorithms (such as DBSCAN or GMM) may offer better results.

2.7.1.2 K-Medoids (PAM—Partitioning Around Medoids)

The K-medoids method is a partitional clustering technique that, unlike K-means, represents each cluster by a medoid (a real point in the data set) instead of a centroid (arithmetic mean) (Kaufman & Rousseeuw, 2009) (Fig. 2.6). This feature makes it more robust to noise and outliers, as medoids are less sensitive to extreme values than averages. However, this advantage comes at a higher computational cost, especially in large data sets. Like K-means, it requires the user to specify the number of k clusters beforehand. The steps are similar to those of k-means as shown below.

1. Initialization: k objects are randomly selected as initial medoids.
2. Assignment: Each point is assigned to the nearest medoid, using metrics such as Euclidean or Manhattan distance (the latter is more resistant to outliers).
3. Optimization: Trade-offs between medoids and non-medoids are evaluated to minimize the sum of dissimilarities (objective function). If a trade-off reduces the total dissimilarity, the medoid is updated.
4. Convergence: Repeat steps 2 and 3 until there are no significant improvements.

k-medoids is more accurate and robust to outliers or non-Gaussian distributions; however, it has a higher computational cost compared to k-means (Subasi, 2020). The standard K-Medoids algorithm selects the initial medoids randomly, which can lead to convergence to local optima, sensitivity to outliers, and inconsistent results between runs. To mitigate this, one can use K-Medoids++, which chooses initial medoids further apart, which generally produces more stable and meaningful clusters, thus avoiding suboptimal solutions. It is analogous to K-Means++, but adapted to work

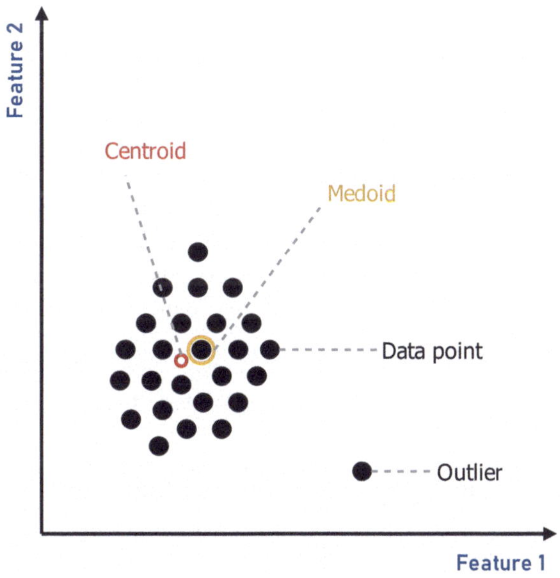

Fig. 2.6 Difference between centroids and medoids

with medoids (actual dataset points) instead of centroids (averages). To implement K-Medoids (including K-Medoids++ initialization) in Python, the library used is scikit-learn-extra, which is officially compatible with scikit-learn.

Medoids are a powerful tool for identifying real, representative examples within a temporal data set, such as daily or weekly records. Unlike averages (which may not exist in real data), medoids select the most typical point in each group, making them ideal for applications such as identifying typical days in electricity consumption, web traffic or sales; finding representative weeks in weather or industrial production data; and reducing temporal data without losing key patterns.

2.7.2 Density-Based Algorithms

2.7.2.1 DBSCAN (Density-Based Spatial Clustering of Applications with Noise)

DBSCAN is an algorithm that identifies clusters as regions of high density separated by areas of low density, allowing the detection of arbitrarily shaped groups (Ester et al., 1996). Unlike methods such as k-means, DBSCAN does not require prior specification of the number of clusters, which makes it a valuable tool for data exploration. The algorithm is based on two key parameters: ε (epsilon), which defines the neighborhood radius, and MinPoints, the minimum number of points required to form a dense region (Fig. 2.7). Points are classified into core points that have at least MinPoints within their ε neighborhood (located in dense areas), border points that have at least one core point within their ε neighborhood (at the boundaries of a cluster), or noise points that have neither core points nor border points in their neighborhood (isolated in low-density areas). DBSCAN iterates by connecting reachable core points by density, thus forming clusters. The DBSCAN algorithm is available in the sklearn.cluster.DBSCAN module of the scikit-learn Python library. A notable advantage is its ability to handle noise and outliers, although the proper choice of ε and MinPoints is critical, as incorrect values may subdivide valid clusters or join them incorrectly. Moreover, its scalability and flexibility to work with multidimensional data make it ideal for applications in data mining and pattern recognition (Theodoridis & Koutroumbas, 2009). However, its performance depends on the relative density of the data and can be affected by significant variations in density.

2.7.2.2 OPTICS (Ordering Points to Identify the Clustering Structure)

The OPTICS algorithm is an advanced extension of DBSCAN designed to overcome its main limitation: the dependence on a fixed distance parameter (ε) (Ankerst et al., 1999). Unlike DBSCAN, which requires a single value of ε to define dense regions, OPTICS processes multiple distances simultaneously, generating a reachability order that captures the hierarchical structure of the clusters. This order is complemented

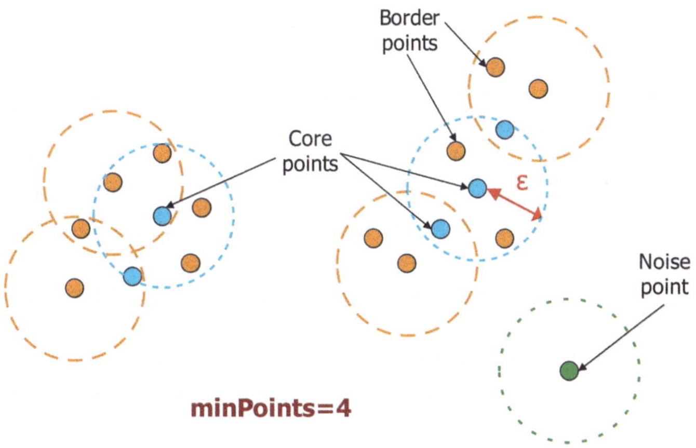

Fig. 2.7 Types of points in DBSCAN

by two key metrics: the core distance, which indicates the local density around each point, and the reachability distance, which reflects the minimum proximity for a point to be considered part of a cluster.

OPTICS operates in three main stages: (1) it calculates the processing order of points, prioritizing those with higher densities (small ε values); (2) it generates a reachability diagram, where valleys represent clusters and ridges separate regions of different density; and (3) it allows to extract clusters a posteriori for any ε' less than a predefined maximum ε, offering flexibility to adjust granularity. This capability makes it ideal for datasets with varying densities, such as in spatial analysis or anomaly detection, where DBSCAN would fail using a single threshold. In addition, OPTICS maintains similar computational complexity to DBSCAN, albeit with minimal additional cost for distance storage. However, its output does not map clusters directly, but requires diagram interpretation or auxiliary methods such as cluster_optics_dbscan to obtain specific partitions. In summary, OPTICS generalizes DBSCAN by providing a hierarchical and adaptive view of density, eliminating the need to specify ε rigidly and facilitating exploratory analysis of complex data. In Python, the OPTICS algorithm is available in the sklearn.cluster.OPTICS module of the scikit-learn library.

2.7.2.3 HDBSCAN (Hierarchical Density-Based Spatial Clustering of Applications with Noise)

HDBSCAN is an advanced extension of DBSCAN and OPTICS that overcomes the limitation of assuming homogeneous density in the data. From DBSCAN it uses the principle of local density (core points, borders and noise) but eliminates the need for a fixed ε, as it explores multiple density scales (Campello et al., 2013). While de

OPTICS adopts the idea of hierarchical reachability, but instead of generating a reachability graph, it builds a hierarchy of clusters using a minimum spanning tree (MST). This tree models density relationships at different levels, followed by a condensation that extracts the most stable clusters. The main advantage of HDBSCAN is its ability to automatically detect clusters with varying densities, eliminating the need to specify critical parameters such as ε (although it retains min_samples to define local density). In addition, it assigns each point a cluster membership score (soft clustering), which allows distinguishing between central and peripheral elements, as well as identifying noise in a more robust way. Its output includes a hierarchical structure that can be analyzed at different scales, making it ideal for complex datasets where natural groups vary in density and shape. However, its computational cost is higher than that of DBSCAN due to the construction of the hierarchy, although it is still efficient for practical applications. So HDBSCAN combines the best of DBSCAN and OPTICS, offering flexible, adaptive and less parameter-dependent clustering, ideal for domains such as bioinformatics, spatial analysis or anomaly detection. The HDBSCAN algorithm is available in the sklearn.cluster.HDBSCAN module of the scikit-learn library in the Python language.

2.7.3 Hierarchical Algorithms

Hierarchical clustering is a technique that organizes data into nested structures using two main approaches: agglomerative (bottom-up) and divisive (top-down) (Fig. 2.8). In the agglomerative method, each data point starts as an individual cluster, and in each iteration the two most similar clusters are merged until a single cluster encompassing all the data is formed. In contrast, the divisive approach starts with a single cluster containing all the points and divides them recursively until each point forms its own cluster. Both methods generate a dendrogram which is a binary tree visualizing the cluster hierarchy, where the height of the branches represents the dissimilarity between groups. This representation allows cutting the tree at different levels to obtain partitions with different numbers of clusters, which provides flexibility in the exploration of underlying structures.

Agglomerative algorithms differ in how they calculate the similarity between clusters (Wilks, 2011) (Fig. 2.9):

- Simple linkage (nearest neighbor): uses the minimum distance between points in different clusters. Tends to produce elongated clusters.
- Full linkage (farthest neighbor): Uses the maximum distance, favoring compact but small clusters.
- Average link: Calculates the average distance between all pairs of inter-cluster points, balancing between the above methods.
- Centroid link: Calculates the distance between their centroids (centers of gravity).
- Ward's method: Minimizes the intra-cluster variance, similar to K-means but in a hierarchical framework, ideal for spherical, compact and well separated clusters.

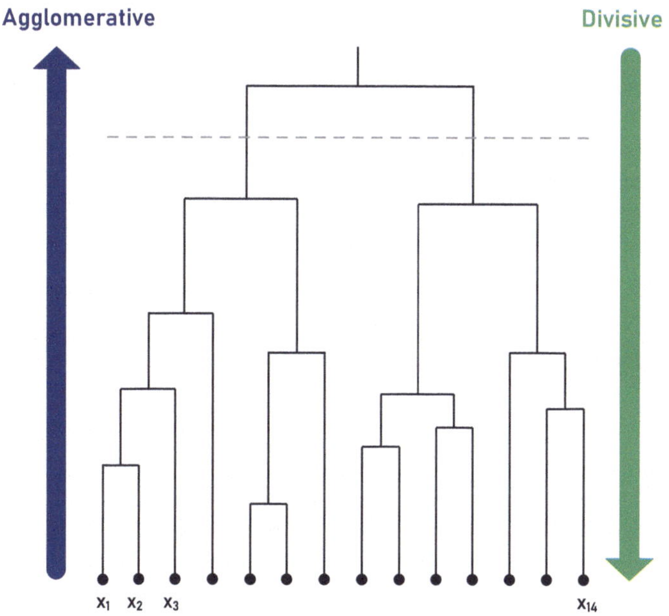

Fig. 2.8 Agglomerative and divisive clustering

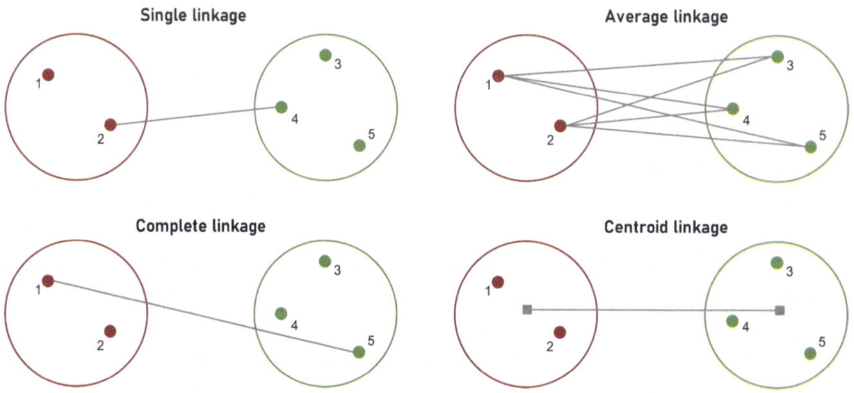

Fig. 2.9 Types of links in hierarchical clustering

One of its strengths is that it is not required to specify the number of clusters a priori. It also provides an intuitive visualization through dendrograms and captures hierarchical relationships in data with nested structures (e.g., biological taxonomies). However, it faces limitations such as having a high computational cost for large datasets; sensitivity to outliers especially when using single linking; and Irreversibility as merges or splits are definitive, which can lead to suboptimal solutions. In the Python language the agglomerative hierarchical algorithm is available in the sklearn.cluster.AgglomerativeClustering module of the Scikit-learn library, where Euclidean, Manhattan or cosine distance can be used, in addition to offering linking criteria such as Ward, average, complete or single. In practice, agglomerative methods are more popular due to their efficient implementation (e.g., using distance matrices). For large sets, it is recommended to combine hierarchical clustering with partitional techniques (e.g., initial K-means) to reduce computational costs. While divisive hierarchical methods are less common due to their complexity, but useful when priority is given to identify initial macro-clusters.

2.7.4 BIRCH

The BIRCH (Balanced Iterative Reducing and Clustering using Hierarchies) algorithm, developed by Zhang, Ramakrishnan and Livny in 1996, is a highly efficient solution for hierarchical clustering of large volumes of data, especially designed for environments with limited memory resources. Its main strength lies in its ability to process information in an incremental and scalable manner, making it an ideal choice for big data applications and continuous data streams. The algorithm operates in two fundamental phases: in the first phase, it builds a CF-Tree (Clustering Feature Tree), a hierarchical structure that summarizes the data into compact subgroups called Clustering Features (CF). Each CF stores three key statistics: the number of points in the group (N), the linear sum of their coordinates (LS) and the sum of their squares (SS), allowing the distribution and dispersion of the data to be represented efficiently without the need to store each point individually. This compression phase drastically reduces the size of the dataset, while preserving its essential structure.

In the second phase, the algorithm applies traditional clustering techniques (such as k-means or hierarchical methods) on the CF-Tree instead of the original data, significantly speeding up the process without sacrificing accuracy. In addition, BIRCH includes an optional refinement phase to readjust clusters and remove outliers. Among its advantages are its speed (linear complexity in many cases), low memory consumption, and robustness to noise. However, it has limitations: it is only applicable to numerical data, its performance depends critically on parameters such as the distance threshold (T) and the branching factor (B), and it does not handle irregularly shaped clusters well. Despite these restrictions, BIRCH is widely used in market segmentation, anomaly detection, and as a preprocessing stage for other algorithms, demonstrating its versatility in large-scale data analysis. Its innovative approach to intelligent information compression positions it as a valuable tool in the field of machine learning and data mining.

2.7.5 Algorithms Based on Probabilistic Models

2.7.5.1 Gaussian Mixture Models

The Gaussian Mixture Model (GMM) is a probabilistic clustering algorithm that assumes that data are generated from a combination of multiple Gaussian distributions, each representing a cluster (Pearson, 1894; Shanmugam, 2009). Unlike methods such as K-means, which assign each point to a single cluster rigidly, GMM uses a soft approach, calculating the probability that a point belongs to each cluster using parameters such as the mean (μ), the covariance matrix (Σ, which defines the shape of the cluster) and the mixing coefficients (π, which indicate the relative weight of each distribution). The algorithm is iteratively optimized by means of Expectation–Maximization (EM): in the E-step, the membership probabilities of each point to the Gaussians are estimated, and in the M-step, the parameters are updated to maximize the likelihood of the data. This flexibility allows modeling clusters with elliptical or overlapping shapes, overcoming the limitations of methods based on Euclidean distances. However, GMM requires a priori specification of the number of clusters, and its performance depends on the data following an approximately Gaussian distribution. In addition, its computational complexity is higher due to the inversion of covariance matrices, especially in high-dimensional spaces. The GMM algorithm is available in the sklearn.mixture.GaussianMixture module in the Scikit-learn Python language library.

2.7.6 Fuzzy Clustering Algorithms (Fuzzy)

2.7.6.1 Fuzzy C-Means

Fuzzy clustering is a flexible approach that, unlike traditional rigid clustering methods (such as K-means), allows each data object to belong to multiple clusters with varying degrees of membership, represented by values in the interval [0, 1] (Bezdek et al., 1984), as shown in Fig. 2.10. This approach is especially useful when the boundaries between clusters are not well defined or when the data have natural overlaps (Saxena et al., 2017). The best-known algorithm in this category is Fuzzy C-Means (FCM), developed by Bezdek in 1981, which generalizes K-means by minimizing a cost function based on the weighted sum of distances between points and cluster centers, where the weights reflect degrees of membership. Unlike traditional K-means, which assigns each point exclusively to a cluster (binary membership), FCM uses a fuzzy partition matrix, where the sum of the memberships of a point to all clusters must be equal to 1. This allows modeling ambiguous situations, such as points located in intermediate zones between clusters. However, FCM requires a priori knowledge of the number of clusters and is sensitive to the initialization of the centers, and can converge to local minima. The Fuzzy C-Means algorithm is available in the

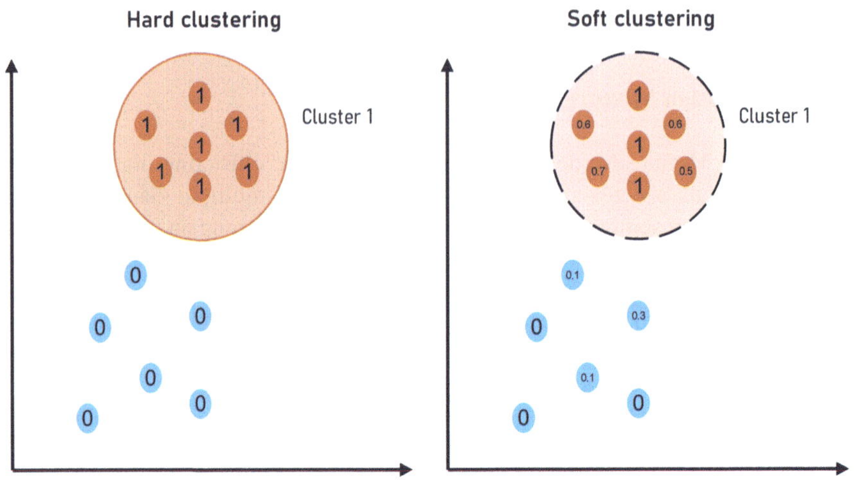

Fig. 2.10 Difference between hard and soft clustering

skfuzzy.cluster.cmeans module of the skfuzzy library in the Python language. Other variants, such as Fuzzy C-Medoids (FCMdd), select medoids (real points in the data set) as centers, which improves robustness against outliers. Although these methods overcome the limitations of rigid algorithms by handling non-compact cluster shapes (such as shells), their performance depends critically on the choice of parameters and the suitability of the fuzzy membership model to the underlying structure of the data.

2.7.7 *Network/Graph-Based Algorithms*

2.7.7.1 Spectral

Spectral Clustering is an advanced clustering algorithm based on graph theory and linear algebra, especially effective for identifying clusters with non-convex shapes or complex structures (such as rings or spirals) where traditional methods such as K-means fail (Von Luxburg, 2007). It is available in the sklearn.cluster.SpectralClustering module of the scikit-learn library in Python. Its operation is based on four key steps: (1) construction of an affinity matrix (using kernels such as the Radial Basis Function) that quantifies the similarities between points; (2) computation of the Laplacian matrix of the graph to capture the underlying connectivity; (3) spectral decomposition to extract the k smallest eigenvectors, which project the data to a low-dimensional space where the clusters are linearly separable; and (4) application of K-means on these eigenvectors to assign the final clusters. Unlike approaches based on Euclidean distances, Spectral Clustering exploits global relationships between points. However, its high computational cost (due to the computation of eigenvectors in $n \times n$ matrices) limits its use to medium-sized datasets or those preprocessed with techniques such as PCA. Moreover, it requires a priori specification of the number of clusters, and its performance depends critically on the choice of the affinity kernel. To improve efficiency in sparse matrices, the Algebraic Multigrid (AMG) solver that requires the pyamg library in Python can be used, although it is not recommended for cases with a large number of clusters. This method combines the flexibility to capture nonlinear structures with the accuracy of spectral methods, but sacrifices scalability in favor of clustering quality.

2.7.8 Other Clustering Algorithms

2.7.8.1 Affinity Propagation

The Affinity Propagation (AP) algorithm represents an innovative approach in the field of clustering, standing out for its ability to automatically determine both the optimal number of clusters and their representative centers (called exemplars), without requiring this parameter to be specified a priori as in traditional methods (K-means, DBSCAN). Developed by Frey and Dueck in 2007, AP is based on a sophisticated "message passing" mechanism between data points, where each evaluates its suitability to act as an exemplar through two critical components: (1) responsibility (R), which quantifies how suitable a point is to represent another compared to alternative candidates, and (2) availability (A), which reflects the willingness of a point to become an exemplar based on the preferences of other points. This iterative process, which updates similarity matrices until convergence is reached, consists of the following key stages: similarity calculation (where metrics such as negative Euclidean distance define relationships between points), sequential update of responsibilities and availabilities, net responsibility calculation, final selection of exemplars, and cluster assignment. Negative Euclidean distance is a key adaptation to transform dissimilarity (distance) measures into similarity values, which is a fundamental requirement of the algorithm. Since AP operates with similarities where larger values indicate higher affinity, the negative Euclidean distance is applied to invert the relationship: closer points (smaller distance) generate fewer negative values, interpreted as higher similarity. This transformation preserves geometric proximity in Euclidean spaces in an efficient manner, although other metrics such as cosine similarity can also be used when the nature of the data requires it.

Among the main advantages of AP are its autonomy to define the cluster structure (ideal for data with unknown patterns), its robustness to noise and outliers, and its flexibility to handle non-spherical or convex shapes. However, the algorithm has significant limitations: its quadratic computational complexity ($O(N^2)$) in time and memory restricts it to small to medium data sets, and its performance depends critically on parameters such as the damping factor (to avoid oscillations) and preference (which influences the number of exemplars). For contexts with large data volumes, alternatives such as HDBSCAN or OPTICS are usually more efficient. Despite these limitations, AP remains a powerful tool in scenarios demanding automatic pattern discovery, offering a unique balance between algorithmic autonomy and the ability to reveal complex structures in domains with high uncertainty. Its implementation, available in libraries such as scikit-learn, requires careful tuning of hyperparameters to maximize its potential in specific problems.

2.8 Future Perspectives and Challenges in Clustering Algorithms

Clustering algorithms face critical challenges that must be addressed to improve their application in various fields, where scalability, dimensionality and robustness are priorities. Key properties to evaluate these algorithms include: (1) Scalability, essential to handle exponential data growth; (2) High dimensionality handling, which requires identifying relevant features to capture intrinsic structure; (3) Robustness against noise and outliers, inevitable in measurement and storage processes; (4) Automatic determination of the number of clusters, since dependence on predefined parameters (as in K-means or Spectral Clustering) limits their usefulness in the absence of prior knowledge; (5) Flexibility for arbitrary shapes, overcoming the limitations of methods that assume spherical clusters; (6) Visualization and interpretability, crucial for applications where decision making depends on understandable results; and (7) Adaptability to mixed data (continuous and categorical), since real data are rarely homogeneous.

In addition, the computational complexity in clustering optimization problems grows dramatically with data size, although local methods can offer practical solutions when the underlying clusters are well defined. Moving forward, interdisciplinary collaboration is needed not only to optimize existing parameters and algorithms, but also to develop techniques that integrate scalability with accuracy, such as the use of deep learning or hybrid models. The next generation of algorithms will need to balance computational efficiency with the ability to discover complex patterns in heterogeneous data, thus ensuring their applicability in a variety of critical scenarios.

Clustering is taking on a key role in the promotion of sustainability by allowing the identification of patterns and groupings in large volumes of information. For example, in urban areas, clustering helps to classify zones according to their energy consumption, CO_2 emissions, or waste generation, thus facilitating the implementation of more efficient public policies. In the industrial sector, this technique makes it possible to optimize supply chains, reduce waste, and improve resource efficiency, contributing to more circular production models. Furthermore, in agriculture, cluster analysis is used to maximize crop yields with the lowest environmental impact. However, its effectiveness depends on data quality and integration with other technologies, such as IoT and artificial intelligence, to achieve a truly sustainable approach. Thus, clustering is positioned as a promising tool for achieving the Sustainable Development Goals (SDGs), although its success requires multidisciplinary collaboration and access to reliable information.

References

Ankerst, M., Breunig, M. M., Kriegel, H. P., & Sander, J. (1999). OPTICS: Ordering points to identify the clustering structure. *ACM Sigmod Record, 28*(2), 49–60. https://doi.org/10.1145/304181.304187

Arthur, D., & Vassilvitskii, S. (2006). *k-means++: The advantages of careful seeding*. Stanford.

Bezdek, J. C., Ehrlich, R., & Full, W. (1984). FCM: The fuzzy c-means clustering algorithm. *Computers & Geosciences, 10*(2–3), 191–203. https://doi.org/10.1016/0098-3004(84)90020-7

Caliński, T., & Harabasz, J. (1974). A dendrite method for cluster analysis. *Communications in Statistics-Theory and Methods, 3*(1), 1–27.

Campello, R. J., Moulavi, D., & Sander, J. (2013). Density-based clustering based on hierarchical density estimates. In Pacific-Asia conference on knowledge discovery and data mining (pp. 160–172). Berlin, Heidelberg: Springer Berlin Heidelberg. https://doi.org/10.1007/978-3-642-37456-2_14

Davies, D. L., & Bouldin, D. W. (2009). A cluster separation measure. *IEEE Transactions on Pattern Analysis and Machine Intelligence, 2*, 224–227.

Ester, M., Kriegel, H. P., Sander, J., & Xu, X. (1996, August). A density-based algorithm for discovering clusters in large spatial databases with noise. In *kdd* (Vol. 96, No. 34, pp. 226–231).

Han, J., Pei, J., & Tong, H. (2022). Data mining: concepts and techniques. Morgan Kaufmann.

Kaufman, L., & Rousseeuw, P. J. (2009). *Finding groups in data: An introduction to cluster analysis*. Wiley.

Keogh, E., & Kasetty, S. (2002). On the need for time series data mining benchmarks: a survey and empirical demonstration. In *Proceedings of the eighth ACM SIGKDD international conference on Knowledge discovery and data mining* (pp. 102–111).

Keogh, E., & Ratanamahatana, C. A. (2005). Exact indexing of dynamic time warping. *Knowledge and information systems, 7*, 358–386.

Kotu, V., & Deshpande, B. (2019). Clustering. *Data Science*, 221–261. https://doi.org/10.1016/B978-0-12-814761-0.00007-1

Little, R. J., & Rubin, D. B. (2019). *Statistical analysis with missing data*. Wiley.

Pearson, K. (1894). Contributions to the mathematical theory of evolution. *Philosophical Transactions of the Royal Society of London., 185*, 71–110. https://doi.org/10.1098/rsta.1894.0003

Rousseeuw, P. J. (1987). Silhouettes: A graphical aid to the interpretation and validation of cluster analysis. *Journal of Computational and Applied Mathematics, 20*, 53–65.

Rubin, D. B. (1976). Inference and missing data. *Biometrika, 63*(3), 581–592.

Saxena, A., Prasad, M., Gupta, A., Bharill, N., Patel, O. P., Tiwari, A., Er, M. J., Ding, W., & Lin, C. T. (2017). A review of clustering techniques and developments. *Neurocomputing, 267*, 664–681. https://doi.org/10.1016/j.neucom.2017.06.053

Shanmugam, R. (2009). Finite mixture models. *Technometrics, 44*(1), 82. https://doi.org/10.1198/tech.2002.s651

Subasi, A. (2020). Clustering examples. *Practical Machine Learning for Data Analysis Using Python*, 465–511. https://doi.org/10.1016/b978-0-12-821379-7.00007-2

Tibshirani, R., Walther, G., & Hastie, T. (2001). Estimating the number of clusters in a data set via the gap statistic. *Journal of the Royal Statistical Society: Series B (statistical Methodology), 63*(2), 411–423.

Theodoridis, S., & Koutroumbas, K. (2009). Clustering algorithms I: Sequential algorithms. *Pattern Recognition*, 627–652. https://doi.org/10.1016/b978-1-59749-272-0.50014-1

Theodoridis, S., Pikrakis, A., Koutroumbas, K., & Cavouras, D. (2010). *Introduction to pattern recognition: a matlab approach*. Academic Press.

Von Luxburg, U. (2007). A tutorial on spectral clustering. *Statistics and Computing, 17*, 395–416. https://doi.org/10.1007/s11222-007-9033-z

Wierzchoń, S. T., & Kłopotek, M. A. (2018). Cluster analysis. In *Modern Algorithms of Cluster Analysis*. Studies in Big Data, vol. 34. Springer, Cham. https://doi.org/10.1007/978-3-319-69308-8_2

Witten, I. H., & Frank, E. (2005). *Data mining: Practical machine learning tools and techniques 2nd ed*. Ser. Morgan Kaufmann Series in Data Management Sys.

Wilks, D. S. (2011). Cluster Analysis. *International Geophysics, 100*, 603–616. https://doi.org/10.1016/B978-0-12-385022-5.00015-4

Supplementary Online Elements Scikit-learn

https://scikit-learn.org/stable/

Chapter 3
Programming for Clustering: Python, R, and MATLAB

3.1 Introduction to Clustering Tools and Computational Environments

Clustering is an unsupervised machine learning technique that groups similar data points into clusters, revealing underlying patterns without predefined labels (Chander & Gopalakrishnan, 2023; Usama et al., 2019). In the context of environmental and energy research, clustering can help identify patterns such as climate zone groupings, consumer energy usage profiles, sensor readings patterns, and more (Wang & Biljecki, 2022; Quatrosi, 2022; Alomari et al., 2022). Effective use of clustering requires not only understanding the algorithms conceptually, but also the practical skills to implement these algorithms with programming. This chapter bridges that gap by focusing on programming for clustering, using three widely used languages/tools in data science: Python, R, and MATLAB.

Modern data science work often involves a complex software stack with many libraries and dependencies. To manage this complexity, the Anaconda distribution has become a popular solution. Anaconda is a free, open-source distribution that includes the Python (and R) language along with a package manager (conda) and hundreds of pre-installed packages commonly used in data science (download Anaconda from *anaconda.com*). In other words, Anaconda provides a ready-to-use environment where libraries for clustering (and other machine learning tasks) are easily accessible and can be installed or updated in a controlled manner. By creating isolated conda environments, researchers can ensure that the specific versions of Python/R and libraries they use for a clustering project remain consistent, which improves reproducibility of results across different machines and over time. Anaconda Navigator is a graphical interface that simplifies the management of virtual environments and the launch of applications like Jupyter Notebook, RStudio, and Spyder within the appropriate settings. The installation process is straightforward across all major operating

A. Y. Raya-Tapia et al., *Machine Learning and Clustering for a Sustainable Future*,
Studies in Computational Intelligence 1233,
https://doi.org/10.1007/978-3-032-03876-0_3

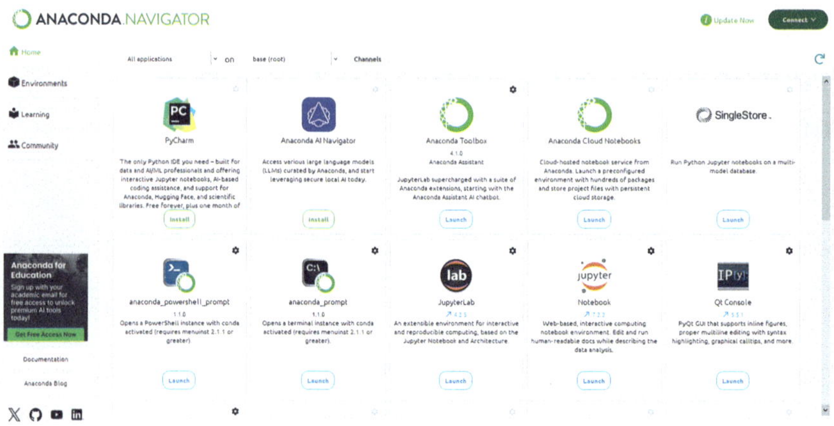

Fig. 3.1 Overview of Anaconda Navigator

systems. Once installed, Navigator offers an integrated platform for package management and application launching, streamlining Python project workflows, particularly in data science and machine learning.

Figure 3.1 illustrates the Anaconda Navigator interface and highlights its main features. In the "Home" section, users can browse available Python environments and access application launchers, documentation, and tutorials. Notably, Jupyter Notebook and Spyder come pre-installed, and users can begin working with them simply by clicking the "Launch" button.

Jupyter Notebook, developed by the Jupyter Project, a non-profit initiative founded by Fernando Pérez, aims to promote open-source tools that support interactive computing across multiple programming languages. Its name reflects its support for Julia, Python, and R.

It is a web-based application that allows users to write, execute, and document code interactively. Designed for literate programming, Jupyter Notebooks combine executable code with rich text explanations, mathematical equations, visualizations, and outputs in a single, shareable document. This makes it especially valuable in research, education, and collaborative programming.

Each notebook operates with a kernel responsible for executing code and a user interface (shell) that displays inputs and outputs. Users can mix code cells with Markdown cells, allowing for seamless integration of code and documentation. The interactive environment is particularly useful for iterative experimentation, such as model tuning in machine learning. Additionally, Jupyter's extensibility allows for integration with other languages and functionalities through plugins and extensions. To launch Jupyter Notebook, users can either click "Launch" from Anaconda Navigator or run jupyter notebook from the terminal or Anaconda Prompt. This action opens the Jupyter dashboard in the default web browser (see Fig. 3.2).

Once opened, the dashboard appears at a local server address (e.g., http://localhost:8888/tree) and displays the contents of the user's working directory. From

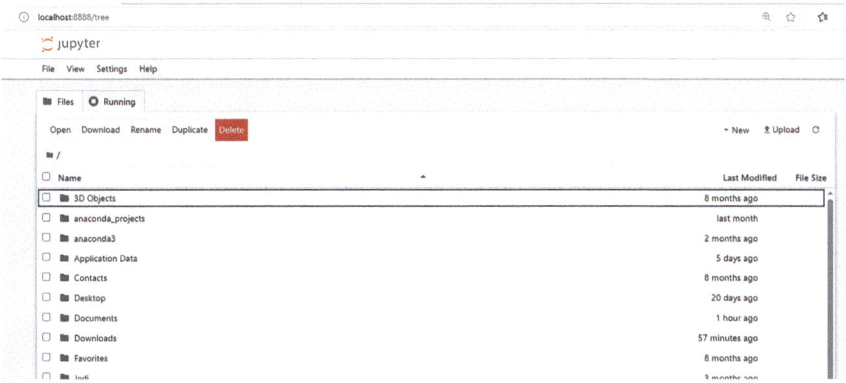

Fig. 3.2 The Jupyter Notebook dashboard

here, users can create new notebooks, upload existing files, or open folders and terminals. The interface provides tools for sorting files by name or modification date, and uses icons to distinguish between file types such as notebooks (.ipynb) and text files (.txt).

Creating a new notebook involves clicking "New" and selecting the desired kernel, such as Python 3. This opens a new browser tab containing the notebook interface. The top of the interface displays the notebook's title (see Fig. 3.3), initially "Untitled", which can be renamed by clicking on it. Below the title is the last checkpoint indicator, showing the most recent autosave time. The menu bar provides access to functions like saving, editing, and changing kernel settings, while the toolbar offers quick access to common actions. The main body of the notebook is where users write and execute content.

In this interface, the default cell type is a code cell, which allows users to enter and execute code. Executed cells display their output immediately below. Users can run cells using the "Run" button or the Shift + Enter shortcut. Markdown cells are used to add text explanations, headings, or equations using formatting syntax. When executed, these cells render as formatted text, improving the readability and structure of the notebook (see Fig. 3.3).

On the other hand, RStudio is a powerful integrated development environment (IDE) specifically designed for R, widely used in data analysis, statistical computing,

Fig. 3.3 Visualization of a new notebook

Fig. 3.4 Exploring the RStudio environment on Windows

and data science. Its user-friendly interface is divided into four main panes, each offering distinct functionalities that enhance the coding workflow and streamline project management (see Fig. 3.4).

The Script Editor is where users write and edit R scripts. It includes features such as syntax highlighting, code auto-completion, and automatic indentation, all of which contribute to a smoother and more efficient coding experience. The Console serves as the primary interaction point with the R interpreter. It allows users to enter commands directly and view immediate outputs, making it ideal for quick calculations, testing code snippets, or running entire scripts line by line. On the top right, the Environment Pane provides tools for managing the current R session. It displays all active objects, such as variables and data frames, along with their structures and values. It also includes a command history tab, where users can review previously executed commands, and a connections tab to manage links with external databases or data sources. The Files pane is a multi-purpose area that integrates several essential tools. The Files tab allows users to navigate the project directory and manage files. The Plots tab displays visualizations generated during the session and provides options to export them. The Packages tab shows installed R packages and facilitates the installation or loading of additional libraries. This pane may also include other useful tabs depending on the configuration, such as Help for documentation, Viewer for web content, Git for version control, and Markdown preview for formatted documents. Understanding this layout is essential for taking full advantage of RStudio's capabilities (Fig. 3.4) in both beginner and advanced data science workflows.

3.1.1 Why Focus on Python, R, and MATLAB?

These three platforms cover a broad range of user preferences and legacy code in scientific research:

- Python is a general-purpose programming language that has become dominant in artificial intelligence (AI), machine learning, and data science due to its readable syntax and an enormous ecosystem of libraries. It excels in integration with workflows, automation, and has de facto standard libraries for machine learning (e.g., scikit-learn, TensorFlow) (Dhruv et al., 2021; Downey & Elkner, 2008; Porter & Zingaro, 2024). Python's libraries include efficient implementations of all major clustering algorithms, making it a powerful tool for environmental data analysis.
- R is a language developed by and for statisticians. It is especially strong in statistical modeling and has a rich collection of packages for data analysis and visualization. R was one of the first languages to provide user-friendly clustering implementations (dating back to the 1990s with libraries like cluster). Many researchers in ecology, environmental science, and energy economics continue to use R for its statistical rigor and plotting capabilities (Tucker et al., 2023; Widemann et al., 2013).
- MATLAB (Matrix Laboratory) is a high-level environment widely used in engineering and scientific computing, including fields such as renewable energy systems and climate modeling. MATLAB's strengths lie in numerical linear algebra (ideal for matrix-heavy computations in clustering) and an interactive environment with powerful built-in visualizations (MathWorks, Inc., 2022). MATLAB's toolboxes (such as the Statistics and Machine Learning Toolbox) provide ready-to-use clustering algorithms optimized in C/C++ behind the scenes. This is valuable for researchers who prefer an integrated environment and for educational settings.

Each of these tools has its strengths and weaknesses for clustering tasks, which are discussed in their respective sections. For instance, Python's advantage is its versatility and production-readiness, whereas R provides specialized techniques and ease of statistical analysis, and MATLAB offers speed and an all-in-one platform at the cost of being proprietary. The goal is not to declare one "best" tool, but to equip the reader with knowledge of how to perform clustering in each, so they can choose the tool that best fits their project requirements and personal or institutional preferences.

Before diving into each language, it's important to note how Anaconda can be leveraged for all three:

- For Python, Anaconda is almost a standard choice; it comes with core libraries such as NumPy, pandas, and scikit-learn pre-installed, and additional packages can be added via conda install or pip. It is assumed that the reader has a working Python environment (preferably via Anaconda) with the libraries needed.
- For R, Anaconda can manage R through the r-essentials package (which includes R and commonly used packages). However, many R users install R separately and

use RStudio. Both approaches are fine. In this chapter, when we use R, you can either run the code in RStudio or even in a Jupyter Notebook by installing the R kernel (using *conda install -c r r-irkernel* or via *R's IRkernel package*). Anaconda's environment management is helpful if you want to maintain a consistent R package environment or if you want to switch between R and Python in the same project (for example, using R's strong plotting with Python's data manipulation, or vice versa).

- For MATLAB, Anaconda does not apply directly (MATLAB is obtained from MathWorks and not managed through conda). However, you can still use Anaconda to manage a Python environment that calls MATLAB engine API or to organize projects that include calling MATLAB scripts. In the context of clustering, MATLAB is treated separately, as it is self-contained. Just ensure you have the Statistics and Machine Learning Toolbox (or relevant toolbox) installed in MATLAB, as that provides most of the clustering functions discussed in this chapter.

By setting up your computational environment properly (e.g., using Anaconda for Python/R), you ensure that you have access to all the necessary libraries and that code examples run smoothly.

3.2 Python for Clustering

3.2.1 Python for Data Science and Clustering: Overview, Strengths, and Weaknesses

Python is a high-level, interpreted programming language known for its clean syntax and enormous versatility. Created by Guido van Rossum and first released in 1991, Python has evolved into one of the most popular languages in both general software development and scientific computing. In the realm of AI, data science, and machine learning, Python stands out because it allows rapid development and has a vast ecosystem of libraries contributed by an active community. For example, scikit-learn (sklearn) is a widely used Python library (Pedregosa et al., 2011) that provides "a wide range of state-of-the-art machine learning algorithms for medium-scale supervised and unsupervised problems". This implies that both simple k-means clustering and advanced density-based methods can be readily implemented using existing Python libraries.

Strengths in clustering

- It has libraries such as NumPy for efficient numerical computations and pandas for data manipulation, which are often used to prepare data before clustering.

- For the clustering algorithms themselves, scikit-learn offers all the classic methods (k-means, hierarchical clustering, DBSCAN, mean-shift, spectral clustering, etc.) under a consistent and user-friendly API.
- Python also supports specialized clustering libraries or extensions beyond scikit-learn. For instance, the SciPy library contains functions for hierarchical clustering (with dendrogram plotting capabilities) and k-means (via scipy.cluster module). There are also domain-specific libraries; for example, scikit-learn-contrib has implementations of algorithms such as HDBSCAN (a more advanced density clustering), and scikit-fuzzy provides fuzzy c-means clustering for soft clustering needs.
- Python is highly extensible and can integrate with big data frameworks (such as PySpark) if one needs to cluster very large datasets distributed across clusters of machines, although this is beyond our scope here, it's worth noting that the Python interface is available in such contexts too.
- Another strength is ease of use and integration. Python code tends to be concise and readable, which is beneficial for prototyping clustering analyses and iterating quickly. The language's general-purpose nature means that after clustering, you can easily integrate the results into web applications, dashboards, or workflows written in Python.

Weaknesses

- Performance: Being an interpreted language, pure Python could be slow for heavy computations, but in practice, libraries such as NumPy and scikit-learn are implemented in optimized C/C++ or Fortran under the hood, so the heavy lifting is done in low-level code. For moderate data sizes (which cover many environmental datasets), Python's performance is usually sufficient. However, extremely large datasets might require more specialized solutions (such as incremental clustering algorithms, out-of-core processing, or using Python interfaces to faster languages).
- Statistical specialization: Compared to R, Python historically had fewer built-in statistical modeling features. While this gap has closed over time (with libraries for almost every statistical method now available in Python), some very specific statistical clustering techniques or nuances might have more straightforward support in R. For example, methods like clustering validation via specific statistical tests or some specialized biostatistics clustering might be slightly less convenient in Python, though not impossible.
- Dependency management: Without tools like Anaconda, managing Python library versions can become messy (the so-called "dependency hell"). But if you use conda environments properly, this is largely mitigated. It's worth noting that the Python scientific stack can sometimes have installation issues (such as compiling libraries) on certain systems, but again, Anaconda solves many of these by providing precompiled binaries.

In summary, Python is an excellent choice for clustering, especially if you value a balanced platform that lets you do data preprocessing, clustering, and post-clustering analysis in one place. It's particularly strong in productionizing clustering analysis (e.g., integrating it into a pipeline or application) and in handling data workflows end-to-end.

3.2.2 Key Libraries and Tools for Clustering in Python

Python's capabilities for clustering come mostly from its libraries. Here we highlight the most relevant libraries/tools and the clustering methods they offer.

Scikit-learn (sklearn): This is the primary machine learning library in Python and a cornerstone for clustering. Scikit-learn provides a well-documented and consistent interface for a variety of clustering algorithms:

- K-Means Clustering: This is the classic partitioning algorithm that divides data into k clusters by minimizing within-cluster variance. Scikit-learn's implementation includes options for different initialization methods (such as k-means++), specifying the number of iterations, etc.
- Hierarchical Agglomerative Clustering: You can choose different linkage criteria (ward, complete, average) and distance metrics. This algorithm builds a hierarchy of clusters (which can be visualized with a dendrogram using SciPy).
- DBSCAN (Density-Based Spatial Clustering of Applications with Noise): This algorithm finds arbitrarily shaped clusters based on density (points packed together) and can mark points as noise if they don't belong to any cluster. One advantage of DBSCAN is that it does not require specifying the number of clusters in advancemathworks.com, and can detect outliers as noise.
- OPTICS: An extended density-based method similar to DBSCAN, but can find clusters in data with varying density.
- Mean-Shift: (sklearn.cluster.MeanShift) which can automatically discover the number of clusters by finding peaks in the data distribution (often used in image processing).
- Spectral Clustering: (sklearn.cluster.SpectralClustering) which uses graph theory (eigenvalues of a similarity matrix) to cluster data that may not be well-separated in input space but is separable in a transformed spectral space.
- Affinity Propagation: (sklearn.cluster.AffinityPropagation) which is another clustering method that automatically determines the number of clusters based on message passing between points.
- Birch: (sklearn.cluster.Birch) a scalable clustering method that incrementally builds a tree (Clustering Feature tree)—useful for large datasets.
- Gaussian Mixture Models: While technically in sklearn.mixture (e.g., GaussianMixture class), these provide soft clustering (each point has a probability of belonging to each cluster) and can be seen as a form of clustering (model-based clustering).

Scikit-learn's extensive collection makes it a one-stop shop for most clustering needs. It's also continually updated; for example, recent versions added Bisecting

K-Means (a variant of k-means that recursively splits clusters, useful for hierarchical k-means approach) and improved performance in various algorithms.

SciPy (Gommers et al., 2024): Python's scientific computing library, which scikit-learn builds upon in some cases. For clustering, the SciPy module provides:

- Functions like linkage to perform hierarchical clustering (agglomerative) given a distance matrix or data.
- A dendrogram function to visualize the hierarchical cluster tree. This is useful for determining a reasonable number of clusters by "cutting" the dendrogram at a certain level.

SciPy also has k-means clustering and vector quantization (these are somewhat lower-level than scikit-learn's KMeans, and less commonly used now, but they exist).

Pandas: While not a clustering library, pandas is often used to prepare data before clustering. For instance, one might use pandas to load a CSV of environmental data, handle missing values, filter or transform features (such as normalizing pollutant concentrations or scaling energy usage values), then convert the DataFrame to a NumPy array to pass to scikit-learn (McKinney, 2017). Pandas also makes it easy to add the cluster labels back into a dataset for further analysis or visualization (e.g., grouping the data by cluster to compute summary statistics per cluster).

NumPy: Again, not for clustering algorithms per se, but all numeric data will likely be in NumPy arrays. Libraries such as scikit-learn and SciPy expect data as NumPy arrays (or arrays embedded in pandas DataFrames). Understanding NumPy (for example, how to slice data, how to compute basic statistics, etc.) is helpful when doing clustering because you might need to manipulate arrays (McKinney, 2017) (for example, extracting a subset of features to cluster on, or initializing your own centroids for k-means in some advanced usage).

Matplotlib (and Seaborn or other plotting libraries): Visualization is a critical part of clustering analysis. Matplotlib is the standard plotting library in Python and is often used to create scatter plots of clustered data, line plots of cluster centroids, or dendrograms (via SciPy as mentioned) (Hetland & Nelli, 2024). Seaborn is a higher-level plotting library based on Matplotlib that can also make attractive scatter plots with relatively little code, and it can color points by cluster label easily if you pass the labels as a hue.

Python also includes several advanced and specialized libraries:

- Tslearn is a dedicated library for time-series clustering. It supports K-Means with Dynamic Time Warping (DTW), DBA-KMeans, and soft-DTW-based clustering (Tavenard et al., 2020). These techniques are particularly valuable in time-dependent environmental processes, such as water quality monitoring or energy demand profiles, where the shape of the time series and their temporal misalignments are critical. Tslearn enables clustering of sequential data based on similarity in temporal patterns rather than static values.
- The kmodes package provides clustering techniques for categorical and mixed-type data (de Vos, 2015–2024). The K-Modes algorithm extends K-Means to handle categorical variables using a simple matching dissimilarity metric and

modes instead of centroids. The K-Prototypes algorithm combines numerical and categorical distance measures to enable clustering of mixed-type datasets. This is highly relevant in process engineering, where datasets may include both numerical sensor readings and categorical attributes such as equipment types or operational states.
- HDBSCAN: An advanced clustering library that implements the HDBSCAN algorithm (Hierarchical DBSCAN), which can find clusters of varying densities and produces a hierarchy of clusters with a stability score for extracting a flat clustering (Campello et al., 2013). HDBSCAN is useful for complex clustering in environmental data where cluster density isn't uniform (e.g., distinguishing both dense and sparse regions of data).
- Scikit-Fuzzy: A scikit-contrib library for fuzzy logic. It includes Fuzzy C-Means clustering, which is a form of clustering where each data point can belong to multiple clusters with certain membership weights. This can be relevant in environmental science when boundaries between clusters are not crisp (e.g., a location might be partially in one climate regime and partially in another).
- Yellowbrick: A library for machine learning visualizations that includes tools for clustering, such as silhouette plots for assessing cluster quality and intercluster distance maps (Bengfort & Bilbro, 2019).

In practice, for 95% of use cases, scikit-learn (possibly combined with SciPy for some additional functionality) will cover all clustering needs in Python. It's also well-documented and widely used, so community support is abundant. Throughout this section, we will primarily demonstrate using scikit-learn for clustering.

3.2.3 Example: Clustering in Python

In this example, a real-world environmental dataset derived from a municipal water treatment plant is explored, containing measurements collected over time from various stages of the water treatment process. The dataset comprises a rich set of physicochemical variables, such as pH, conductivity, temperature, suspended solids, chemical oxygen demand (COD), biological oxygen demand (BOD), and concentrations of several chemical compounds. These variables characterize the water quality at different phases, including inflow, sedimentation, biological treatment, and outflow.

The dataset, publicly available from the UCI Machine Learning Repository (Poch, 1993), contains measurements from 38 operational variables across over 500 daily observations. Due to missing entries, preprocessing is essential before conducting any meaningful analysis. After filtering and normalizing the data, we retain a clean dataset of 380 complete observations, each representing a snapshot of the plant's condition for a particular day.

Our goal is to apply unsupervised learning techniques to this environmental dataset to uncover natural groupings or operational patterns within the water treatment process. These groupings can provide insight into typical plant conditions, detect operational modes (e.g., normal operation vs. transitional states), or highlight anomalies such as equipment malfunction or contamination spikes, all without any labeled output.

To this end, the following steps are performed:

1. Preprocessing the data: Cleaning missing values, converting variables to numerical form, and standardizing features to achieve zero mean and unit variance.
2. Applying multiple clustering algorithms: K-means, DBSCAN, Agglomerative Clustering, and Spectral Clustering will be implemented and compared. These algorithms are based on different assumptions regarding the shape, density, and separability of clusters, thereby offering complementary insights.
3. Evaluating clustering performance: The coherence and separation of the discovered clusters are assessed using internal metrics such as the Silhouette Score and Davies–Bouldin Index (DBI).
4. Visualizing the clusters: Through Principal Component Analysis (PCA), the high-dimensional data are projected into two dimensions for visual interpretation of the cluster structure.

This example demonstrates how clustering can be used to discover hidden structure in environmental and engineering datasets. Although the original dataset includes many features, the approach can be extended to other applications, such as air quality monitoring, industrial process analysis, or ecosystem health classification, where detecting natural groupings is essential for decision-making and anomaly detection.

It is important to note that the analysis is presented in two complementary stages. In the first, each section of the code is presented step by step and with an educational approach, including detailed explanations of the function of each tool used, the methodological justification behind its application, and the interpretation of the results obtained. This structure allows the reader to understand not only how clustering is implemented in Python, but also why certain technical decisions are made at each stage of the process. Subsequently, the complete and functional code is provided without intermediate comments, ready to be copied and executed in a Python programming environment. This condensed version facilitates its practical reuse in personal, educational, or professional projects.

1. **Library import**: Before starting the clustering analysis, it is necessary to import a series of libraries that provide essential functionalities for data management, algorithm application, and results visualization:

```
**********************************************************************

# Importing essential libraries
import pandas as pd
import numpy as np
import matplotlib.pyplot as plt
import seaborn as sns
from sklearn.preprocessing import StandardScaler
from sklearn.cluster import KMeans
from sklearn.metrics import silhouette_score
from sklearn.cluster import DBSCAN, AgglomerativeClustering, SpectralClustering
from sklearn.metrics import davies_bouldin_score
from sklearn.decomposition import PCA

**********************************************************************
```

The *pandas* and *numpy* libraries are essential for handling structured data; *pandas* simplifies the loading, cleaning, and transformation of datasets, while *numpy* provides tools for performing efficient mathematical operations on numerical arrays.

For generating visualizations, *matplotlib.pyplot* and *seaborn* are used. The former enables the creation of basic plots such as line, bar, or scatter plots, while the latter extends these capabilities with predefined styles and specialized functions to represent data more clearly and aesthetically.

During data preprocessing, *StandardScaler* from the *sklearn.preprocessing* module is applied to standardize variables to a common scale with zero mean and unit variance. This normalization is a crucial step, as many clustering algorithms are sensitive to differences in variable scales.

The *sklearn.cluster* module contains various clustering algorithms implemented in this study. These include: *KMeans*, which clusters data around optimized centroids; *DBSCAN*, which detects densely populated regions and identifies outliers; *AgglomerativeClustering*, which builds cluster hierarchies using a bottom-up approach; and *SpectralClustering*, which leverages similarity graphs and spectral decomposition to detect complex groupings. To evaluate the quality of the results, tools from the *sklearn.metrics* module are employed; particularly *silhouette_score*, which measures internal cohesion and separation between clusters, and *davies_bouldin_score*, which assesses cluster compactness and relative distance.

Finally, *PCA* from the *sklearn.decomposition* module is used, a dimensionality reduction technique that projects multivariate data onto a two-dimensional plane. This transformation is especially useful for visualizing the distribution of clusters and analyzing their structure in a more accessible manner.

2. **Data Loading and Description**: Once the necessary libraries have been imported, the next step is to load the dataset into memory and perform an initial inspection to become familiar with its structure and content. This allows us to identify potential issues, such as missing values or incorrect data types, and to plan the appropriate preprocessing steps.

****

```
# Load data
df_water = pd.read_csv('water-treatment.csv')

# Initial inspection
print("Initial view of the dataset:")
print(df_water.head())
print(df_water.info())
```

****

The *pd.read_csv()* function is used to load the CSV file containing the measurements collected at the water treatment plant, allowing the data to be stored in a DataFrame for subsequent manipulation. Once the dataset is loaded, *df_water.head()* displays the first few rows, which is useful for getting a general impression of the dataset's structure and content.

On the other hand, *df_water.info()* provides a detailed technical summary that includes the number of records, column names, and data types, as well as the number of non-null values per variable. This information is essential for identifying missing data, data type errors, or columns that require cleaning or transformation before analysis.

3. **Data Preprocessing**: Before applying clustering techniques, it is essential to properly prepare the dataset. This process includes handling missing values, converting data types, and normalizing the variables. These steps ensure that the clustering algorithms operate on a clean, numerical, and uniformly scaled dataset.

****

```
# Handle missing values represented by'?'
df_water.replace('?', np.nan, inplace=True)

# Convert all columns to numeric type
for col in df_water.columns:
    df_water[col] = pd.to_numeric(df_water[col], errors='coerce')

# Eliminate columns with more than 30% missing values and remaining rows with NaNs
df_water.dropna(axis=1, thresh=int(0.7 * len(df_water)), inplace=True)
df_water.dropna(inplace=True)

# Normalization of data for clustering
scaler = StandardScaler()
df_scaled_water = scaler.fit_transform(df_water)
```

****

In many environmental datasets, such as the one used in this analysis, missing values are often encoded using special characters, such as the question mark "?". To allow Python to handle them properly, these values are replaced with NaN, a standard notation for missing data. Subsequently, all columns are converted to numeric format using the *pd.to_numeric()* function with the parameter *errors* = '*coerce*', which enforces the conversion and replaces any conversion errors with NaN. Once the missing values are identified, columns with more than 30% missing data are removed, as they are unlikely to contribute useful information to the analysis. Then, all remaining rows containing missing values are also deleted to ensure that the clustering algorithms can run without errors related to incomplete data.

4. **Optimal Cluster Determination (K-means)**: One of the key challenges in clustering is selecting how many groups (k) the algorithm should identify. To address this, heuristic methods are applied to analyze how clustering quality varies with different numbers of clusters. In this step, two complementary approaches are used: *the Elbow Method* (inertia) and the *Silhouette Index*.

```
************************************************************************

# Elbow and silhouette method to identify optimum number
inertia, silhouette_scores = [], []
k_range = range(2, 10)

for k in k_range:
    kmeans = KMeans(n_clusters=k, random_state=42)
    labels = kmeans.fit_predict(df_scaled_water)
    inertia.append(kmeans.inertia_)
    silhouette_scores.append(silhouette_score(df_scaled_water, labels))

# Elbow and silhouette graphics
fig, ax = plt.subplots(1, 2, figsize=(12, 5))
ax[0].plot(k_range, inertia, 'o-', color='blue')
ax[0].set_title('Elbow Method')
ax[0].set_xlabel('Number of clusters')
ax[0].set_ylabel('Inertia')

ax[1].plot(k_range, silhouette_scores, 'o-', color='green')
ax[1].set_title('Silhouette Score')
ax[1].set_xlabel('Number of clusters')
ax[1].set_ylabel('Silhouette Score')
plt.tight_layout()
plt.show()

************************************************************************
```

In this analysis, two commonly accepted criteria were applied: *inertia* and the *Silhouette Score*, whose results are shown in Fig. 3.5. Inertia represents the total sum of squared distances between data points and the centroid of their respective cluster. This value decreases as the number of clusters increases, but at a certain point, the marginal improvements drop sharply. This inflection point, known as the "elbow", indicates an optimal value of *k*, beyond which the model begins to overfit without

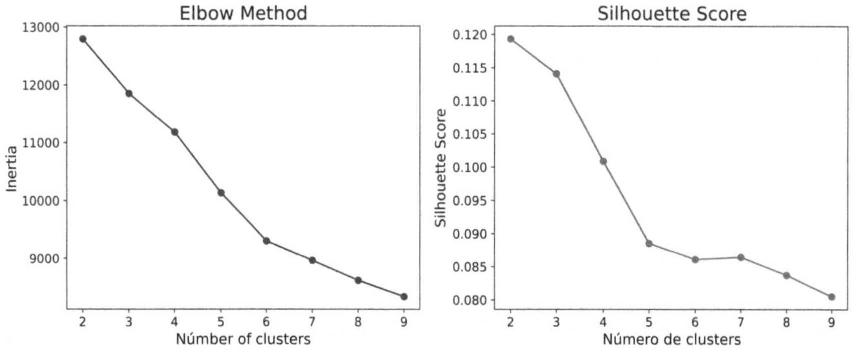

Fig. 3.5 Optimal number of clusters determined by the Elbow Method and Silhouette Score (Python)

significant gains in compactness. Figure 1a illustrates how the slope of the curve visibly flattens around $k = 3$ or $k = 4$.

On the other hand, the Silhouette Score evaluates how similar a point is to its own cluster compared to other clusters. Its values range from -1 to 1, with scores close to 1 indicating strong intra-cluster cohesion and good separation from other groups. Figure 3.5b shows that this index also reaches its maximum at $k = 3$, which supports the suggestion obtained from the elbow method.

This procedure is essential to ensure that clustering algorithms operate with a structurally coherent configuration. In this example, both methods consistently indicate that the most natural grouping lies between three and four clusters, which will be used in the subsequent analyses.

5. **Application of Clustering Algorithms**: Once the optimal number of clusters has been determined, the next step is to apply different clustering algorithms to the preprocessed dataset. The goal is to compare how each technique segments the observations from the water treatment plant, considering that each algorithm is based on different mathematical principles and responds differently to the distribution of the data.

```
****************************************************************************
# K-means with k=3
kmeans = KMeans(n_clusters=3, random_state=42)
labels_kmeans = kmeans.fit_predict(df_scaled_water)

# DBSCAN with adjusted parameters
dbscan = DBSCAN(eps=5, min_samples=5)
labels_dbscan = dbscan.fit_predict(df_scaled_water)

# Agglomerative Clustering with k=3
agglo = AgglomerativeClustering(n_clusters=3)
labels_agglo = agglo.fit_predict(df_scaled_water)

# Spectral Clustering with k=3
spectral = SpectralClustering(n_clusters=3, random_state=42, affinity='nearest_neighbors')
labels_spectral = spectral.fit_predict(df_scaled_water)
****************************************************************************
```

6. **Evaluation and Comparison of Algorithm Performance**: After applying the different clustering algorithms, it is essential to assess the quality of the resulting groupings. To this end, internal metrics are used to quantify cluster cohesion and separation without the need for true labels, since the learning process is unsupervised.

```
****************************************************************************
results = pd.DataFrame({
    'Algorithm': ['K-means', 'DBSCAN', 'Agglomerative', 'Spectral'],
    'Silhouette Score': [
        silhouette_score(df_scaled_water, labels_kmeans),
        silhouette_score(df_scaled_water, labels_dbscan),
        silhouette_score(df_scaled_water, labels_agglo),
        silhouette_score(df_scaled_water, labels_spectral)
    ],
    'Davies-Bouldin Index': [
        davies_bouldin_score(df_scaled_water, labels_kmeans),
        davies_bouldin_score(df_scaled_water, labels_dbscan),
        davies_bouldin_score(df_scaled_water, labels_agglo),
        davies_bouldin_score(df_scaled_water, labels_spectral)
    ]
})

print("Comparative evaluation of the algorithms:")
print(results)
****************************************************************************
```

Table 3.1 Summarizes the values obtained by each algorithm

Algorithm	Silhouette score	Davies-Bouldin index
K-means	0.114075	2.292117
DBSCAN	0.265065	4.237612
Agglomerative	0.087210	2.014509
Spectral	0.111055	2.265010

The Silhouette Score ranges from −1 to 1. Values close to 1 indicate that the data points are well-clustered and well-separated; values near 0 suggest overlapping clusters. In this evaluation, DBSCAN achieved the highest Silhouette Score (0.265), indicating strong internal cohesion within the detected groups (see Table 3.1).

On the other hand, the DBI, with an ideal value close to 0, reflects a combination of intra-cluster dispersion and inter-cluster separation. In this case, Agglomerative Clustering obtained the lowest DBI (2.01), although it had the lowest Silhouette Score, indicating that its groups were less internally consistent. These metrics should be interpreted together rather than in isolation (see Table 3.1). For example, DBSCAN showed strong cohesion but higher inter-group dispersion (high DBI), while Spectral Clustering and K-means showed intermediate values on both metrics, which may indicate a more balanced structure. This type of evaluation provides an objective basis for comparing methods, although it should always be complemented by a visual and contextual interpretation of the problem under study.

7. **Visualization of Results Using PCA**: Since the original data from the water treatment process contains multiple variables, it is difficult to graphically represent the distribution of clusters in the full feature space. To address this, PCA is applied, a dimensionality reduction technique that projects the data onto a lower-dimensional space (in this case, 2D) while maximizing the explained variance. The following code block reduces the normalized data to two principal components and generates a visual comparison of the results obtained by each clustering algorithm:

****

```
# Reduce dimensions for clear display
pca = PCA(n_components=2)
X_pca = pca.fit_transform(df_scaled_water)

# Comparative graphics
fig, axes = plt.subplots(2, 2, figsize=(16, 12))

sns.scatterplot(x=X_pca[:,0], y=X_pca[:,1], hue=labels_kmeans, palette='Set1', ax=axes[0,0])
axes[0,0].set_title('K-means Clusters')

sns.scatterplot(x=X_pca[:,0], y=X_pca[:,1], hue=labels_dbscan, palette='Set2', ax=axes[0,1])
axes[0,1].set_title('DBSCAN Clusters')

sns.scatterplot(x=X_pca[:,0], y=X_pca[:,1], hue=labels_agglo, palette='Set3', ax=axes[1,0])
axes[1,0].set_title('Agglomerative Clusters')

sns.scatterplot(x=X_pca[:,0], y=X_pca[:,1], hue=labels_spectral, palette='Set1', ax=axes[1,1])
axes[1,1].set_title('Spectral Clusters')

for ax in axes.flat:
    ax.set_xlabel('PCA Component 1')
    ax.set_ylabel('PCA Component 2')

plt.tight_layout()
plt.show()
```

****

Figure 3.6 illustrates how each algorithm segments the data based on the two principal components extracted through PCA. The results provide key visual insights for comparing the clustering structures:

- In the top-left panel, K-means produces well-defined clusters aligned along the first principal component, highlighting its effectiveness in scenarios where groups have approximately spherical shapes.
- In the top-right panel, DBSCAN reveals a density-based segmentation, with a significant portion of the data labeled as noise (label -1). This outcome suggests the presence of outliers or atypical behaviors within the treatment process.
- In the bottom-left panel, Agglomerative Clustering forms more diffuse groupings with greater overlap among points, which may hinder straightforward interpretation.
- Finally, in the bottom-right panel, Spectral Clustering produces more defined clusters than Agglomerative Clustering, although with some internal dispersion. The visual quality of the clusters is moderate, consistent with its performance metrics.

PCA visualizations, such as the one shown in Fig. 3.6, are powerful tools for analyzing clustering results in multivariate problems, enabling analysts to more clearly identify cohesion, separation, and the presence of potential outliers in the data.

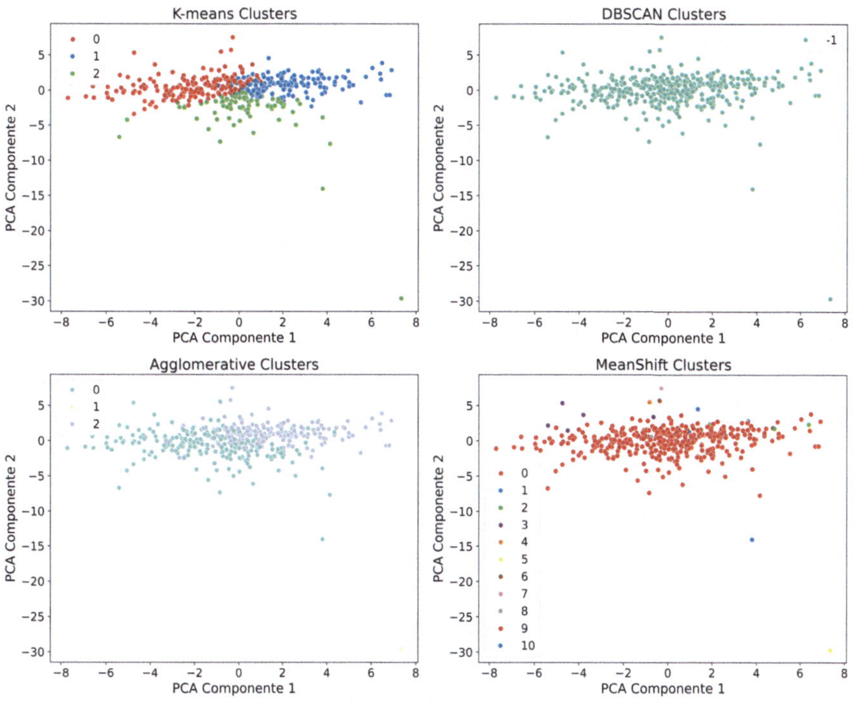

Fig. 3.6 PCA-Based Visualization of Clustering Results Using K-means, DBSCAN, Agglomerative, and Spectral Algorithms

The complete code is presented below:

```
************************************************************************

#==========================================================================
========
# Water Treatment Plant Dataset
#==========================================================================
========

# -----------------------------------------
# Step 1: Data upload and description
# -----------------------------------------

# Import of essential libraries
import pandas as pd
import numpy as np
import matplotlib.pyplot as plt
```

```
import seaborn as sns
from sklearn.preprocessing import StandardScaler
from sklearn.cluster import KMeans
from sklearn.metrics import silhouette_score
from sklearn.cluster import DBSCAN, AgglomerativeClustering, SpectralClustering
from sklearn.metrics import davies_bouldin_score
from sklearn.decomposition import PCA

# Load data
df_water = pd.read_csv('water-treatment.csv')

# Initial inspection
print("Initial view of the dataset:")
print(df_water.head())
print(df_water.info())

# ------------------------------------------
# Step 2: Data Preprocessing
# ------------------------------------------

# Handle missing values represented by'?'
df_water.replace('?', np.nan, inplace=True)

# Convert all columns to numeric type
for col in df_water.columns:
    df_water[col] = pd.to_numeric(df_water[col], errors='coerce')

# Eliminate columns with more than 30% missing values and remaining rows with NaNs
df_water.dropna(axis=1, thresh=int(0.7 * len(df_water)), inplace=True)
df_water.dropna(inplace=True)

# Normalization of data for clustering
scaler = StandardScaler()
df_scaled_water = scaler.fit_transform(df_water)

# ------------------------------------------
# Step 3: Optimal cluster determination (K-means)
# ------------------------------------------
# Elbow and silhouette method to identify optimum number
inertia, silhouette_scores = [], []
k_range = range(2, 10)

for k in k_range:
    kmeans = KMeans(n_clusters=k, random_state=42)
    labels = kmeans.fit_predict(df_scaled_water)
    inertia.append(kmeans.inertia_)
    silhouette_scores.append(silhouette_score(df_scaled_water, labels))

# Elbow and silhouette graphics
fig, ax = plt.subplots(1, 2, figsize=(12, 5))
ax[0].plot(k_range, inertia, 'o-', color='blue')
```

```
ax[0].set_title('Elbow Method')
ax[0].set_xlabel('Number of clusters')
ax[0].set_ylabel('Inertia')

ax[1].plot(k_range, silhouette_scores, 'o-', color='green')
ax[1].set_title('Silhouette Score')
ax[1].set_xlabel('Number of clusters')
ax[1].set_ylabel('Silhouette Score')
plt.tight_layout()
plt.show()

# -------------------------------------------
# Step 4: Application of clustering algorithms
# -------------------------------------------

# K-means with k=3
kmeans = KMeans(n_clusters=3, random_state=42)
labels_kmeans = kmeans.fit_predict(df_scaled_water)

# DBSCAN with adjusted parameters
dbscan = DBSCAN(eps=5, min_samples=5)
labels_dbscan = dbscan.fit_predict(df_scaled_water)

# Agglomerative Clustering with k=3
agglo = AgglomerativeClustering(n_clusters=3)
labels_agglo = agglo.fit_predict(df_scaled_water)

# Spectral Clustering with k=3
spectral = SpectralClustering(n_clusters=3, random_state=42, affinity='nearest_neighbors')
labels_spectral = spectral.fit_predict(df_scaled_water)

# -------------------------------------------
# Step 5: Performance evaluation and comparison
# -------------------------------------------

results = pd.DataFrame({
    'Algorithm': ['K-means', 'DBSCAN', 'Agglomerative', 'Spectral'],
    'Silhouette Score': [
        silhouette_score(df_scaled_water, labels_kmeans),
        silhouette_score(df_scaled_water, labels_dbscan),
        silhouette_score(df_scaled_water, labels_agglo),
        silhouette_score(df_scaled_water, labels_spectral)
    ],
    'Davies-Bouldin Index': [
        davies_bouldin_score(df_scaled_water, labels_kmeans),
        davies_bouldin_score(df_scaled_water, labels_dbscan),
        davies_bouldin_score(df_scaled_water, labels_agglo),
        davies_bouldin_score(df_scaled_water, labels_spectral)
    ]
})

print("Comparative evaluation of the algorithms:")
print(results)
```

```
# ------------------------------------------
# Step 6: Results display (PCA)
# ------------------------------------------

# Reduce dimensions for clear display
pca = PCA(n_components=2)
X_pca = pca.fit_transform(df_scaled_water)

# Gráficos comparativos
fig, axes = plt.subplots(2, 2, figsize=(16, 12))

sns.scatterplot(x=X_pca[:,0], y=X_pca[:,1], hue=labels_kmeans, palette='Set1', ax=axes[0,0])
axes[0,0].set_title('K-means Clusters')

sns.scatterplot(x=X_pca[:,0], y=X_pca[:,1], hue=labels_dbscan, palette='Set2', ax=axes[0,1])
axes[0,1].set_title('DBSCAN Clusters')

sns.scatterplot(x=X_pca[:,0], y=X_pca[:,1], hue=labels_agglo, palette='Set3', ax=axes[1,0])
axes[1,0].set_title('Agglomerative Clusters')

sns.scatterplot(x=X_pca[:,0], y=X_pca[:,1], hue=labels_spectral, palette='Set1', ax=axes[1,1])
axes[1,1].set_title('Spectral Clusters')

for ax in axes.flat:
    ax.set_xlabel('PCA Component 1')
    ax.set_ylabel('PCA Component 2')

plt.tight_layout()
plt.show()
```

****

3.3 R for Clustering

3.3.1 R for Data Science and Clustering: Overview, Strengths, and Weaknesses

R is a programming language and environment specifically designed for statistical computing and graphics. It has been a favorite tool of statisticians and data analysts for decades, particularly in academia. R was released in the mid-1990s (as an open-source implementation inspired by the S language) and since then has grown an expansive library of packages (available through CRAN, the Comprehensive R Archive Network) that cover nearly every statistical or data analysis technique imaginable, including a rich set of clustering methods.

When it comes to clustering, R holds a special place historically. Many clustering algorithms, especially those used in exploratory data analysis, were first made easily available through R packages in the 2000s. For example, methods like Partitioning Around Medoids (PAM) and agglomerative hierarchical clustering were implemented in R's cluster package (Kaufman and Rousseeuw, 1990). This underscores one of R's strengths: it often provides cutting-edge or specialized statistical techniques sooner and in a more specialized form than other languages, because it is developed by the statistical community itself.

Strengths in clustering

- Statistical richness: R has a wide variety of clustering algorithms implemented, often more than available in base libraries of other languages. Beyond standard k-means and hierarchical clustering, R offers a variety of advanced methods through its packages. These include robust approaches such as PAM and CLARA, fuzzy clustering with FANNY, model-based clustering with Mclust, and density-based methods such as DBSCAN. Enhanced hierarchical clustering and graph-based clustering techniques are also available, providing flexible tools for diverse clustering needs.
- Built-in support and ease of use: R provides essential clustering tools without the need for additional packages. The *kmeans()* function in the stats package implements the k-means algorithm and allows easy cluster assignment. The *hclust()* function, combined with *dist()* and *plot()*, enables hierarchical clustering and dendrogram visualization. Basic plotting functions such as *plot()* can create scatter or line plots, using colors and symbols to distinguish clusters.
- Visualization: R is particularly known for its strong graphics capabilities. *ggplot2*, a widely-used plotting package, allows creating complex multi-layered visuals. For clustering, one might use *ggplot2* to create scatter plots with different colors for clusters, add cluster centroids, etc. There are also specialized visualization functions in some clustering packages; for instance, the *factoextra* package provides *fviz_cluster* to automatically create a scatter plot of clusters and even draw cluster ellipses, etc., and *fviz_dend* for dendrograms.
- Interactive data analysis: Tools such as RStudio (a popular IDE for R) and Jupyter notebooks with an R kernel, provide an interactive environment to run clustering analysis step by step, view immediate results, adjust parameters, and so on. RStudio in particular is tailored for R, providing conveniences like viewing data frames, plotting panes, etc., which can make the clustering workflow efficient for an analyst.
- Community and documentation: Because clustering is a well-trodden path in R, one can find many tutorials, examples, and help from the community. Whether it's how to interpret a dendrogram or how to choose a clustering algorithm, there are R documentation pages and vignettes. For example, the cluster package comes with a vignette explaining the algorithms and how to use them.

Weaknesses in clustering

- Speed and scalability: Historically, R has had issues with very large datasets because data is often held in memory and operations can be slower if not vectorized. Clustering can be memory intensive (especially hierarchical clustering which

requires an $O(n^2)$ distance matrix). For moderate data sizes (hundreds or thousands of observations), R is perfectly fine, but if you had, say, a million observations to cluster, base R functions might struggle or become slow. There are packages and strategies to address this, but out-of-the-box Python or certain Java-based tools might handle large scale slightly more gracefully. However, for most environmental datasets (which might be tens of thousands of points at most, or if larger, often aggregated or preprocessed), R's performance is usually not an issue.
- Integration and deployment: R is fantastic for analysis, but integrating R code into production systems is often considered trickier than Python. For example, if you want to deploy a clustering model into a web service, Python's ecosystem is more straightforward. R does have solutions, such as deploying via R Shiny or Plumber APIs, or even calling R from Python using *rpy2*, but this is something to consider if one's project might need to operationalize the clustering. In a pure research or analysis context, this is not a problem at all (R shines there).
- Learning curve: If a user is not already familiar with R, learning R might be slightly more challenging if they come from other languages, because R has its own idiosyncrasies (e.g., how indexing works, factors vs numeric types, etc.). But for someone already comfortable with R's basics, clustering functions are relatively easy to pick up.

In summary, R's primary advantage for clustering is depth and ease of statistical analysis. It's often the go-to for exploring data and trying out different clustering approaches quickly. It also produces publication-quality graphs, which is a reason many researchers use R to create figures for papers (for example, a nicely annotated cluster plot or dendrogram).

3.3.2 Clustering Libraries and Functions in R

Unlike Python's scenario (where one library, scikit-learn, covers most algorithms), in R the functionality is a bit more spread out between base functions and several packages. Here are the key tools:

- Base R (stats package), through the stats package, provides core clustering functions (R Core Team, 2025). The *kmeans()* function performs k-means clustering using the classical Lloyd algorithm (defaulting to the Hartigan-Wong algorithm) and returns cluster assignments, centroids, and within-cluster sum of squares. Users can specify either the number of clusters or the initial centers. The *hclust()* function carries out hierarchical clustering on a distance matrix, with linkage methods like complete, single, average, or Ward's method, producing an object that can be visualized as a dendrogram and segmented into clusters using *cutree()*. The *dist()* function computes distance matrices, essential for hierarchical clustering, supporting various distance measures such as Euclidean or Manhattan. Additionally, the *silhouette()* function from the *cluster* package, often available by default, helps evaluate clustering quality by computing silhouette widths.

- The *cluster* package, often pre-installed with R or easily added via install.packages("cluster"), provides several important clustering functions (Maechler et al., 2025). It includes *pam()* for Partitioning Around Medoids, offering a robust alternative to k-means by minimizing dissimilarities based on representative data points rather than means. For large datasets, *clara()* applies PAM on random samples to improve efficiency. Fuzzy clustering is supported through *fanny()*, which assigns membership coefficients to observations across clusters. Hierarchical clustering methods are also available, with *agnes()* for agglomerative clustering and *diana()* for divisive clustering, each producing detailed cluster objects that can be printed or plotted. Although the plotting capabilities are basic compared to modern standards, they are functional for visualizing results such as dendrograms and cluster plots. Additionally, the package offers silhouette analysis through *silhouette()* and *plot.silhouette()*, aiding in the evaluation of clustering quality.
- The *factoextra* package, developed by Alboukadel Kassambara, is a helpful tool for extracting and visualizing clustering results, although it is not strictly necessary (Kassambara & Mundt, 2020). It simplifies the creation of visualizations, such as cluster scatter plots with *fviz_cluster()*, dendrograms with *fviz_dend()*, and silhouette plots with *fviz_silhouette()*. For high-dimensional data, *factoextra* often uses the first two principal components to generate clear scatter plots, while for two-dimensional data, it plots the original features. The package also offers *get_dist()* to compute distance matrices with additional options, and *fviz_dist()* to visualize them as heatmaps. Due to its simplicity and integration with *ggplot2*, *factoextra* is widely used for teaching and quick clustering analyses.
- The *dbscan* package provides an implementation of the DBSCAN algorithm in R (Hahsler & Piekenbrock, 2025), along with other methods such as OPTICS. The *dbscan()* function performs clustering by assigning cluster labels, with noise typically labeled as 0. To assist in selecting the *eps* parameter, the package includes *kNNdistplot()*, which plots the distance to each point's *k-th* nearest neighbor, helping users identify an appropriate value based on the plot's knee.
- The *mclust* package implements Gaussian Mixture Model clustering in R, offering an interface that automatically evaluates different numbers of clusters and covariance structures (Scrucca et al., 2023). Using *Mclust()*, it fits the best model based on the Bayesian Information Criterion (BIC) and estimates the optimal number of clusters. This package is particularly useful when clustering is suspected but a statistically principled method is needed to determine the number of groups, assuming multivariate normality within clusters.

Several other R packages are worth mentioning for specialized clustering tasks. *NbClust* helps determine the optimal number of clusters by computing multiple validity indices. *Clustertend* assesses clustering tendency, checking whether the data structure is suitable for clustering using measures like the Hopkins statistic. For advanced dendrogram operations, *ape* and *dendextend* offer tools for comparing and customizing dendrograms. Graph-based clustering can be performed with *tidygraph* and *igraph*, while spatial clustering tasks might use *spdep* or *sf*, though these are more niche. Given our focus, we will primarily use base R functions, the *cluster* package, and possibly *factoextra* to simplify visualization and interpretation.

3.3.3 Example: Clustering in R

In this example, we replicate the analysis previously presented in Python, but now using the R programming language (Rstudio). The same environmental dataset from a municipal water treatment plant is used, which includes physicochemical variables such as pH, conductivity, temperature, suspended solids, COD, BOD, among others.

The objective remains the same: to apply unsupervised learning techniques (K-means, DBSCAN, Agglomerative Clustering, and Spectral Clustering) in order to uncover natural groupings or patterns within the water treatment process. These clusters may provide insights into operational conditions, anomalies, or transitional states.

The structure of the R-based analysis includes the following stages:

- Data Preprocessing
- Optimal Number of Clusters Determination
- Clustering Algorithm Implementation
- Performance Evaluation
- Visualization using PCA

The following sections explain each step in detail, with an educational focus and clear code commentary. At the end, a complete version of the code is presented, ready to be copied and executed in R (RStudio or Jupyte Notebook).

1. **Libraries**: Before starting the clustering analysis, it is necessary to import a series of libraries that provide essential functionalities for data management, algorithm application, and results visualization:

```
*************************************************************************

# Required libraries
library(dbscan)
library(kernlab)
library(cluster)
library(clusterCrit)
library(ggplot2)
library(gridExtra)

*************************************************************************
```

This block loads the necessary libraries for the analysis. The *dbscan* package is used to perform density-based clustering, allowing the identification of core points, border points, and noise. The *kernlab* package provides kernel-based algorithms, including spectral clustering, which is useful for capturing non-convex cluster shapes. The *cluster* package includes functions for hierarchical clustering and for computing internal validation measures such as the silhouette width. The *clusterCrit* package offers access to a range of internal validation indices, including the *Silhouette* Score and the *Davies–Bouldin Index*, which are commonly used to evaluate clustering quality. The *ggplot2* package is a widely used visualization library that enables the creation of high-quality, aesthetically pleasing plots. Lastly, the *gridExtra* package allows for the arrangement of multiple *ggplot2* plots in a single visual layout, facilitating side-by-side comparison of clustering results.

2. **Data Preparation and Preprocessing**: Before applying any clustering algorithms, it is crucial to ensure that the dataset is clean, complete, and properly formatted. The following R code demonstrates the essential preprocessing steps applied to the water treatment dataset.

```
***********************************************************************

# Load dataset
df_water <- read_csv("BOOK CLUSTERING/water-treatment.csv")

# Replace missing values marked as '?'
df_water[df_water == "?"] <- NA

# Convert all values to numeric
df_water <- as.data.frame(lapply(df_water, function(x) as.numeric(x)))

# Remove columns with more than 30% missing values and drop remaining incomplete rows
threshold <- 0.7 * nrow(df_water)
df_water <- df_water[ , colSums(!is.na(df_water)) >= floor(threshold)]
df_water <- na.omit(df_water)

# Normalize the dataset
df_scaled_water <- scale(df_water)
***********************************************************************
```

The dataset is loaded using *read_csv()* from the *readr* package. It contains environmental measurements from a municipal water treatment plant, including variables such as pH, temperature, and chemical concentrations. In this dataset, missing values are encoded as question marks ("?"). These are replaced with NA, which is R's standard way of representing missing data. This step is essential for ensuring compatibility with functions that handle missing values. The dataset is then converted entirely to numeric format using *lapply()*. Any non-numeric data (such as strings or characters) will be coerced to numeric values, and invalid conversions will result in NA, which will be handled in the next step. Columns with more than 30% missing values are removed, as they likely lack sufficient information to contribute meaningfully to the analysis. After this, any rows still containing missing values are also removed to ensure that all observations are complete. This is necessary because many clustering algorithms in R do not accept missing data. Finally, the dataset is normalized using the *scale()* function. This step standardizes each variable to have a mean of 0 and a standard deviation of 1. Normalization is particularly important for clustering algorithms, as it ensures that all features contribute equally to distance calculations, regardless of their original units or scales.

3. **Optimal Number of Clusters (K-means)**: To determine the most appropriate number of clusters for the dataset, two widely accepted internal validation methods are applied: the Elbow Method and the Silhouette Score. These approaches help evaluate how well the data points are grouped for different values of k, the number of clusters, without relying on external labels.

```
**********************************************************************

set.seed(42)
k_range <- 2:9
inertia <- numeric(length(k_range))
silhouette_scores <- numeric(length(k_range))

d_euclidean <- dist(df_scaled_water)

for (i in seq_along(k_range)) {
 km <- kmeans(df_scaled_water, centers = k_range[i], nstart = 10)
 inertia[i] <- km$tot.withinss
 sil <- silhouette(km$cluster, d_euclidean)
 silhouette_scores[i] <- mean(sil[, "sil_width"])
}
 # Plots
 df_eval <- data.frame(Clusters = k_range, Inertia = inertia, Silhouette = silhouette_scores)

 p1 <- ggplot(df_eval, aes(x = Clusters, y = Inertia)) +
  geom_point(color = "blue") + geom_line(color = "blue") +
  ggtitle("Elbow Method") + xlab("Number of Clusters") + ylab("Inertia")

 p2 <- ggplot(df_eval, aes(x = Clusters, y = Silhouette)) +
  geom_point(color = "green") + geom_line(color = "green") +
  ggtitle("Silhouette Score") + xlab("Number of Clusters") + ylab("Silhouette Score")

 grid.arrange(p1, p2, ncol = 2)

**********************************************************************
```

The code begins by setting a random seed for reproducibility using *set.seed(42)*. A range of cluster values from 2 to 9 is defined and stored in *k_range*. Two empty vectors, *inertia* and *silhouette_scores*, are initialized to store the evaluation results for each value of k. A Euclidean distance matrix (*d_euclidean*) is calculated from the normalized data, which will be used for computing silhouette scores.

Within the loop, for each k value, the K-means algorithm is applied using *kmeans()* with 10 random initializations ($nstart = 10$) to ensure stability in the results. The total within-cluster sum of squares is stored in *inertia[i]*, which measures how compact the clusters are. A smaller value indicates that points are closer to their assigned cluster centers. The silhouette values, calculated using the *silhouette()* function, reflect the cohesion within clusters and the separation between clusters. The mean silhouette

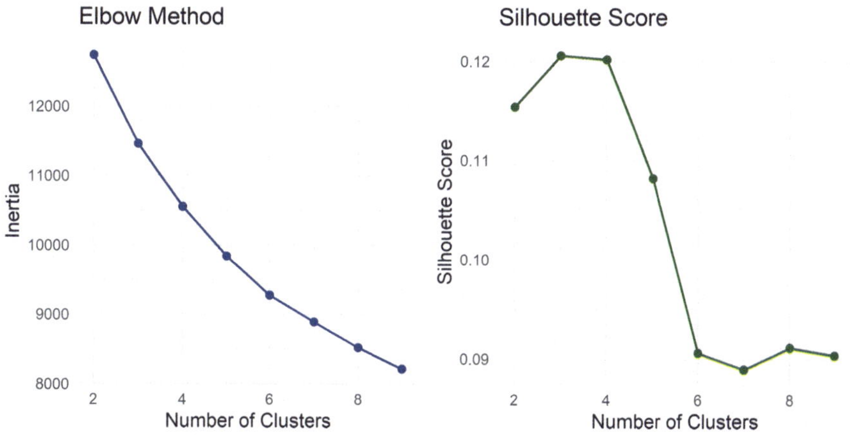

Fig. 3.7 Optimal number of clusters determined by the Elbow Method and Silhouette Score (Rstudio)

width is stored in *silhouette_scores[i]* as an overall measure of clustering quality for that k.

After computing both metrics for all values of k, the results are compiled into a data frame and visualized with two side-by-side plots (Fig. 3.7) using *ggplot2*. The first plot (Elbow Method) shows how inertia decreases as the number of clusters increases, with a noticeable change in slope indicating a potential optimal number of clusters. The second plot (Silhouette Score) shows how the average silhouette width varies with k, helping to identify the number of clusters that yields the highest cohesion and separation.

According to the visual inspection of these plots (Fig. 3.7), the optimal number of clusters lies around 3 or 4. In this case, $k = 3$ is selected as the most balanced and consistent choice based on both metrics. This step ensures that subsequent clustering analyses are performed under a configuration that reflects meaningful structure in the data.

4. **Clustering Algorithm Implementation**: In this step, four different clustering algorithms are applied to the preprocessed dataset to explore various grouping structures. Using these diverse methods provides a robust basis for understanding the data's underlying patterns from multiple analytical perspectives.

```
****************************************************************************

# K-means
set.seed(42)
kmeans_result <- kmeans(df_scaled_water, centers = 3, nstart = 10)
labels_kmeans <- kmeans_result$cluster

# DBSCAN
dbscan_result <- dbscan(df_scaled_water, eps = 5, minPts = 5)
labels_dbscan <- dbscan_result$cluster

# Agglomerative (Hierarchical)
d <- dist(df_scaled_water)
hc <- hclust(d, method = "ward.D2")
labels_agglo <- cutree(hc, k = 3)

# Spectral Clustering
spectral_model <- specc(as.matrix(df_scaled_water), centers = 3)
labels_spectral <- spectral_model@.Data

****************************************************************************
```

5. **Performance Evaluation**: In this step, clustering performance is evaluated using the Silhouette Score and the DBI, two internal metrics that assess cluster cohesion and separation. The Silhouette Score indicates how well points fit within their assigned clusters, with higher values reflecting better-defined groupings. The DBI measures how distinct the clusters are, with lower values indicating better separation. DBSCAN achieved the highest Silhouette Score, suggesting strong internal cohesion, but also showed the highest DBI, indicating greater overlap between clusters. This highlights the need to consider multiple metrics when comparing clustering results (see Table 3.2).

Table 3.2 Summarizes the values obtained by each algorithm in R

Algorithm	Silhouette score	Davies-Bouldin index
K-means	0.09744102	1.854991
DBSCAN	0.08515396	4.237612
Agglomerative	NaN	2.014509
Spectral	0.09568897	2.434893

```
*************************************************************************

# Adjust DBSCAN labels
labels_dbscan_adj <- as.integer(factor(labels_dbscan, levels = sort(unique(labels_dbscan))))

score_km <- intCriteria(as.matrix(df_scaled_water), labels_kmeans, c("Silhouette", "Davies_Bouldin"))
score_db <- intCriteria(as.matrix(df_scaled_water), labels_dbscan_adj, c("Silhouette", "Davies_Bouldin"))
score_ag <- intCriteria(as.matrix(df_scaled_water), labels_agglo, c("Silhouette", "Davies_Bouldin"))
score_sp <- intCriteria(as.matrix(df_scaled_water), labels_spectral, c("Silhouette", "Davies_Bouldin"))

results <- data.frame(
  Algorithm = c("K-means", "DBSCAN", "Agglomerative", "Spectral"),
  Silhouette_Score = c(score_km$silhouette, score_db$silhouette, score_ag$silhouette, score_sp$silhouette),
  Davies_Bouldin_Index = c(score_km$davies_bouldin, score_db$davies_bouldin, score_ag$davies_bouldin,
score_sp$davies_bouldin)
)

print(results)

*************************************************************************
```

6. **Visualization using PCA**: To clearly interpret the clustering structure, we apply PCA to reduce dimensionality. The PCA visualization provides a clear and interpretable 2D representation of the clusters. DBSCAN stands out by identifying outliers, while K-means and Spectral Clustering produce more balanced groupings (see Fig. 3.8).

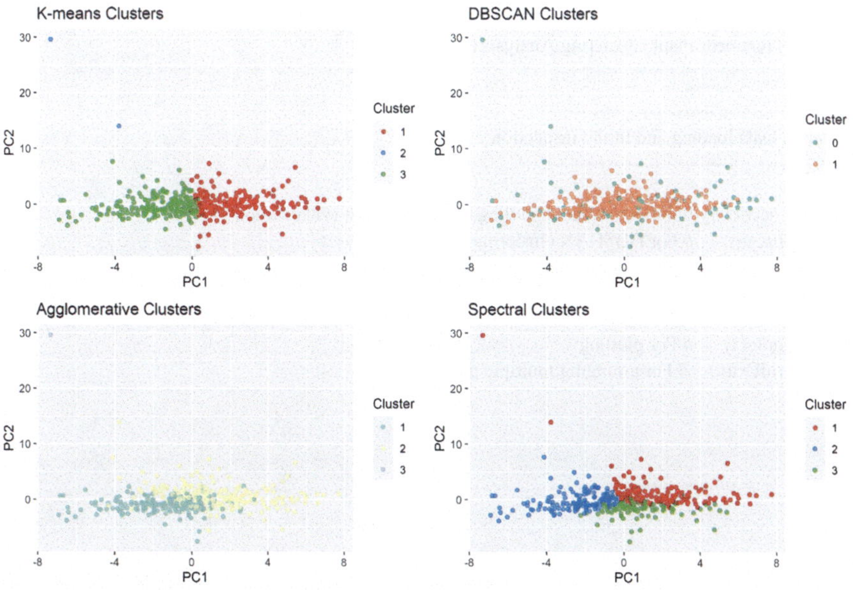

Fig. 3.8 PCA-Based visualization of clustering results using K-means, DBSCAN, Agglomerative, and Spectral algorithms (Rstudio)

****

```
pca_model <- prcomp(df_scaled_water, rank. = 2)
pca_coords <- pca_model$x[, 1:2]

plot_clusters <- function(coords, labels, title, palette){
  df <- data.frame(PC1 = coords[,1], PC2 = coords[,2], Cluster = factor(labels))
  ggplot(df, aes(x = PC1, y = PC2, color = Cluster)) +
    geom_point(alpha = 0.7) +
    scale_color_brewer(palette = palette) +
    labs(title = title, x = "PC1", y = "PC2")
}

p_km <- plot_clusters(pca_coords, labels_kmeans, "K-means Clusters", "Set1")
p_db <- plot_clusters(pca_coords, labels_dbscan, "DBSCAN Clusters", "Set2")
p_ag <- plot_clusters(pca_coords, labels_agglo, "Agglomerative Clusters", "Set3")
p_sp <- plot_clusters(pca_coords, labels_spectral, "Spectral Clusters", "Set1")

grid.arrange(p_km, p_db, p_ag, p_sp, ncol = 2, nrow = 2)
```

****

Finally, the full code is provided below, ready to copy and execute in R (RStudio or Jupyter Notebook). This version includes all code blocks above, without additional explanation.

****

```
#=======================================================================
# Water Treatment Plant Clustering Analysis (R)
#=======================================================================

# -----------------------------------------
# Step 1: Data loading and initial inspection
# -----------------------------------------

# Import necessary libraries for data handling, clustering, and visualization
library(dbscan)      # For DBSCAN clustering
library(kernlab)     # For spectral clustering (specc function)
library(cluster)     # For silhouette function and hierarchical clustering
library(clusterCrit) # For internal validation indices (Silhouette, Davies-Bouldin)
library(ggplot2)     # For plotting
library(gridExtra)   # For arranging multiple plots
```

```
# Load the dataset (CSV) into a data frame
# The dataset uses "?" to denote missing values, read as strings for now

df_water <- read_csv("BOOK CLUSTERING/water-treatment.csv")

# Initial inspection of data: view first few rows and structure

print("Initial view of the dataset:")
print(head(df_water))
print(str(df_water))

# The structure reveals multiple columns (operational variables) and their types.
# Many columns are read as character due to the presence of "?" as missing entries.

# -----------------------------
# Step 2: Data preprocessing
# -----------------------------

# Replace missing value placeholders ("?") with NA (R's missing value)
df_water[df_water == "?"] <- NA

# Convert all columns to numeric type (coercing non-numeric to NA)
df_water <- as.data.frame(lapply(df_water, function(x) as.numeric(x)))
# At this point, any values that were non-numeric (including "?") are now NA,
# and all columns are of numeric type.

# Remove columns with more than 30% missing values
threshold <- 0.7 * nrow(df_water)     # minimum non-NA entries required (70% of rows)
df_water <- df_water[ , colSums(!is.na(df_water)) >= floor(threshold)]
# Remove any remaining rows that still contain NA (ensure complete cases)
df_water <- na.omit(df_water)
# After this cleaning, df_water contains only numeric data with no missing values.

# Normalize the data for clustering (zero mean, unit variance for each variable)
df_scaled_water <- scale(df_water)

# ------------------------------------------------------------
# Step 3: Optimal cluster determination using Elbow and Silhouette
# ------------------------------------------------------------

# Use K-means clustering on a range of k values to evaluate inertia and silhouette
set.seed(42)  # for reproducibility of k-means initialization
k_range <- 2:9
inertia <- numeric(length(k_range))          # sum of squared distances (within-cluster)
silhouette_scores <- numeric(length(k_range))  # average silhouette score for each k

# Compute K-means for each k, record inertia and silhouette
d_euclidean <- dist(df_scaled_water)  # pre-compute distance matrix for silhouette calculations
for (i in seq_along(k_range)) {
  k <- k_range[i]
  km <- kmeans(df_scaled_water, centers = k, nstart = 10)
  inertia[i] <- km$tot.withinss  # total within-cluster sum of squares (inertia)
```

```
  # Compute average silhouette score for this clustering
  sil <- silhouette(km$cluster, d_euclidean)
  silhouette_scores[i] <- mean(sil[, "sil_width"])
}

# Plot the Elbow Method (inertia) and Silhouette scores for each k
df_eval <- data.frame(Clusters = k_range, Inertia = inertia, Silhouette = silhouette_scores)
p1 <- ggplot(df_eval, aes(x = Clusters, y = Inertia)) +
  geom_point(color = "blue") + geom_line(color = "blue") +
  ggtitle("Elbow Method") + xlab("Number of clusters") + ylab("Inertia")
p2 <- ggplot(df_eval, aes(x = Clusters, y = Silhouette)) +
  geom_point(color = "green") + geom_line(color = "green") +
  ggtitle("Silhouette Score") + xlab("Number of clusters") + ylab("Silhouette Score")
grid.arrange(p1, p2, ncol = 2)


# --------------------------------------------------
# Step 4: Apply clustering algorithms with chosen k=3
# --------------------------------------------------

# K-means clustering with k = 3 clusters
set.seed(42)  # ensure reproducible clustering
kmeans_result <- kmeans(df_scaled_water, centers = 3, nstart = 10)
labels_kmeans <- kmeans_result$cluster  # cluster assignments (1, 2, 3)

# DBSCAN clustering with eps = 5 and minPts = 5 (density-based clustering)
dbscan_result <- dbscan(df_scaled_water, eps = 5, minPts = 5)
labels_dbscan <- dbscan_result$cluster   # cluster assignments (0 is noise, 1,2,... are clusters)
# Note: In DBSCAN, points labeled "0" are considered noise/outliers not belonging to any cluster.

# Agglomerative Hierarchical Clustering with Ward's method, cut into 3 clusters
d <- dist(df_scaled_water)                 # distance matrix
hc <- hclust(d, method = "ward.D2")          # hierarchical clustering (Ward linkage)
labels_agglo <- cutree(hc, k = 3)           # cut tree to obtain 3 clusters (labels 1,2,3)

# Spectral clustering with k = 3 clusters using Gaussian RBF kernel (default in specc)
spectral_model <- specc(as.matrix(df_scaled_water), centers = 3)
labels_spectral <- spectral_model@.Data       # extract cluster assignments (1,2,3)


# -------------------------------------------------------
# Step 5: Performance evaluation and comparison of clusters
# -------------------------------------------------------

# Calculate Silhouette Score and Davies-Bouldin Index for each clustering result
# (Using the scaled data for internal validation measures)
# Prepare cluster label vectors (ensure no invalid labels for indices)

labels_dbscan_adj <- as.integer(factor(labels_dbscan, levels = sort(unique(labels_dbscan))))
# The above converts DBSCAN labels into a contiguous sequence starting at 1
# (treating noise as its own cluster) for evaluation purposes.

# Compute internal validation indices using clusterCrit
```

```
score_km <- intCriteria(as.matrix(df_scaled_water), labels_kmeans, c("Silhouette", "Davies_Bouldin"))
score_db <- intCriteria(as.matrix(df_scaled_water), labels_dbscan_adj, c("Silhouette", "Davies_Bouldin"))
score_ag <- intCriteria(as.matrix(df_scaled_water), labels_agglo, c("Silhouette", "Davies_Bouldin"))
score_sp <- intCriteria(as.matrix(df_scaled_water), labels_spectral, c("Silhouette", "Davies_Bouldin"))

sil_k <- score_km$silhouette
sil_d <- score_db$silhouette
sil_a <- score_ag$silhouette
sil_s <- score_sp$silhouette

db_k <- score_km$davies_bouldin
db_d <- score_db$davies_bouldin
db_a <- score_ag$davies_bouldin
db_s <- score_sp$davies_bouldin

# Create a summary table of the results
results <- data.frame(
  Algorithm = c("K-means", "DBSCAN", "Agglomerative", "Spectral"),
  Silhouette_Score = c(sil_k, sil_d, sil_a, sil_s),
  Davies_Bouldin_Index = c(db_k, db_d, db_a, db_s)
)
print("Comparative evaluation of the algorithms:")
print(results)


# ---------------------------------------------
# Step 6: Results visualization using 2D PCA plot
# ---------------------------------------------

# Reduce data to 2 principal components for visualization
pca_model <- prcomp(df_scaled_water, rank. = 2)   # PCA to 2 dimensions
pca_coords <- pca_model$x[, 1:2]                  # scores for first two principal components

# Prepare data frames for plotting each clustering result
df_km <- data.frame(PC1 = pca_coords[,1], PC2 = pca_coords[,2], Cluster = factor(labels_kmeans))
df_db <- data.frame(PC1 = pca_coords[,1], PC2 = pca_coords[,2], Cluster = factor(labels_dbscan))
df_ag <- data.frame(PC1 = pca_coords[,1], PC2 = pca_coords[,2], Cluster = factor(labels_agglo))
df_sp <- data.frame(PC1 = pca_coords[,1], PC2 = pca_coords[,2], Cluster = factor(labels_spectral))

# Build scatter plots for each clustering outcome
p_km <- ggplot(df_km, aes(x = PC1, y = PC2, color = Cluster)) +
  geom_point(alpha = 0.7) + scale_color_brewer(palette = "Set1") +
  labs(title = "K-means Clusters", x = "PC1", y = "PC2")
p_db <- ggplot(df_db, aes(x = PC1, y = PC2, color = Cluster)) +
  geom_point(alpha = 0.7) + scale_color_brewer(palette = "Set2") +
  labs(title = "DBSCAN Clusters", x = "PC1", y = "PC2")
p_ag <- ggplot(df_ag, aes(x = PC1, y = PC2, color = Cluster)) +
  geom_point(alpha = 0.7) + scale_color_brewer(palette = "Set3") +
  labs(title = "Agglomerative Clusters", x = "PC1", y = "PC2")
p_sp <- ggplot(df_sp, aes(x = PC1, y = PC2, color = Cluster)) +
  geom_point(alpha = 0.7) + scale_color_brewer(palette = "Set1") +
  labs(title = "Spectral Clusters", x = "PC1", y = "PC2")
```

```
# Arrange the four scatter plots in a 2x2 grid for comparison
grid.arrange(p_km, p_db, p_ag, p_sp, ncol = 2, nrow = 2)
```

****

3.4 MATLAB for Clustering

3.4.1 *MATLAB for Data Science and Clustering: Overview, Strengths, and Weaknesses*

MATLAB is a high-level programming and interactive environment developed by MathWorks, first released in the 1980s. The name stands for Matrix Laboratory because it was originally designed for matrix computations (e.g., solving linear algebra problems). Over the years, MATLAB has grown into a broad platform for technical computing, with toolboxes covering statistics, machine learning, signal processing, optimization, and more. In many academic fields, especially engineering, MATLAB has been a staple due to its ease of use (many find its language and environment straightforward to pick up for numerical tasks) and powerful built-in functions.

In the context of clustering and more broadly machine learning, MATLAB provides functionality primarily through the Statistics and Machine Learning Toolbox (which is an add-on product; in older versions, the statistics toolbox was separate, now they are combined). There are also specialized toolboxes (for example, a Deep Learning Toolbox, a Fuzzy Logic Toolbox, etc.) that include algorithms that can be applied to clustering. For these purposes, the focus will be placed on what is available in the Statistics and Machine Learning Toolbox, as it includes implementations of K-means, hierarchical clustering, DBSCAN, Gaussian mixtures, among others, analogous to those discussed for Python and R.

Strengths in clustering

- Integrated Environment and Ease of Use: MATLAB's IDE is well-suited for iterative experimentation. The workspace browser, variable editor, and plotting tools allow one to quickly load data (from a.mat file, CSV, etc.), run clustering, and visualize results. For someone who is not primarily a software developer, MATLAB can be less intimidating than dealing with package installations or environment variables as in Python/R. Everything in MATLAB is contained in the environment after installation. This makes it popular in academic settings for teaching and prototyping.
- Performance: MATLAB's algorithms are typically implemented in highly optimized native code (C/C++ or Fortran behind the scenes) and make use of MATLAB's optimized math libraries. Clustering algorithms such as k-means or hierarchical clustering in MATLAB are often very fast on moderately large

data because they leverage vectorized operations. Moreover, MATLAB easily takes advantage of multi-core CPUs for many operations. If you have the Parallel Computing Toolbox, some clustering can also be parallelized or run on GPU (e.g., kmeans in recent versions can use parallelism).

- Rich math and visualization library: Once clusters are obtained, MATLAB has strong capabilities for further analysis, computing statistics, plotting results (2D, 3D plots). For instance, MATLAB's plotting functions can produce scatter plots, 3D plots, and even interactive figures (with brushing data points, etc.). Creating a publication-quality figure might sometimes be easier in R or Python with fine control, but MATLAB too can produce high-quality graphs and offers a lot of control over appearance.
- Toolbox algorithms: MATLAB's Stats and ML Toolbox includes not just clustering but also classification, regression, etc. This means you can do a holistic analysis in one environment. For example, after clustering data, you might use classification tools to see if you can predict cluster membership from other features, or use the PCA function for dimensionality reduction prior to clustering, etc., all within MATLAB.
- Specialized methods: Some algorithms are available in MATLAB that might not be as readily available in base Python/R without extra libraries. For example, Fuzzy C-Means (FCM) clustering is included in MATLAB's Fuzzy Logic Toolbox (*fcm* function). Also, MATLAB has interactive tools, e.g., in the Classification Learner / Clustering app where you can perform clustering via a GUI for those who prefer that to coding.

Weaknesses in clustering

- Cost and Accessibility: MATLAB is a commercial product. Not everyone has access to it. This is a major practical limitation. Python and R are free; thus, in contexts where open-source is mandated or for broader accessibility (e.g., sharing code with colleagues who may not have MATLAB licenses), MATLAB is at a disadvantage.
- Community and Extensibility: While MATLAB has a community (MATLAB Central, File Exchange where users share custom functions), it's smaller compared to the open-source communities for Python and R. If you need a very cutting-edge or niche clustering algorithm, you might find a user-contributed MATLAB function on File Exchange, but it might not be as robust or up-to-date as an R/Python package maintained by dozens of contributors. Python/R often get the newest research implementations faster due to the open nature. MATLAB updates rely on MathWorks or enthusiastic users.
- Language and Integration: The MATLAB language, while easy for numerical computing, is not as general-purpose as Python. For instance, file handling or string processing is better now than it used to be, but still not as convenient as Python's. If you need to integrate clustering results into a larger software system (like a web service, or into a database), Python or R might offer more direct routes. MATLAB can interface with other languages (there's a MATLAB engine for Python, you can compile MATLAB code, etc.), but these add complexity.

- Memory consumption: In some cases, MATLAB can be heavy on memory. It often keeps multiple copies of data if you're not careful (though newer versions improved with something called copy-on-write). For large datasets, one must ensure not to run out of memory. Python with libraries like NumPy is also memory heavy at times, but offers ways to handle out-of-core data with other libraries. R likewise might have memory issues if not using special packages, so this weakness is not unique to MATLAB, but worth noting.

Overall, MATLAB is a strong tool for prototyping clustering on moderately large datasets, especially if you already have MATLAB in your workflow (common in engineering projects, e.g., an electrical engineer clustering load profiles might already be simulating circuits in MATLAB, so sticking to one environment is convenient). In a sustainable energy context, MATLAB might be used to cluster time-series data (such as load curves, wind power output patterns, etc.), and then use those clusters in further simulations or optimizations (which MATLAB also handles).

3.4.2 Clustering Functions and Tools in MATLAB

In MATLAB (with the Statistics and Machine Learning Toolbox), the key functions for clustering include:

- kmeans: This function performs k-means clustering. MATLAB's kmeans is highly optimized and can automatically use parallel computing if available. It also supports different algorithms (like using k-means++ initialization via Name-Value pairs).
- Hierarchical Clustering: MATLAB supports this several functions. The *pdist* function computes pairwise distances, similar to R's dist or Python's *pairwise_distances*, returning a vector of distances. The linkage function then constructs the hierarchical clustering tree, producing a matrix that encodes the tree structure, much like R's *hclust* object or Python's linkage matrix. To visualize the clustering, dendrogram plots the hierarchical tree, with the option to display all leaves. For a more streamlined approach, the *clusterdata* function combines distance computation, linkage, and tree cutting into a single step, directly assigning observations to clusters based on the specified method and number of clusters. Additionally, the cluster function can manually cut a linkage result to form a specified number of clusters.
- DBSCAN: Introduced in newer versions (around 2017 or later). It returns cluster indices (with -1 for noise points, or 0 depending on version, but typically -1 is noise). You can specify a distance metric. DBSCAN in MATLAB identifies core points and noise.
- Gaussian Mixture Models: MATLAB offers fitgmdist to fit a Gaussian mixture distribution to data. While primarily for modeling, it can be used for clustering by looking at the component assignments:

MATLAB offers several clustering methods beyond hierarchical clustering. Since R2018b or R2019a, the *kmedoids* function, part of the Statistics and Machine Learning Toolbox, allows clustering around medoids, similar to *PAM* in R. Although MATLAB does not have a dedicated *spectral clustering* function, it can be performed manually through similarity matrix computation and eigen decomposition, or indirectly using graph-related functions. *Agglomerative clustering* with various linkage methods is supported through linkage and cluster, although custom implementations may be needed for specialized constraints. *Fuzzy clustering* is available via the fcm function in the Fuzzy Logic Toolbox. For validation and evaluation, MATLAB provides the silhouette function to visualize and assess clustering quality, and evalclusters to determine the optimal number of clusters using criteria such as *Calinski-Harabasz* and *Silhouette scores.*

3.4.3 Example: Clustering in MATLAB

In this example, we replicate the clustering analysis previously conducted in Python and R, now using MATLAB. The dataset remains the same: a real-world environmental dataset obtained from a municipal water treatment plant, consisting of physicochemical variables such as pH, conductivity, temperature, suspended solids, COD, BOD, among others.

The goal is to apply unsupervised learning techniques in MATLAB to uncover natural groupings in the data, identify typical operational states, transitional phases, or detect anomalies. Due to differences in algorithm availability across platforms, the following clustering methods are used in MATLAB: K-means, DBSCAN, Agglomerative Hierarchical Clustering, and Gaussian Mixture Models (GMM) (as an alternative to Spectral Clustering).

The analysis includes the following stages:

1. Data Preprocessing
2. Optimal Number of Clusters Determination
3. Clustering Algorithm Implementation
4. Performance Evaluation
5. PCA-Based Visualization

Each step is explained in detail, followed by a complete MATLAB script ready to be executed.

1. **Data Preprocessing**: Before performing any clustering analysis, it is essential to ensure that the dataset is clean, consistent, and formatted appropriately for mathematical operations. This step is crucial to avoid errors during model execution and to improve the quality of clustering results. It is important to mention that the first column with “D-1/3/90” type format has been eliminated from the database (watertreatment).

```
*********************************************************************

% Load the dataset and treat '?' as missing values
data = readtable('watertreatment.csv');

% Convert to numeric matrix and remove columns with more than 30% missing values
data_numeric = table2array(data);
num_thresh = round(0.7 * size(data_numeric,1));
data_numeric(:, sum(~isnan(data_numeric)) < num_thresh) = [];

% Remove any remaining rows with missing values
data_numeric = rmmissing(data_numeric);

% Normalize the data (zero mean and unit variance)
data_scaled = normalize(data_numeric);

*********************************************************************
```

The readtable command loads the dataset from a CSV file into a MATLAB table. If the dataset contains missing values represented by special characters such as '?', these should be treated as missing during import (via the 'TreatAsEmpty' parameter, which may be added as needed). MATLAB stores the dataset as a table to preserve variable names and heterogeneous data types. Once loaded, the table is converted into a numerical matrix using *table2array*, which assumes that all columns contain numeric or convertible data. Then, columns with more than 30% missing entries are removed. This threshold ensures that columns with insufficient or unreliable information do not influence the clustering process. After filtering the columns, any remaining rows that still contain missing values (NaN) are removed using *rmmissing*. This guarantees that the dataset is complete and ready for the mathematical operations required by clustering algorithms, which generally do not tolerate missing values. Finally, the data is standardized using *normalize*, which scales each feature to have zero mean and unit variance. This step is essential because clustering algorithms such as K-means and *Gaussian Mixture Models* are sensitive to the scale of variables. Without normalization, features with larger ranges may dominate the distance calculations and bias the clustering outcome.

2. **Optimal Number of Clusters (K-means)**: Before applying K-means clustering, it is important to determine the most appropriate number of clusters, denoted as k, to partition the data. Selecting too few clusters may cause underfitting, while too many may lead to overfitting or meaningless segmentation. In this step, we use two widely accepted heuristic methods: the Elbow Method and the Silhouette Score.

```
************************************************************************

% Determine optimal number of clusters using Elbow and Silhouette methods
k_range = 2:9;
inertia = zeros(length(k_range),1);
silhouette_scores = zeros(length(k_range),1);

for k = k_range
   [idx, ~, sumd] = kmeans(data_scaled, k, 'Replicates', 10, 'Display', 'off');
   inertia(k-1) = sum(sumd);
   silhouette_avg = mean(silhouette(data_scaled, idx));
   silhouette_scores(k-1) = silhouette_avg;

end
************************************************************************
```

This section tests different values of k (from 2 to 9) to identify the optimal number of clusters using the Elbow Method and Silhouette Score. For each k, K-means is applied, and metrics are computed to assess cluster compactness and separation. The Elbow plot helps identify the point where adding more clusters yields diminishing returns, while the Silhouette plot highlights the most coherent clustering structure. Both metrics suggest that $k = 3$ provides a balanced and meaningful partitioning of the data, which is used in the next clustering steps (see Fig. 3.9).

3. **Clustering Algorithm Implementation**: This section applies four different clustering algorithms to the preprocessed and normalized dataset, allowing comparison between methods that rely on different assumptions about the structure and distribution of the data.

```
************************************************************************

% K-means Clustering (k=3)
[idx_kmeans, ~] = kmeans(data_scaled, 3, 'Replicates', 10);

% DBSCAN
epsilon = 2;  % Tuned after exploratory analysis
MinPts = 5;
labels_dbscan = dbscan(data_scaled, epsilon, MinPts);

% Agglomerative Hierarchical Clustering
Z = linkage(data_scaled, 'ward');
labels_agglom = cluster(Z, 'maxclust', 3);

% Gaussian Mixture Model (GMM) – Alternative to Spectral Clustering
gmModel = fitgmdist(data_scaled, 3, 'RegularizationValue', 1e-5);
labels_gmm = cluster(gmModel, data_scaled);

************************************************************************
```

K-means partitions the data into three clusters by minimizing intra-cluster variance, using multiple initializations to improve stability. DBSCAN identifies clusters

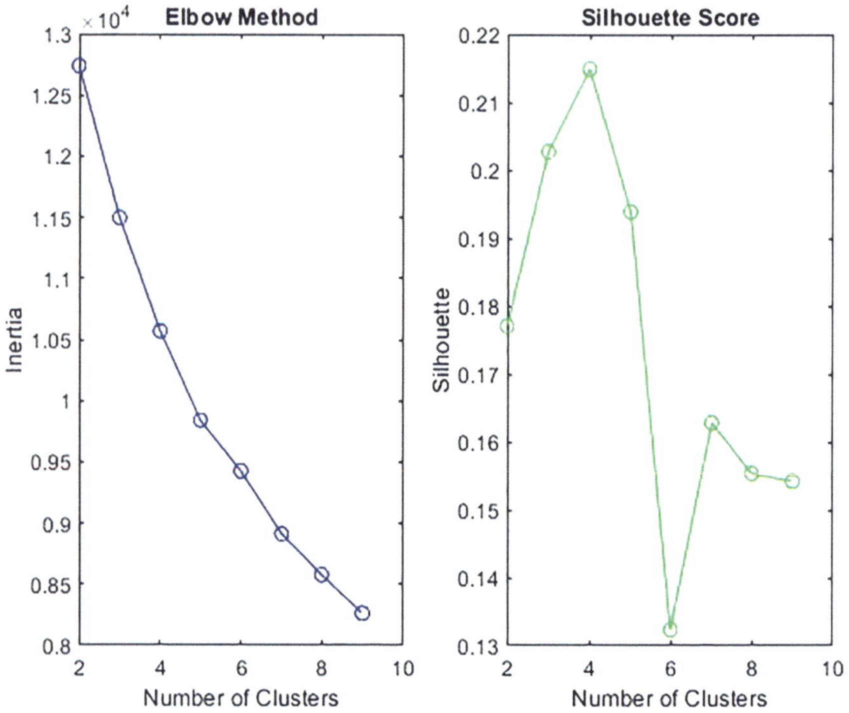

Fig. 3.9 Optimal number of clusters determined by the Elbow Method and Silhouette Score (MATLAB)

based on data density, allowing for arbitrary shapes and detecting outliers, without needing to predefine the number of clusters. Agglomerative clustering builds a hierarchy of clusters using Ward's linkage and then cuts the dendrogram to form three groups. Gaussian Mixture Models assume the data comes from overlapping Gaussian distributions and assign cluster membership probabilistically, with regularization ensuring numerical stability. Together, these methods provide complementary insights into the data's structure.

4. **Performance Evaluation**: After applying the clustering algorithms, it is essential to assess how well each method partitions the data. In this step, we use two internal validation metrics: Silhouette Score and DBI, both of which require cluster labels to be formatted correctly.

```
************************************************************************

% Set DBSCAN labels to be consecutive positive integers
labels_dbscan_adj = labels_dbscan;
labels_dbscan_adj(labels_dbscan == 0) = -1;
labels_dbscan_adj = grp2idx(labels_dbscan_adj);

% Silhouette Score
eva_kmeans   = evalclusters(data_scaled, idx_kmeans, 'silhouette');
eva_dbscan   = evalclusters(data_scaled, labels_dbscan_adj, 'silhouette');
eva_agglom   = evalclusters(data_scaled, labels_agglom, 'silhouette');
eva_gmm      = evalclusters(data_scaled, labels_gmm, 'silhouette');

% Davies–Bouldin Index
eva_kmeans_db = evalclusters(data_scaled, idx_kmeans, 'DaviesBouldin');
eva_dbscan_db = evalclusters(data_scaled, labels_dbscan_adj, 'DaviesBouldin');
eva_agglom_db = evalclusters(data_scaled, labels_agglom, 'DaviesBouldin');
eva_gmm_db    = evalclusters(data_scaled, labels_gmm, 'DaviesBouldin');

% Comparative table
Algorithms = {'K-means','DBSCAN','Agglomerative','GMM'};
Results = table(Algorithms', ...
  [eva_kmeans.CriterionValues; eva_dbscan.CriterionValues; eva_agglom.CriterionValues;
eva_gmm.CriterionValues], ...
  [eva_kmeans_db.CriterionValues; eva_dbscan_db.CriterionValues; eva_agglom_db.CriterionValues;
eva_gmm_db.CriterionValues], ...
  'VariableNames', {'Algorithm','Silhouette','DaviesBouldin'});
disp('Comparative Evaluation:');
disp(Results);

************************************************************************
```

To ensure compatibility with MATLAB's evalclusters function, DBSCAN labels are adjusted to be consecutive positive integers, since the algorithm labels noise as 0. The Silhouette Score, ranging from -1 to 1, is used to assess cluster cohesion and separation for each method. The final results are summarized in a comparative table, where higher Silhouette Scores and lower DBI values indicate better clustering performance.

5. **PCA-Based Visualization**: To facilitate visual interpretation of the clustering results, we apply PCA to reduce the high-dimensional dataset to just two dimensions. This allows us to plot the observations in a 2D space while preserving as much of the original variance as possible.

```
**********************************************************************
[~, score, ~] = pca(data_scaled);

% Visualization
titles = {'K-means','DBSCAN','Agglomerative','GMM'};
cluster_labels = {idx_kmeans, labels_dbscan, labels_agglom, labels_gmm};

figure;
for i = 1:4
    subplot(2,2,i);
    gscatter(score(:,1), score(:,2), cluster_labels{i});
    title(titles{i});
    xlabel('PC1');
    ylabel('PC2');
    grid on;
end

**********************************************************************
```

PCA is applied to reduce the dataset to two principal components (PC1 and PC2), which capture the most variance and allow for visual representation. Four scatter plots are then generated, one for each clustering algorithm, showing how the data points are grouped in this reduced space (see Fig. 3.10). These visualizations provide a clear comparison of clustering behavior, helping to identify which methods yield distinct, compact clusters and which reveal overlapping groups or outliers.

Finally, the full code is provided below, ready to copy and execute in MATLAB. This version includes all code blocks above, without additional explanation.

```
**********************************************************************
% Load the dataset and treat '?' as missing values
data = readtable('watertreatment.csv');

% Convert to numeric matrix and remove columns with more than 30% missing values
data_numeric = table2array(data);
num_thresh = round(0.7 * size(data_numeric,1));
data_numeric(:, sum(~isnan(data_numeric)) < num_thresh) = [];

% Remove any remaining rows with missing values
data_numeric = rmmissing(data_numeric);

% Normalize the data (zero mean and unit variance)
data_scaled = normalize(data_numeric);


%Optimal Number of Clusters (K-means)
```

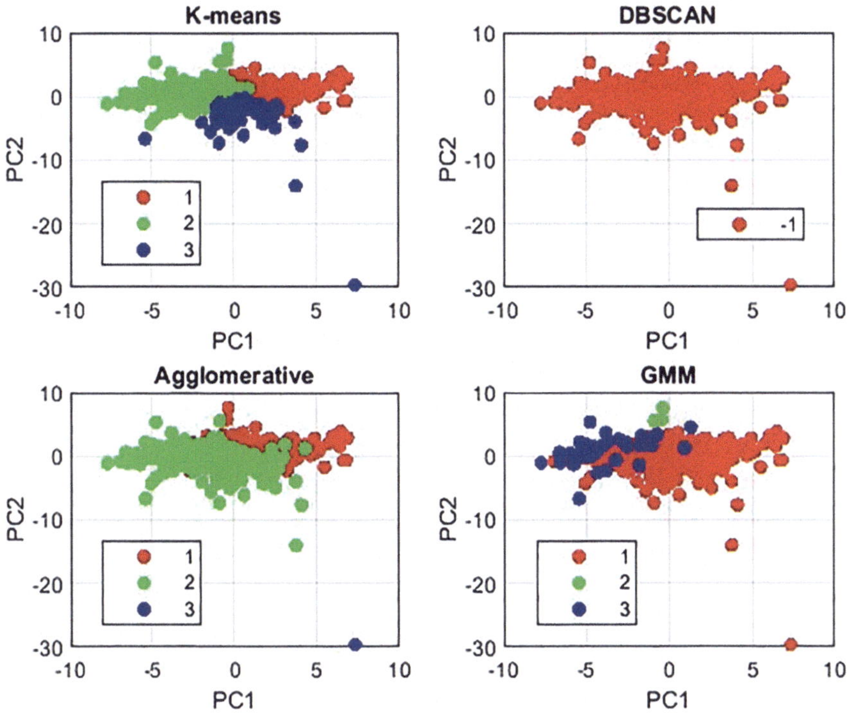

Fig. 3.10 PCA-Based visualization of clustering results using K-means, DBSCAN, Agglomerative, and Spectral algorithms (Matlab)

```
% Determine optimal number of clusters using Elbow and Silhouette methods
k_range = 2:9;
inertia = zeros(length(k_range),1);
silhouette_scores = zeros(length(k_range),1);

for k = k_range
    [idx, ~, sumd] = kmeans(data_scaled, k, 'Replicates', 10, 'Display', 'off');
    inertia(k-1) = sum(sumd);
    silhouette_avg = mean(silhouette(data_scaled, idx));
    silhouette_scores(k-1) = silhouette_avg;
end

% Plotting
figure;
subplot(1,2,1);
plot(k_range, inertia, '-ob');
title('Elbow Method');
xlabel('Number of Clusters');
ylabel('Inertia');
```

```
subplot(1,2,2);
plot(k_range, silhouette_scores, '-og');
title('Silhouette Score');
xlabel('Number of Clusters');
ylabel('Silhouette');

%Clustering Algorithm Implementation
% K-means Clustering (k=3)
[idx_kmeans, ~] = kmeans(data_scaled, 3, 'Replicates', 10);

% DBSCAN
epsilon = 2;  % Tuned after exploratory analysis
MinPts = 5;
labels_dbscan = dbscan(data_scaled, epsilon, MinPts);

% Agglomerative Hierarchical Clustering
Z = linkage(data_scaled, 'ward');
labels_agglom = cluster(Z, 'maxclust', 3);

% Gaussian Mixture Model (GMM) – Alternative to Spectral Clustering
gmModel = fitgmdist(data_scaled, 3, 'RegularizationValue', 1e-5);
labels_gmm = cluster(gmModel, data_scaled);


% Performance Evaluation
% Set DBSCAN labels to be consecutive positive integers
labels_dbscan_adj = labels_dbscan;
labels_dbscan_adj(labels_dbscan == 0) = -1;
labels_dbscan_adj = grp2idx(labels_dbscan_adj);

% Silhouette Score
eva_kmeans     = evalclusters(data_scaled, idx_kmeans, 'silhouette');
```

```
eva_dbscan   = evalclusters(data_scaled, labels_dbscan_adj, 'silhouette');
eva_agglom   = evalclusters(data_scaled, labels_agglom, 'silhouette');
eva_gmm      = evalclusters(data_scaled, labels_gmm, 'silhouette');

% Davies–Bouldin Index
eva_kmeans_db  = evalclusters(data_scaled, idx_kmeans, 'DaviesBouldin');
eva_dbscan_db  = evalclusters(data_scaled, labels_dbscan_adj, 'DaviesBouldin');
eva_agglom_db  = evalclusters(data_scaled, labels_agglom, 'DaviesBouldin');
eva_gmm_db     = evalclusters(data_scaled, labels_gmm, 'DaviesBouldin');

% Comparative table
Algorithms = {'K-means','DBSCAN','Agglomerative','GMM'};
Results = table(Algorithms', ...
  [eva_kmeans.CriterionValues; eva_dbscan.CriterionValues; eva_agglom.CriterionValues;
eva_gmm.CriterionValues], ...
  [eva_kmeans_db.CriterionValues; eva_dbscan_db.CriterionValues; eva_agglom_db.CriterionValues;
eva_gmm_db.CriterionValues], ...
  'VariableNames', {'Algorithm','Silhouette','DaviesBouldin'});

disp('Comparative Evaluation:');
disp(Results);


%PCA-Based Visualization

[~, score, ~] = pca(data_scaled);

% Visualization
titles = {'K-means','DBSCAN','Agglomerative','GMM'};
cluster_labels = {idx_kmeans, labels_dbscan, labels_agglom, labels_gmm};

figure;
for i = 1:4
  subplot(2,2,i);
  gscatter(score(:,1), score(:,2), cluster_labels{i});
  title(titles{i});
  xlabel('PC1');
  ylabel('PC2');
  grid on;
end
```

****

3.5 Conclusions

This chapter has demonstrated the practical implementation of clustering techniques in Python, R, and MATLAB using a real-world environmental dataset. Through methodical data preprocessing, algorithmic application, and comprehensive evaluation, we observed that each language offers unique strengths: Python excels in workflow integration and library versatility, R provides rich statistical tools and visualization capabilities, while MATLAB offers an efficient and self-contained environment for engineering applications. Among the algorithms tested, DBSCAN revealed the highest internal cohesion (Silhouette Score), though with higher inter-cluster overlap (DBI), whereas K-means and Spectral Clustering showed more balanced performance. PCA-based visualizations effectively complemented numerical metrics by providing intuitive cluster structure representations. The comparative approach presented here not only illustrates the versatility of clustering tools across platforms but also underscores the importance of selecting algorithms and environments based on the specific context, data characteristics, and analytical goals of each project.

References

Alomari, M. F., Mahmoud, M. A., & Ramli, R. (2022). A systematic review on the energy efficiency of dynamic clustering in a heterogeneous environment of wireless sensor networks (WSNs). *Electronics, 11*(18), 2837. https://doi.org/10.3390/electronics11182837

Bengfort, B., & Bilbro, R. (2019). Yellowbrick: Visualizing the Scikit-Learn model selection process. *Journal of Open Source Software, 4*(35), 1075. https://doi.org/10.21105/joss.01075

Campello, R. J. G. B., Moulavi, D., & Sander, J. (2013). Density-based clustering based on hierarchical density estimates. In J. Pei, V. S. Tseng, L. Cao, H. Motoda & G. Xu (Eds.), *Advances in Knowledge Discovery and Data Mining. PAKDD 2013* (Vol. 7819, pp. 160–172). Springer. https://doi.org/10.1007/978-3-642-37456-2_14

Chander, B., & Gopalakrishnan, K. (2023). *Data clustering using unsupervised machine learning.* In *Statistical Modeling in Machine Learning* (pp. 179–204). Academic Press. https://doi.org/10.1016/B978-0-323-91776-6.00015-4

de Vos, N. J. (2015–2024). *KMODES categorical clustering library* (latest version) [Software]. GitHub. https://github.com/nicodv/kmodes

Dhruv, A. J., Patel, R., & Doshi, N. (2021). Python: The most advanced programming language for computer science applications. *Science and Technology Publications*, 292–299. https://doi.org/10.5220/0010307902920299

Downey, A., & Elkner, J. (2008). *How to think like a computer scientist: Learning with Python.* Green Tea Press. https://www.greenteapress.com/thinkpython/thinkCSpy.pdf

Gommers, R., Virtanen, P., Haberland, M., Burovski, E., Reddy, T., Weckesser, W., ... & Striega, K. (2024). *scipy/scipy: SciPy 1.15.0* [Software]. Zenodo. https://doi.org/10.5281/zenodo.14593523

Hahsler, M., & Piekenbrock, M. (2025). *dbscan: Density-Based Spatial Clustering of Applications with Noise (DBSCAN) and Related Algorithms* (Version 1.2.2) [R package]. https://CRAN.R-project.org/package=dbscan

Hetland, M. L., & Nelli, F. (2024). Activity 1: Data Analysis with Pandas, Matplotlib, and Seaborn. In *Beginning Python* (pp. 1–22). Apress. https://doi.org/10.1007/979-8-8688-0196-9_25

Kassambara, A., & Mundt, F. (2020). *factoextra: Extract and visualize the results of multivariate data analyses* [R package]. https://cran.r-project.org/web/packages/factoextra/readme/README.html

Maechler, M., Rousseeuw, P., Struyf, A., Hubert, M., & Hornik, K. (2025). *Cluster: Cluster analysis basics and extensions* (Version 2.1.8.1) [R package]. CRAN. https://CRAN.R-project.org/package=cluster

McKinney, W. (2017). *Python for Data Analysis: Data Wrangling with Pandas, NumPy, and IPython.* O'Reilly Media. https://www.oreilly.com/library/view/python-for-data/9781491957653/

Pedregosa, F., Varoquaux, G., Gramfort, A., Michel, V., Thirion, B., Grisel, O., ... & Vanderplas, J. (2011). Scikit-learn: Machine learning in Python. *Journal of Machine Learning Research, 12*, 2825–2830. https://www.jmlr.org/papers/volume12/pedregosa11a/pedregosa11a.pdf

Poch, M. (1993). *Water Treatment Plant*. UCI Machine Learning Repository. https://doi.org/10.24432/C5FS4C

Porter, L., & Zingaro, D. (2024). *Learn AI-Assisted Python Programming: With Github Copilot and ChatGPT*. Simon and Schuster. https://books.google.es/books?hl=es&id=fZwhEQAAQBAJ

Quatrosi, M. (2022). Clustering environmental performances, energy efficiency and clean energy patterns: A comparative static approach across EU Countries (No. 0722). *SEEDS Working Paper Series*. https://www.sustainability-seeds.org/papers/RePec/srt/wpaper/0722.pdf

R Core Team. (2025). *R: A language and environment for statistical computing*. R Foundation for Statistical Computing. https://cran.r-project.org/doc/manuals/r-release/fullrefman.pdf

Scrucca, L., Fraley, C., Murphy, T. B., & Raftery, A. E. (2023). *Model-based clustering, classification, and density estimation using mclust in R*. Chapman and Hall/CRC. https://doi.org/10.1201/9781003277965

Tavenard, R., Faouzi, J., Vandewiele, G., Divo, F., Androz, G., Holtz, C., ... & Woods, E. (2020). Tslearn, a machine learning toolkit for time series data. *Journal of Machine Learning Research, 21*, Article 118. https://jmlr.csail.mit.edu/papers/v21/20-091.html

The MathWorks, Inc. (2022). *MATLAB version: 9.13.0 (R2022b)*. The MathWorks, Inc. https://www.mathworks.com

Tucker, M. C., Shaw, S. T., Son, J. Y., & Stigler, J. W. (2023). Teaching statistics and data analysis with R. *Journal of Statistics and Data Science Education, 31*(1), 18–32. https://doi.org/10.1080/26939169.2022.2089410

Usama, M., Qadir, J., Raza, A., Arif, H., Yau, K. L. A., Elkhatib, Y., ... & Al-Fuqaha, A. (2019). Unsupervised machine learning for networking: Techniques, applications and research challenges. *IEEE Access, 7*, 65579–65615. https://doi.org/10.1109/ACCESS.2019.2916648

Wang, J., & Biljecki, F. (2022). Unsupervised machine learning in urban studies: A systematic review of applications. *Cities, 129*, Article 103925. https://doi.org/10.1016/j.cities.2022.103925

Widemann, B. T. Y., Bolz, C. F., & Grelck, C. (2013). The functional programming language R and the paradigm of dynamic scientific programming. In H. W. Loidl & R. Peña (Eds.), *Trends in Functional Programming. TFP 2012* (Vol. 7829, pp. 175–190). Springer. https://doi.org/10.1007/978-3-642-40447-4_12

Chapter 4
Clustering Applications in Process Systems Engineering

4.1 Introduction

The discipline of chemical engineering centers on producing essential goods that support everyday life. These production activities depend on structured frameworks and operational strategies that guide choices impacting cost, environmental impact, and safety. Central to this field is the application of mathematical models, which play a vital role in understanding and designing chemical operations. As a result, the subfield known as process systems engineering (PSE) emerged, originally focused on evaluating the economic viability of chemical processes; especially concerning energy use and output efficiency (Bakshi, 2019; Reis & Saraiva, 2021). Initially dealing with the performance of single units, PSE has since expanded its scope to encompass entire plants and global supply networks (Vassiliadis et al., 2024). More recently, Bakshi (2019) has also found applications at microscopic scales, such as atomic, molecular, and particle levels, driving progress in techniques for designing molecules.

Process Systems Engineering (PSE) has evolved into a distinct and recognized methodical area. It encompasses a broad range of activities such as computational modeling, simulation, process design and innovation, product refinement, scheduling, planning, real-time monitoring, fault detection, predictive analysis, control systems, and optimization, along with other forms of systems-level evaluation (Reis & Saraiva, 2021). PSE, as a multidisciplinary discipline, brings together elements of pragmatic mathematics, engineering principles, and algorithmic computer science to solve intricate, multi-faceted challenges and support decision-making informed (see for examples Biegler & Grossmann, 2004). Because of its wide applicability, PSE holds a prominent place in academic research and serves as a powerful tool within industrial settings. In recent years, the chemical sector is now addressing a new wave of goals related to environmental protection, economic performance, social responsibility, safety, and long-term sustainability (Ortiz-Espinoza

A. Y. Raya-Tapia et al., *Machine Learning and Clustering for a Sustainable Future*,
Studies in Computational Intelligence 1233,
https://doi.org/10.1007/978-3-032-03876-0_4

et al., 2019; Teh et al., 2019). A notable example is the push to reduce ecological impact through more efficient operation and management of chemical facilities (Khan et al., 2023; Pan et al., 2016; Vooradi et al., 2018). In response to these evolving demands, PSE provides the theoretical groundwork and computational tools necessary to meet both existing and upcoming tasks (Grossmann & Harjunkoski, 2019). PSE methodologies have been applied across various sectors, including pharmaceuticals, agrochemicals, food processing, and biochemical production, for purposes such as process design, system optimization, and operational planning (Pistikopoulos et al., 2021). As a result, the field continues to grow in both scope and relevance.

In its early stages, PSE relied on rule-of-thumb strategies and pre-established expert knowledge to guide process design efforts (Douglas, 1985). Subsequently, mathematical programming methodologies were employed as process models, and numerical resolution techniques were developed (El-Halwagi & Manousiouthakis, 1989; Papoulias & Grossmann, 1983). Currently, data-driven procedures are being implemented in response to the substantial volume of created data, alongside hybrid approaches that integrate many techniques (López-Flores et al., 2023). Physics-informed modeling techniques can be employed to construct data-driven approaches that allow for a deeper investigation of complex systems (Gkinis et al., 2019; Wang et al., 2022). Compared to mathematical programming strategies, typically restricted to scenarios that satisfy material and energy conservation laws, data-driven approaches provide greater adaptability and enable the exploration of more intricate and dynamic phenomena (López-Flores et al., 2022).

The Industry 4.0 (i.e., the fourth industrial revolution) brings together digital and physical technologies. This digital transformation is producing massive volumes of data (Sajid et al., 2021), prompting a fast and transformative evolution in process systems engineering, and driving accelerated growth in innovative tools and methods (Reis & Saraiva, 2021). These changes are fueled by breakthroughs in technologies such as cloud computing, smart sensing devices, and high-speed data processing. Combined with the availability of large-scale data (big data) and sophisticated techniques for interpreting diverse datasets, these developments are unlocking new possibilities for capturing, storing, and analyzing information. The volume of data generated within process industries is escalating due to the proliferation of inexpensive equipment and sensors (Thibault et al., 2023). Companies gather extensive data everyday via sensor networks, lab testing, financial documentation, weather data, digital simulations, scientific literature, and patent databases (Dobbelaere et al., 2021). This presents a significant potential and an urgent necessity to use data-driven solutions for extracting insights and supporting informed decision-making. Relying solely on the expertise, instincts, and experience of personnel is no longer sufficient for achieving optimal operational performance (Choi et al., 2022).

When applying data-driven techniques, it is essential to handle the data with care, as it may be compromised by various errors and inaccuracies. Faulty data can mask real changes in processes, reduce the effectiveness of control systems, lead to equipment wear or failure, compromise product standards, and cause unsafe or economically inefficient operations (Thibault et al., 2023). Relying on such flawed data can result in misleading or inaccurate outcomes. Therefore, a critical stage in data research is cleaning and correcting the data to ensure its reliability and accuracy for informed decision-making. This process involves identifying anomalies, distinguishing between operational and steady-state conditions, and minimizing noise. Properly processed data is essential for downstream applications like optimization, cost or energy assessments, system monitoring, fault detection, and troubleshooting (Xu et al., 2015).

This research examines and critically assesses clustering approaches utilized in the domain of process systems engineering, emphasizing their role in enhancing intricate systems. It offers an extensive overview, ranging from the theoretical underpinnings of clustering to its sophisticated applications across several domains of process systems engineering. This paper highlights the role of clustering in integrating predictive models, data analysis techniques, and optimization tools, hence facilitating adaptable resolutions to modern difficulties. Furthermore, it examines nascent applications in sustainable problems, highlighting its significance in constructing resilient frameworks. The research delineates the primary limits of clustering and suggests potential avenues for its incorporation into engineering and sustainable contexts.

This study provides a comprehensive evaluation of clustering approaches applied within PSE, focusing on their role to enhancing complex system performance. It spans from foundational clustering principles to advanced implementations across various PSE fields. The paper underscores how clustering supports the fusion of data analytics, forecasting methods, and optimization strategies, thereby enabling flexible responses to current challenges. In addition, it searches for emergent uses in sustainable technologies, emphasizing its importance in building robust and adaptive systems. The work also identifies the key constraints associated with clustering and outlines promising directions for its broader adoption in industrial and environmentally conscious applications.

4.2 Utilizing Clustering Techniques for Optimization Purposes

Process optimization is an essential field in applied and engineering sciences, focused on system performance and sustainability, and resource efficiency by mathematical modeling, simulation tools, and unconventional analytical procedures. In PSE, this inludes a wide range of applications, extending from the development of chemical production facilities to the strategic planning and coordination of supply chains, aiming to optimize benefits, minimize costs, and mitigate environmental concerns.

Accomplishing these goals necessitates instruments proficient in evaluating extensive data sets, discerning pertinent patterns, and facilitating strategic decision-making. Between the various technologies available, clustering has proven to be an extremely effective method for unsupervised learning (López-Flores et al., 2024).

An unsupervised learning technique called clustering was developed to categorize and organize data based on similarities, density patterns, or probabilistic attributes, exclusive of relying on predefined categories (Mehta et al., 2020; Oyewole et al., 2023). The core principle involves categorizing data into subsets or "clusters," guaranteeing that items inside a single group are homogenous relative to one another, whereas differing significantly from those in other groups. Such a characteristic renders it a potent instrument for analyzing heterogeneous and complex datasets, such as those frequently observed in process systems engineering.

Initial clustering methodologies, exemplified by the k-means algorithm, were grounded in basic mathematical ideas, specifically the minimization of Euclidean distance among data points and their corresponding centroids. Over time, advanced techniques were established, including density-based algorithms like DBSCAN and hierarchical clustering, which may discern intricate features, such as irregular forms or nested group relationships (Fahim, 2023; Ran et al., 2023). These advances expanded their usefulness across several fields (Ikotun et al., 2023).

In PSE, clustering has become an essential instrument for process optimization from several viewpoints. Its usefulness resides in its capacity to arrange extensive industrial data, thereby enabling strategic decision-making. Key implementations are outlined as follows:

- Data Classification for System Enhancement: In the development of chemical and energy infrastructures, clustering techniques are used to sort both historical and real-time data into meaningful segments. This helps uncover recurring operational patterns that can be applied to improve system performance, reduce material waste, and limit environmental pollutants (Maestri et al., 2010; Rastgoo et al., 2024; Rovira et al., 2022).
- System Fault Detection and Operational Monitoring: Algorithms such as DBSCAN and k-means are widely used to distinguish between various system states and detect anomalies within complex processes. This contributes to higher reliability and reduce operational interruptions (Rahoma et al., 2023; Zheng & Zhao, 2020).
- Modeling of Time-Dependent Processes: Clustering helps organize dynamic data into structured classes, offering a basis for building forecasting and decision-making models essential for real-time process optimization and control (Muñoz López et al., 2020).
- Designing for Sustainability: Clustering provides eco-conscious system design by classifying balanced high-performance solutions. An important example is the use of illustrative time periods, which simplify long-term simulations by reducing computational effort while accurately reflecting changes in resource availability—beneficial for systems like renewable energy grids and integrated energy-water-food networks (Morlet-Espinosa et al., 2023; Raya Tapia et al., 2025).
- Optimization with Multiple Goals: Grouping Pareto-optimal outcomes, clustering allows users to analyze trade-offs among objectives such as cost efficiency, energy savings, and environmental impact. This provides both visual and quantitative insights to guide strategic decisions (Sendín et al., 2009; Valencia-Marquez et al., 2023).

Clustering acts as a conduit among raw data and optimization efforts by presenting an organized depiction of the systems under analysis. Within process systems engineering, this enables the amalgamation of data attributes with computational representations, simulation techniques, and optimization methods. Clustering improves the effectiveness of conventional procedures by simplifying intricate data and offering essential insights, hence facilitating the creation of innovative adaptable and sustainable solutions. This cross-disciplinary approach has the potential to revolutionize how process systems are designed and operated, offering impactful solutions to worldwide challenges such as sustainable development and the shift toward cleaner energy sources.

4.3 Summary of Clustering Methods

Clustering techniques often function in the following manner: The process begins by mapping data points within a high-dimensional space, where each axis corresponds to a specific attribute or variable. A similarity measure, commonly Euclidean distance or a probabilistic metric is then applied to evaluate how closely related the data points are (Kalgotra et al., 2020; Niwattanakul et al., 2013; Saxena et al., 2017; Xu & Wunsch, 2010). Clustering algorithms group these points into clusters based on defined principles such as spatial closeness or local data concentration. Many clustering methods function through iterative refinement, gradually adjusting groupings to minimize internal variability within clusters. However, some algorithms bypass iteration entirely, relying instead on structural relationships like density regions or hierarchical levels to form clusters. The quality and validity of the clustering outcomes are typically verified using numerical evaluation criteria or validated by domain experts to ensure they accurately reflect the context of the problem being analyzed.

As of now, academics have suggested and examined over 100 distinct clustering approaches (Ghosal et al., 2020). The approaches may be classified into five principal categories (Fahad, 2014): hierarchical, partitioning-based, density-oriented, model-driven, and grid-structured clustering approaches.

Partitioning-based is a sequential method that categorizes data according to their closeness to cluster centroids, frequently assessed using metrics like Euclidean or Manhattan distance (N'Cir et al., 2015; Fahad, 2014). It presumes that each cluster has a minimum of one data point, as well as every data point is allocated exclusively to a single cluster. This procedure starts with the initialization of cluster centroids, followed by the computation of distances and the allocation of data points to the bordering cluster. Then, the centroids are adjusted, and this cycle continues until the results stabilize or a specified stopping condition is satisfied. Popular techniques such as K-Medoids, Fuzzy K-Means, K-Modes, and K-Means are extensively implemented because of their straightforward implementation and ability to handle large datasets. However, they may struggle with identifying clusters that overlap or have irregular shapes. These methodologies have been utilized across diverse domains, tumor identification and analysis of medical imagery (Patel et al., 2013; Shanker et al., 2017), infrastructure development (Molokomme et al., 2021; Otok et al., 2022), studies on energy usage monitoring (Chévez et al., 2017; Ruiz et al., 2020), urban gentrification (Raya-Tapia et al., 2024a), route planning and traffic pattern evaluation (Song et al., 2020), air quality assessment and climate modeling (Govender & Sivakumar, 2020), and the evaluation of countries' progress towards the 17 Sustainable Development Goals (Raya-Tapia et al., 2024b).

Hierarchical arranges data into a tiered framework via two primary methodologies: divisive (top-down) and agglomerative (bottom-up). The agglomerative method starts with each data point treated as a separate cluster, then progressively merges the closest pairs based on a selected distance measure until all points are assembled into a single overarching cluster. In contrast, the divisive method begins with the entire dataset as one large cluster and incrementally splits it into smaller subgroups as dissimilarities increase. The procedure produces a tree-like representation of the data, known as a dendrogram, which visually captures the hierarchy and relationships among data points (Ran et al., 2023; Xie et al., 2020). Representative approaches in this group comprise Echidna, Chameleon, ROCK, BIRCH, and CURE. These techniques have been applied in various areas, such as evaluating student performance (Rana & Garg, 2016), conducting air quality and climate studies (Govender & Sivakumar, 2020), monitoring water quality (Warsito et al., 2021), analyzing consumption patterns for energy (Wang et al., 2020), optimizing industrial operations (Ahn & Chang, 2019), and enhancing Wireless Sensor Networks (Jan et al., 2017).

Density- oriented delineates groups according to the concentration of data inside the data space. Regions characterized by a high concentration of data points are grouped into clusters, while sparsely populated regions serve as boundaries, effectively isolating noise and handling outliers (Fahim, 2023). The method starts by selecting an unvisited data point and assessing its surrounding region inside a given radius 'ε'. If the quantity of adjacent points reaches or above a certain threshold, a cluster is established; otherwise, the point is marked as noise. This procedure is executed recursively on all remaining unvisited spots. Notable density- oriented methods in this group encompass DENCLUE, DBCLASD, OPTICS, and DBSCAN, all known for their very proficient at detecting clusters of irregular form and handling noise. These methodologies have been utilized across several domains, such as earthquake event occurrences (Nicolis et al., 2024), anomaly detection (Retiti Diop Emane et al., 2024), waste detection (Fauzan et al., 2024), renewable energy systems (Hou et al., 2024), and food products (Nurulhikmah & Abdi, 2024).

Ultimately, Grid-structured and Model-Based employ distinct but successful methodologies for cluster formation. Grid-structured systematically arranges data points into a consistent grid framework with statistical techniques instead of directly querying the database. It assesses the density of each grid cell, eliminates cells with low point concentration, and combines neighboring high-density cells to form clusters (Nalawadę et al., 2023). The performance of this approach is influenced by the chosen grid resolution, making it especially suitable for processing extensive datasets efficiently. Conversely, Model-Based depends on established probabilistic or statistical approaches to determine clusters within data. This technique can accomplish complex patterns and can or can not need the number of clusters to be given beforehand. Representative algorithms encompass STING, CLIQUE, and WaveCluster for grid-structured, and Expectation Maximization (derived from Gaussian Mixture Models), COBWEB, and neural network-based models for model-based clustering (Gormley et al., 2023).

4.4 Clustering Applications in PSE

4.4.1 System Fault Detection and Operational Monitoring

Clustering approaches now play a crucial role in Diagnosis and Fault Detection, especially within engineering settings marked by complex operations, nonlinear behaviors, and systems with multiple modes. Clustering improves the resilience and accuracy of diagnostic systems by facilitating the identification, categorization, and analysis of problems in demanding environments, ultimately supporting safer and more efficient processes. This section analyzes noteworthy research that demonstrates the novel use of clustering in Fault Detection and Diagnosis, highlighting their contributions throughout the optimization lifecycle, from initial analysis to implementation and post-process evaluation.

Ardali et al. (2024) established an extensive fault diagnostic system that integrates NSGAII for selecting relevant features, t-SNE for reducing data dimensionality, and clustering techniques such as DBSCAN and k-means. In the initial stage prior to optimization, NSGAII diminished input variables by 18%, preserving diagnostic performance while lowering computational demands. During optimization, t-SNE outperformed PCA in key evaluation measures, such as Adjusted Mutual Information (AMI) and Adjusted Rand Index (ARI), offering improved data visualization and clustering precision. The final post-optimization phase focused on refining fault identification, achieving a minimal classification error rate of 0.0133 on the Tennessee Eastman dataset, and successfully handling eight distinct fault scenarios with notable scalability.

Zheng and Zhao (2020) devised an approach that combines DBSCAN, stacked autoencoders, and t-SNE to detect faults in industrial processes. In the initial phase before optimization, stacked autoencoders were employed to reduce dimensionality and extract meaningful features, laying the groundwork for effective clustering. In the optimization process, t-SNE and DBSCAN were utilized, resulting in a clustering purity of 97.8% and a fault detection precision of 93% for the Tennessee Eastman Process dataset. Following optimization, pseudo-supervised approaches were developed using the clustered outputs, showing how clustering-derived knowledge can be leveraged to enhance fault cataloging performance.

Shi et al. (2024) employed transfer and unsupervised learning methodologies to identify defects in high-purity carbonate processing systems. In the initial phase, t-SNE and stacked autoencoders in conjunction with DBSCAN produced pseudo-labeled data, with a clustering accuracy of 98.7%. Optimization initiatives elevated defect identification rates to 96% with better unsupervised approaches. In the final phase, transfer learning strategies, such as Joint Adaptation Network (JAN) and Deep Domain Confusion (DDC), were utilized to harmonize multimodal datasets, leading to a 4% increase in defect detection performance within target environments.

Srinivasan et al. (2004) presented a Dynamic Principal Component Analysis (DPCA -based framework designed to cluster transitional and steady states in flexible manufacturing environments. In the early phase, dimensionality reduction using PCA helped isolate important aspects that distinguished various operating conditions. During optimization, DPCA categorized stable and shifting modes, markedly decreasing fault detection errors to just 2.36 samples. In the final phase, research yielded insights into the dynamics of transitional periods, ultimately improving system supervision and control in adaptable chemical processes.

Maestri et al. (2010) utilized a rigorous clustering methodology to oversee systems exhibiting several steady states, tackling issues of data contamination and outliers during the early phase. The optimization stage initiatives employed a trimmed determinant criterion (TDC) to maintain fundamental variable associations, resulting in a 50% enhancement in missing alert detection rates and a reduction of false alarm rates to just 1%. The final phase focused on refining clustering analysis to enhance fault detection capabilities across both simulated scenarios and real-world applications, including methanol production facilities.

Rahoma et al. (2023) introduced a combined technique that integrates PCA for feature reduction with k-means algorithm to detect faults in gas treatment systems and distillation units. In the initial phase, data dimensionality was reduced, taking 85% of the variance among main process variables. Clustering successfully segregated operating modes during optimization, resulting in a fault detection accuracy of 93%. The final phase confirmed the method's scalability and suitability for real-time defect diagnosis in engineering environments.

Ziaei-Halimejani et al. (2021) proposed a defect diagnostic system that combines reappearance quantification analysis (RQA) with clustering aproaches to address the complexities of nonlinear chemical systems. In the preliminary phase, RQA features were used to highlight critical data patterns, thereby increasing the effectiveness of subsequent clustering. During optimization, DBSCAN was applied to detect dense data groupings, achieving ARI values exceeding 0.88.

Tang and Zhang (2021) employed the K-WD-KPCA technique for fault identification in the feedwater treatment operations of coal-fired power facilities. The initial stage initiatives classified operational states by k-means algorithm, facilitating the creation of tailored KPCA models for each condition. Optimization of integrated wavelet denoising (WD) to improve sensitivity and reduce variations caused by process variability. The final phase process confirmed the concept using real-world data, markedly decreasing false alarm rates and improving across systems operating under multiple conditions.

4.4.2 Renewable Energy Systems

Clustering techniques are important for enhancing the efficiency and effectiveness of renewable energy system design and operation. They address issues including resource unpredictability, temporal volatility, and scalability, providing data-driven solutions that are both economical and resilient.

Rastgoo et al. (2024) formulated a GAN-LSTM-based model to identify characteristic time segments in renewable energy systems. This approach utilizes adaptive feature learning via GANs in the pre-optimization phase, improving data representation with loss terms appropriate to clustering. During the optimization phase, LSTM networks effectively managed sequence-based data and spatiotemporal correlations, markedly enhancing accuracy by decreasing errors such as MAPE (15.1%) and RMSE (0.108). Post-optimization concentrated on temporal aggregation, grouping records into representative days, resulting in a 74.58% decrease in MAE relative to traditional approaches, and showcasing its efficacy in automating decision-making while managing extensive energy data.

Otnes et al. (2023) employed k-means algorithm to optimize thermomechanical fatigue models in photovoltaic (PV) modules. This pre-optimization method grouped 5–20 exemplary daily temperature profiles from comprehensive datasets covering 5–10 years. The approach decreased computing expenses by as much as 70-fold while preserving essential statistical characteristics, like temperature variability and climatic distinctions. The clustering-based representative datasets facilitated effective and scalable numerical modeling, enhancing developments in photovoltaic system dependability and degradation research.

Balasubramanian and Balachandra (2021) utilized methodologies to classify regular load curves in power networks integrated with renewable energy. In the pre-optimization phase, twelve sample load classes were determined based on temporal and seasonal trends, which were subsequently combined into seven final load curves for optimization. This technique enabled dynamic demand-response planning, aligning 90% of hourly energy loads within a ±12% margin of the actual recorded data, demonstrating the significance of clustering in improving grid dependability and operational flexibility.

Pérez-Uresti et al. (2023) applied time-series clustering techniques to enhance the planning and development of renewable energy utility facilities. Pre-optimization clustering found representative periods (RPs), balancing variable representation with computational expenses. The optimization utilized a mixed-integer linear programming (MILP) formulation to ascertain resilient plant designs, whilst post-optimization assessed out-of-sample robustness. A comparative analysis elucidated the trade-offs of biomass boilers and gasification systems, focusing on investment costs and leakage levels. This thorough methodology highlighted the effectiveness of clustering in optimizing cost, resilience, and structure in energy systems.

Wang and Li (2020) employed hierarchical algorithm inside temporal convolution networks to enhance process modeling and prediction precision in methanol production systems. Pre-optimization utilized clustering to partition operational data into six discrete sections, therefore augmenting the receptive field of their modified temporal convolution network (MRO-ARFTCN). The optimization phase utilized this structured data to derive significant causal characteristics, attaining enhanced performance with a testing error of 0.2631, surpassing conventional RNN and LSTM techniques.

Raya-Tapia et al. (2025) assessed three methodologies for identifying representative weeks in power system planning utilizing a MILP model. Pre-optimization for the selection of sample weeks was conducted utilizing k-means with an Euclidean metric, k-means with a dynamic temporal warping (DTW) metric, in conjunction with a combinatorial approach. The combinatorial technique exhibited the lowest root mean square error (RMSE) in the net load duration curve representation, registering a value of 2.80, succeeded by k-means with dynamic time warping (DTW) at 3.21, and k-means utilizing the Euclidean metric at 5.49. In terms of overall system cost, the combinatorial technique was more costly (\$4.4274 $\times$ 10^{10}), although k-means utilizing DTW and k-means employing the Euclidean metric decreased costs by 0.1% and 0.2%, respectively.

4.4.3 Chemical Process Optimization

Clustering algorithms are essential for segmenting and analyzing intricate chemical processes, facilitating informed decision-making by streamlining data interpretation and revealing significant trends. This section analyzes significant research that underscore the importance of clustering in enhancing chemical process engineering throughout various analytical phases.

Morlet-Espinosa et al. (2023) established a framework that combines DBSCAN with variational autoencoders to enhance the design of energy-water-food nexus networks. This method facilitated the study of datasets comprising more than 438,000 individual entries, with a clustering duration of roughly 2007.89 s. Pre-optimization entailed dimensionality reduction by variational autoencoders, markedly decreasing processing requirements. The optimization phase employed DBSCAN to enhance the analysis of models with 60,000 constraints, facilitating the decision-making process in resource management and nexus design.

Rovira et al. (2022) illustrated the effectiveness of HDBSCAN and UMAP in detecting non-globular clusters within reactive flow data. In comparison to t-SNE, UMAP demonstrated execution speeds that were 65% quicker while more successfully maintaining both global and local structures. The application of HDBSCAN during the optimization phase improved cluster granularity by independently identifying noise and anomalous patterns. This integration yielded actionable data into reactive flow areas, enabling targeted actions and enhancing operational efficiency.

Beaver et al. (2007) utilized k-PCA algorithm on cyclic chemical process data, achieving the capture of up to 95% of process variability and successfully identifying several steady states and transitions. During the pre-optimization phase, the study utilized dynamic windowing techniques to differentiate high- and low-frequency data patterns. Optimization improved PCA models to attain effective partitioning of process states, augmenting monitoring and fault detection capacities in cyclic processes.

Seghers et al. (2023) employed the PaCMAP method to improve the clarity and separation of clusters within chemical process datasets, particularly in pyrolysis reactors. This strategy effectively balanced the preservation of global and local data structures, allowing high-resolution identification of operational events across various scales. Pre-optimization encompassed data pretreatment for dimensionality reduction, whereas optimization maintained essential patterns during embedding. The methodology offered comprehensive insights into macro-level operational trends and micro-level variations, greatly improving process monitoring and problem diagnostics.

Muñoz López et al. (2020) used clustering and manifold learning methods to automate time detection in batch processes. The research utilized t-SNE and unsupervised techniques for phase identification, attaining a fault detection accuracy of 92.3%. Pre-optimization discerned essential properties in multi-phase data, whereas optimization employed regression and classification models to synchronize phases across differing batch lengths. This framework exhibited strong efficacy in managing irregular and multi-phase process data, enhancing diagnostic accuracy.

4.4.4 Supply Chain Optimization

Clustering approaches have transformed supply chains by optimizing logistics, enhancing the effectiveness of decision-making frameworks, and facilitating the development of environmentally responsible network architectures. By tackling critical hurdles at each phase, these strategies offer resilient solutions to intricate supply chain problems, facilitating more effective and adaptable procedures. This section analyzes pivotal lessons that underscore the transformational impact of clustering on supply chain optimization, demonstrating its many applicability throughout different stages of deployment.

Mahesh Prabhu et al. (2020) utilized a hierarchical algorithm to categorize providers based on criteria such as cost, delivery distance, and lead time. This methodology established 3 distinct supplier groups emphasizing cost-efficiency. This led to a 20% decrease in logistics costs and a 10% improvement in delivery times. The preliminary stage entailed data standardization to guarantee impartial cluster creation, whilst optimization concentrated on categorizing providers to refine supply chain processes and improve global performance.

Geiger (2024) employed clustering algorithm to address manufacture routing issues, combining clustering results with metaheuristic approaches. This combined method resulted in a 41% reduction in relative gaps and significantly improved routing efficiency. During the pre-optimization phase, clustering played a key role in grouping production sites, whereas the optimization phase employed these clusters to fine-tune routing procedures and reduce operating expenses.

Ng and Lam (2013) employed fuzzy clustering to develop environmentally conscious supply chains, with a material recovery rate of 88.9%. In the initial stage, clustering organized facilities based on their material exchange patterns, promoting engineering symbiosis and minimizing releases. In the optimization process, a mixed-integer nonlinear programming problem reduced transportation expenses and environmental effects, illustrating how clustering can be effectively integrated into sustainable supply network planning.

Seng Ng and Srinivasan (2010) presented a cooperative multi-agent architecture that incorporated clustering methods for identifying defects and supporting decisions within supply chain systems. In the initial stage, dimensionality reduction and pattern recognition were achieved using techniques like Self-Organizing Maps (SOM) and Principal Component Analysis (PCA). Optimization utilized Bayesian decision fusion to expand the precision of fault identification and classification. The final phase techniques enhanced decision-making using historical data.

4.4.5 Real-Time Monitoring and Decision Intelligence Platforms

Clustering approaches are crucial for augmenting monitoring frameworks by strengthening fault detection precision, enabling effective classification of operating conditions, and facilitating the interpretation of multifaceted process datasets.

Liu et al. (2023) proposed a clustering-driven architecture intended to track transitional phases within engineering operations. In the initial stage, the methodology employed neighbor inconsistent pairs (NIPs) to reduce dynamic attributes, enabling efficient separation of operational data into stable and transitional modes. In the optimization process, clustering techniques effectively recognized transitional phases, frequently misidentified as faults, attaining fault detection accuracy exceeding 99%. During the final phase, this strategy sustained false alarm rates around 2.5%, yielding substantial perceptions into process variability and enhancing supervising dependability.

Thomas et al. (2018) utilized dimensionality reduction and clustering methods, including PCA and DBSCAN, to categorize various operational modes in manufacture processes. The initial phase entailed decreasing data dimensionality while maintaining the structure of meaningful clusters. Clustering techniques effectively distinguished between normal and abnormal operating states during optimization, attaining ARI scores greater than 0.8. The final phase study demonstrated dynamic shifts and time-based relationships within the data, highlighting clustering's efficacy in managing high-dimensional information.

Bo and Hao (2015) devised a qualitative trend clustering (QTC) technique, utilizing arrangement alignment to classify operative abnormalities. In the initial phase, the technique utilized the Smith-Waterman approach to align qualitative trends and discover patterns within operational data. Optimization aimed at effectively differentiating between abnormal and normal trends, with the subsequent phase enhancing the fault knowledge base for real fault diagnosis in engineering environments.

4.4.6 Optimization Under Multiple Objectives

Clustering techniques are crucial in multi-objective optimization by effectively handling intricate trade-offs, structuring Pareto-optimal outcomes, and communication and involvement with stakeholders. These strategies improve decisions in multi-criteria systems where objectives may be at odds, as evidenced by several applications in engineering fields.

Valencia-Marquez et al. (2023) employed fuzzy clustering to assess competing objectives in the interconnection among water, energy, and waste systems. This methodology uncovered essential synergies between regulatory strategies for biomass utilization and implementation of thermal energy storage, establishing an organized methodology for reconciling water resource management, energy optimization, and waste minimization. The study elucidated optimal techniques to facilitate the design of environmentally sustainable systems by clustering Pareto solutions, emphasizing the effectiveness of clustering in harmonizing ecological goals with operational performance.

Sendín et al. (2009) broadened the use of clustering techniques with multiple initialization points to enhance limited systems in chemical engineering. The strategy enhanced solution resilience and minimized unnecessary assessments by detecting local minima during pre-optimization. Post-optimization research shown its capacity to ease the resolution of advanced problems in nonlinear systems, providing a pragmatic solution to large-scale optimization issues.

García-Martínez et al. (2022) illustrated the significance of clustering in simulating coalitions across involved parties during the development of community-based energy infrastructures. The initial phase facilitated the grouping of stakeholders according to ideological distances, allowing for the discovery of common interests. Post-optimization tactics enabled equitable solutions addressing multiple objectives, reducing disagreements and securing widespread acceptance, demonstrating clustering's capacity to integrate technological objectives with societal needs.

Tang et al. (2024) utilized kernel fuzzy c-means algorithm to improve predictive control in selective catalytic reduction (SCR) systems used for nitrogen oxide mitigation in coal-fired power facilities. This approach was essential in the initial phase, as clustering divided operational datasets for training prediction approaches. The optimization phase employed these models to attain an 85% decrease with respect to the variability of NOx emission levels, highlighting the significance of clustering in intricate, time-varying industrial processes.

Huang et al. (2006) utilized progressive cluster formation technique in bioremediation design, using projection methods to enhance effectiveness of remediation strategies. The study attained average relative deviations below 12% in forecasting benzene concentrations using clustered outputs from biodegradation models. This technique successfully balanced operating costs and pollution reduction, demonstrating clustering's significance in high-demand decision procedures involving significant resources.

4.5 Examination of Clustering Methodologies Applied in PSE

Clustering approaches have emerged as pivotal instruments in PSE (refer to Table 4.1). This section provides a critical analysis of their functions, limits, and future possibilities, integrating lessons from many applications.

Table 4.1 In-depth analysis of clustering techniques for the advancement of process systems engineering

Disciplinary scope	Initial Phase	Optimization	Final Phase
Fault detection and diagnosis	Choose important data features	Robust clustering	Enhancement of fault categorization accuracy
Renewable energy systems	Temporal clustering for representative period determination	Time-series aggregation coupled with system-level optimization	Analyzing the expansion capacity and operational reliability
Chemical processes	Data preprocessing through autoencoder-based dimensionality reduction	Reactive fluid dynamics clustering integrated with real-time anomaly monitoring	Analysis of evolving operational patterns
Supply chain management	Supplier segmentation and facility classification	Economical logistics planning and resource reclamation	Sustainability diagnostics and optimization-guided decision assistance
Multi-objective optimization	Stakeholder classification alongside structured analysis of trade-off solutions	Model calibration for forecasting purposes and integrative systems analysis	Balancing conflicting parameters through comparative evaluation

4.5.1 Key Analytical Reflections and Interpretations

4.5.1.1 Influence Throughout Distinct Process Phases

- Clustering algorithms excel in the initial phase by diminishing dataset complexity, identifying representative characteristics, and organizing data in a format conducive to optimized processing. Methods including DBSCAN, PCA, and fuzzy clustering have demonstrated efficacy in practical implementations including both sustainable energy system development and analytical approaches in chemical engineering.
- During the optimization stage, clustering improves model precision and operational decision-making. For example, the use of kernel fuzzy c-means clustering in in NOx emission reduction systems and k-means in energy load management illustrates its versatility in managing intricate systems with changeable dynamics.
- The final phase advantages arise from clustering's capacity to enhance solution strategies, ensure model reliability, and deliver practical recommendations. Illustrations encompass enhancing fault diagnostic precision in industrial systems and assessing compromises among multiple goals in optimization problems.

4.5.1.2 Technical Strengths and Functional Benefits

- Clustering techniques are well-suited for managing complex datasets characterized by multiple operating modes and nonlinear behaviors, commonly encountered in energy and industrial domains.
- The ability to integrate clustering with other approaches, like neural network-based models, knowledge transfer techniques, and advanced optimization strategies, greatly increases its applicability across many fields.
- Its function in connecting technical and social dimensions, shown by stakeholder-based strategies in the development of decentralized energy systems, underscores its wider applicability.

4.5.1.3 Critical Issues and Methodological Boundaries

- Extensive utilization of clustering techniques is significantly influenced by initial parameter configurations and the quality of data preparation, hindering consistent replication and broader application. The computational expense of clustering, especially for extensive datasets or time-sensitive applications, remains a constraint notwithstanding improvements in algorithmic performance.
- Understanding clustering results and incorporating them into structured decision analysis models frequently necessitates domain-specific knowledge, impeding wider use in less specialized areas.

4.5.1.4 Next-Generation Approaches and Industry Trajectories

- Integrated approaches compounding clustering with sophisticated machine learning methods, like transfer learning, LSTMs, and GANs, are positioned to tackle complicated, high-dimensional data with greater effectiveness.
- Application of clustering techniques in dynamic monitoring and control frameworks, especially in the context of sustainable power networks and chemical processing plants, constitutes a vital domain for future investigation.
- The importance of clustering in improving sustainability evaluations, especially in the optimization of interconnected water, energy, and food systems and circular economy systems, is becoming increasingly essential in tackling global environmental issues.

The adaptability and scalability of clustering provide it an essential instrument in PSE, stronger, more resource-efficient, and sustainability-driven outcomes. Nonetheless, tackling its computing requirements and improved integration with automation technologies will be crucial for progressing its functionalities. Multidisciplinary approaches combining clustering with real-time analytics and optimization models must be central to future investigations, hence maintaining its significance in addressing increasing difficulties in industrial and environmental systems.

4.6 Outlook and Summary

The incorporation of clustering technologies into PSE signifies not only an optimization tactic but a fundamental transformation in the analysis and management of complex systems. In addition to their primary function of organizing and simplifying extensive information, clustering algorithms enable the move towards data-driven decision-making across several technical fields. Nonetheless, the pivotal inquiry persists: how can these techniques advance to meet the escalating requirements of sustainable performance, system flexibility, and accuracy in progressively integrated environments?

A significant difficulty in PSE is the ever-changing nature of operating settings, particularly within the domains of renewable energy infrastructure and chemical engineering processes. Current clustering systems, while robust, frequently exhibit static characteristics, depending on predetermined parameters and assumptions regarding data distributions. It is essential to design adaptive clustering models that change in accordance with system dynamics. These models would integrate real-time feedback mechanisms, enabling the dynamic reconfiguration of clusters as system circumstances change. This capacity is particularly pertinent in energy networks including variable renewable resources, where operational states can vary significantly over brief intervals.

A further frontier exists in the integration of clustering methodologies with predictive analytics and prescriptive models. The significance of clustering is enhanced when it influences decision-making frameworks, like predictive maintenance plans or real-time control techniques. Utilizing clustering results as inputs for machine learning models may enhance predictions by including subtle operational conditions into forecasts. This integration of descriptive and predictive analytics offers potential for more robust and efficient system designs, particularly in the face of uncertainty.

Ethical concerns and stakeholder engagement require careful attention. Given that clustering-based models impact choices in vital domains like energy access and supply chain priorities, it is imperative to ensure transparency and inclusion. Clustering algorithms must deliver not just technically optimum solutions but also conform to social, economic, and environmental fairness objectives. Future research should concentrate on creating backgrounds that mix multi-criteria assessment measures by harmonizing technical performance with wider societal implications.

The function of clustering in promoting multidisciplinary collaboration represents a significantly underutilized possibility. By integrating disciplines, such as merging chemical engineering skills with data science techniques, clustering might stimulate new ideas and inventions. The exchange of ideas is essential for tackling global issues such as climate change, because solutions need the integration of knowledge from several disciplines.

In conclusion, although clustering has become a fundamental aspect of PSE, its future depends on transcending conventional applications. The forthcoming generation of clustering tools will be characterized by dynamic, integrative, and morally conscious methodologies, assuring their relevance in a landscape that necessitates both technological proficiency and adaptation, as well as equity.

References

Ahn, H., & Chang, T.-W. (2019). A similarity-based hierarchical clustering method for manufacturing process models. *Sustainability, 11*, 2560. https://doi.org/10.3390/su11092560

Ardali, N. R., Zarghami, R., & Gharebagh, R. S. (2024). Optimized data driven fault detection and diagnosis in chemical processes. *Computers & Chemical Engineering, 186*, Article 108712. https://doi.org/10.1016/j.compchemeng.2024.108712

Bakshi, B. R. (2019). Toward sustainable chemical engineering: The role of process systems engineering. *Annual Review of Chemical and Biomolecular Engineering, 10*, 265–288. https://doi.org/10.1146/annurev-chembioeng-060718-030332

Balasubramanian, S., & Balachandra, P. (2021). Characterising electricity demand through load curve clustering: A case of Karnataka electricity system in India. *Computers & Chemical Engineering, 150*, Article 107316. https://doi.org/10.1016/j.compchemeng.2021.107316

Beaver, S., Palazoglu, A., & Romagnoli, J. A. (2007). Cluster analysis for autocorrelated and cyclic chemical process data. *Industrial and Engineering Chemistry Research, 46*, 3610–3622. https://doi.org/10.1021/ie060544v

Biegler, L. T., & Grossmann, I. E. (2004). Retrospective on optimization. *Computers & Chemical Engineering, 28*, 1169–1192. https://doi.org/10.1016/j.compchemeng.2003.11.003

Bo, Z., & Hao, Y. (2015). Qualitative trend clustering of process data for fault diagnosis. In *2015 IEEE International Conference on Automation Science and Engineering (CASE)*. Presented at the 2015 IEEE International Conference on Automation Science and Engineering (CASE), IEEE, Gothenburg, Sweden, pp. 1584–1588. https://doi.org/10.1109/CoASE.2015.7294327

Chévez, P., Barbero, D., Martini, I., & Discoli, C. (2017). Application of the k-means clustering method for the detection and analysis of areas of homogeneous residential electricity consumption at the Great La Plata region, Buenos Aires. *Argentina. Sustainable Cities Society, 32*, 115–129. https://doi.org/10.1016/j.scs.2017.03.019

Choi, Y., An, N., Hong, S., Cho, H., Lim, J., Han, I.-S., Moon, I., & Kim, J. (2022). Time-series clustering approach for training data selection of a data-driven predictive model: Application to an industrial bio 2,3-butanediol distillation process. *Computers & Chemical Engineering, 161*, Article 107758. https://doi.org/10.1016/j.compchemeng.2022.107758

Dobbelaere, M. R., Plehiers, P. P., Van De Vijver, R., Stevens, C. V., & Van Geem, K. M. (2021). Machine learning in chemical engineering: Strengths, weaknesses, opportunities, and threats. *Engineering, 7*, 1201–1211. https://doi.org/10.1016/j.eng.2021.03.019

Douglas, J. M. (1985). A hierarchical decision procedure for process synthesis. *AIChE Journal, 31*, 353–362. https://doi.org/10.1002/aic.690310302

El-Halwagi, M. M., & Manousiouthakis, V. (1989). Synthesis of mass exchange networks. *AIChE Journal, 35*, 1233–1244. https://doi.org/10.1002/aic.690350802

Fahad, A., Alshatri, N., Tari, Z., Alamri, A., Khalil, I., Zomaya, A. Y., Foufou, S., & Bouras, A. (2014). A Survey of clustering algorithms for big data: Taxonomy and empirical analysis. *IEEE Transactions on Emerging Topics in Computing, 2*, 267–279. https://doi.org/10.1109/TETC.2014.2330519

Fahim, A. (2023). A varied density-based clustering algorithm. *Journal of Computer Science, 66*, Article 101925. https://doi.org/10.1016/j.jocs.2022.101925

Fauzan, A., Fadillah, G., Fitria, A., Adriana, H., & Bariklana, M. (2024). Cluster mapping of waste exposure using DBSCAN approach: Study of spatial patterns and potential distribution in Bantul regency. *JOIV International Journal Information. 8*, 751. https://doi.org/10.62527/joiv.8.2.2475

García-Martínez, J., Reyes-Patiño, J. L., López-Sosa, L. B., & Fuentes-Cortés, L. F. (2022). Anticipating alliances of stakeholders in the optimal design of community energy systems. *Sustain. Energy Technology Assessment, 54*, Article 102880. https://doi.org/10.1016/j.seta.2022.102880

Geiger, A. (2024). Enhancing supply chain coordination: A comparative analysis of clustering techniques for the production routing problem. *Computers & Industrial Engineering, 196*, Article 110455. https://doi.org/10.1016/j.cie.2024.110455

Ghosal, A., Nandy, A., Das, A.K., Goswami, S., & Panday, M. (2020). A short review on different clustering techniques and their applications. In: J. K. Mandal, & D. Bhattacharya (Eds.), *Emerging Technology in Modelling and Graphics, Advances in Intelligent Systems and Computing*. Springer Singapore, Singapore, pp. 69–83. https://doi.org/10.1007/978-981-13-7403-6_9

Gkinis, P. A., Koronaki, E. D., Skouteris, A., Aviziotis, I. G., & Boudouvis, A. G. (2019). Building a data-driven reduced order model of a chemical vapor deposition process from low-fidelity CFD simulations. *Chemical Engineering Science, 199*, 371–380. https://doi.org/10.1016/j.ces.2019.01.009

Gormley, I. C., Murphy, T. B., & Raftery, A. E. (2023). Model-based clustering. *Annual Review Statistics Its Application, 10*, 573–595. https://doi.org/10.1146/annurev-statistics-033121-115326

Govender, P., & Sivakumar, V. (2020). Application of k-means and hierarchical clustering techniques for analysis of air pollution: A review (1980–2019). *Atmospheric Pollution Research, 11*, 40–56. https://doi.org/10.1016/j.apr.2019.09.009

Grossmann, I. E., & Harjunkoski, I. (2019). Process systems engineering: Academic and industrial perspectives. *Computers & Chemical Engineering, 126*, 474–484. https://doi.org/10.1016/j.compchemeng.2019.04.028

Hou, G., Wang, J., & Fan, Y. (2024). Wind power forecasting method of large-scale wind turbine clusters based on DBSCAN clustering and an enhanced hunter-prey optimization algorithm. *Energy Conversion Management, 307*, Article 118341. https://doi.org/10.1016/j.enconman.2024.118341

Huang, G. H., Huang, Y. F., Wang, G. Q., & Xiao, H. N. (2006). Development of a forecasting system for supporting remediation design and process control based on NAPL-biodegradation simulation and stepwise-cluster analysis. *Water Resource Research,* 42, 2005WR004006. https://doi.org/10.1029/2005WR004006

Ikotun, A. M., Ezugwu, A. E., Abualigah, L., Abuhaija, B., & Heming, J. (2023). K-means clustering algorithms: A comprehensive review, variants analysis, and advances in the era of big data. *Information Sciences, 622,* 178–210. https://doi.org/10.1016/j.ins.2022.11.139

Jan, B., Farman, H., Javed, H., Montrucchio, B., Khan, M., & Ali, S. (2017). Energy efficient hierarchical clustering approaches in wireless sensor networks: A survey. *Wireless Communications and Mobile Computing, 2017,* 1–14. https://doi.org/10.1155/2017/6457942

Kalgotra, P., Sharda, R., & Luse, A. (2020). Which similarity measure to use in network analysis: Impact of sample size on phi correlation coefficient and Ochiai index. *International Journal of Information Management, 55,* Article 102229. https://doi.org/10.1016/j.ijinfomgt.2020.102229

Khan, W. U., Ahmed, S., Dhoble, Y., & Madhav, S. (2023). A critical review of hazardous waste generation from textile industries and associated ecological impacts. *Journal of the Indian Chemical Society, 100,* Article 100829. https://doi.org/10.1016/j.jics.2022.100829

Liu, J., Zhao, S., Xie, Y., Jahanshahi, H., Wei, S., & Mohammadzadeh, A. (2023). Fault monitoring-oriented transition process identification of complex industrial processes with neighbor inconsistent pair-based attribute reduction. *Journal of Process Control, 121,* 30–49. https://doi.org/10.1016/j.jprocont.2022.11.011

López-Flores, F. J., Lira-Barragán, L. F., Rubio-Castro, E., El-Halwagi, M. M., & Ponce-Ortega, J. M. (2023). Hybrid machine learning-mathematical programming approach for optimizing gas production and water management in shale gas fields. *ACS Sustainable Chemistry Engineering, 11,* 6043–6056. https://doi.org/10.1021/acssuschemeng.3c00569

López-Flores, F. J., Ramírez-Márquez, C., & Ponce-Ortega, J. M. (2024). Process systems engineering tools for optimization of trained machine learning models: Comparative and perspective. *Industrial and Engineering Chemistry Research, 63,* 13966–13979. https://doi.org/10.1021/acs.iecr.4c00632

López-Flores, F. J., Rubio-Castro, E., & Ponce-Ortega, J. M. (2022). Incorporating machine learning for thermal engines modeling in industrial waste heat recovery. *Chemical Engineering Research and Design, 181,* 239–252. https://doi.org/10.1016/j.cherd.2022.03.022

Maestri, M., Farall, A., Groisman, P., Cassanello, M., & Horowitz, G. (2010). A robust clustering method for detection of abnormal situations in a process with multiple steady-state operation modes. *Computers & Chemical Engineering, 34,* 223–231. https://doi.org/10.1016/j.compchemeng.2009.05.012

Mahesh Prabhu, R., Hema, G., Chepure, S., & Guptha, M. N. (2020). Logistics optimization in supply chain management using clustering qlgorithms. *Scalable Computer Practice Exploring,* 21, 107–114. https://doi.org/10.12694/scpe.v21i1.1628

Mehta, V., Bawa, S., & Singh, J. (2020). Analytical review of clustering techniques and proximity measures. *Artificial Intelligence Review, 53,* 5995–6023. https://doi.org/10.1007/s10462-020-09840-7

Molokomme, D. N., Chabalala, C. S., & Bokoro, P. N. (2021). Enhancement of advanced metering infrastructure performance using unsupervised K-means clustering algorithm. *Energies, 14,* 2732. https://doi.org/10.3390/en14092732

Morlet-Espinosa, J., Flores-Tlacuahuac, A., & Fuentes-Cortes, L. F. (2023). A combined variational encoding and optimization framework for design of the water–energy–food nexus. *Computers & Chemical Engineering, 170,* Article 108076. https://doi.org/10.1016/j.compchemeng.2022.108076

Muñoz López, C. A., Bhonsale, S., Peeters, K., & Van Impe, J. F. M. (2020). Manifold learning and clustering for automated phase identification and alignment in data driven modeling of batch processes. *Frontiers Chemical Engineering, 2,* Article 582126. https://doi.org/10.3389/fceng.2020.582126

Nalawade, S., Gokhale, S., Ingale, S., Arora, S., & Jahirabadkar, S. (2023). Hybrid density- grid based clustering algorithms: A review. In *2023 7th International Conference On Computing, Communication, Control And Automation (ICCUBEA)*. Presented at the 2023 7th International Conference On Computing, Communication, Control And Automation (ICCUBEA), IEEE, Pune, India, pp. 1–5. https://doi.org/10.1109/ICCUBEA58933.2023.10392256

N'Cir, C.-E.B., Cleuziou, G., & Essoussi, N. (2015). Overview of overlapping partitional clustering methods. In M. E. Celebi (Ed.), *Partitional Clustering Algorithms*. Springer International Publishing, Cham, pp. 245–275. https://doi.org/10.1007/978-3-319-09259-1_8

Ng, W. P. Q., & Lam, H. L. (2013). Sustainable supply network design through optimisation with clustering technique integration. *Chemical Engineering Transactions, 35*, 661–666. https://doi.org/10.3303/CET1335110

Nicolis, O., Delgado, L., Peralta, B., Díaz, M., & Chiodi, M. (2024). Space-time clustering of seismic events in Chile using ST-DBSCAN-EV algorithm. *Environmental and Ecological Statistics, 31*, 509–536. https://doi.org/10.1007/s10651-023-00594-3

Niwattanakul, S., Singthongchai, J., Naenudorn, E., & Wanapu, S. (2013). *Using of Jaccard Coefficient for Keywords Similarity.*

Nurulhikmah, F., & Abdi, D. N. E. (2024). Classification of foods based on nutritional content using K-means and DBSCAN clustering methods. *Teknika, 13*, 481–486. https://doi.org/10.34148/teknika.v13i3.1067

Ortiz-Espinoza, A. P., Kazantzi, V., Eljack, F. T., Jiménez-Gutiérrez, A., El-Halwagi, M. M., & Kazantzis, N. K. (2019). Framework for design under uncertainty including inherent safety, environmental assessment, and economic performance of chemical processes. *Industrial and Engineering Chemistry Research, 58*, 13239–13248. https://doi.org/10.1021/acs.iecr.9b02349

Otnes, G., Lindholm, D., Fjær, H., Seljom, P., & Foss, S. E. (2023). Machine learning assisted representative period selection as input to modelling of field degradation in photovoltaic modules. *Solar Energy Materials and Solar Cells, 263*, Article 112559. https://doi.org/10.1016/j.solmat.2023.112559

Otok, B. W., Suharsono, A., Purhadi, P., Standsyah, R. E., & Al Azies, H. (2022). Partitional clustering of underdeveloped area infrastructure with unsupervised learning approach: A case study in the island of Java. *Indonesia. Journal Regional City Plan, 33*, 29–48. https://doi.org/10.5614/jpwk.2022.33.2.3

Oyewole, G. J., & Thopil, G. A. (2023). Data clustering: Application and trends. *Artificial Intelligence Review, 56*, 6439–6475. https://doi.org/10.1007/s10462-022-10325-y

Pan, M., Sikorski, J., Akroyd, J., Mosbach, S., Lau, R., & Kraft, M. (2016). Design technologies for eco-industrial parks: From unit operations to processes, plants and industrial networks. *Applied Energy, 175*, 305–323. https://doi.org/10.1016/j.apenergy.2016.05.019

Papoulias, S. A., & Grossmann, I. E. (1983). A structural optimization approach in process synthesis—II. *Computers & Chemical Engineering, 7*, 707–721. https://doi.org/10.1016/0098-1354(83)85023-6

Patel, P. M., Shah, B. N., & Shah, V. (2013). Image segmentation using K-mean clustering for finding tumor in medical application. *International Journal Computer Trends Technology, 4*(5), 1239–1242.

Pérez-Uresti, S. I., Lima, R. M., Martín, M., & Jiménez-Gutiérrez, A. (2023). On the design of renewable-based utility plants using time series clustering. *Computers & Chemical Engineering, 170*, Article 108124. https://doi.org/10.1016/j.compchemeng.2022.108124

Pistikopoulos, E. N., Barbosa-Povoa, A., Lee, J. H., Misener, R., Mitsos, A., Reklaitis, G. V., Venkatasubramanian, V., You, F., & Gani, R. (2021). Process systems engineering—The generation next? *Computers & Chemical Engineering, 147*, Article 107252. https://doi.org/10.1016/j.compchemeng.2021.107252

Rahoma, A., Imtiaz, S., Ahmed, S., & Khan, F. (2023). Detection and diagnosis of process fault using unsupervised learning methods and unlabeled data. *International Journal of Advances in Engineering Sciences and Applied Mathematics, 15*, 24–36. https://doi.org/10.1007/s12572-023-00327-6

Ran, X., Xi, Y., Lu, Y., Wang, X., & Lu, Z. (2023). Comprehensive survey on hierarchical clustering algorithms and the recent developments. *Artificial Intelligence Review, 56*, 8219–8264. https://doi.org/10.1007/s10462-022-10366-3

Rana, S., & Garg, R. (2016). Application of hierarchical clustering algorithm to evaluate students performance of an institute. In *2016 Second International Conference on Computational Intelligence & Communication Technology (CICT)*. Presented at the 2016 Second International Conference on Computational Intelligence & Communication Technology (CICT), pp. 692–697. https://doi.org/10.1109/CICT.2016.143

Rastgoo, R., Amjady, N., & Zareipour, H. (2024). A deep generative model for selecting representative periods in renewable energy-integrated power systems. *Applied Soft Computing, 165*, Article 112107. https://doi.org/10.1016/j.asoc.2024.112107

Raya-Tapia, A. Y., López-Flores, F. J., Tovar-Facio, J., & Ponce-Ortega, J. M. (2025). Comparative framework of representative weeks selection methods for the optimization of power systems. *Computers & Chemical Engineering, 194*, Article 108985. https://doi.org/10.1016/j.compchemeng.2024.108985

Raya-Tapia, A. Y., Ramírez-Márquez, C., & Ponce-Ortega, J. M. (2024a). Clustering of zones according to the level of gentrification by using an unsupervised learning algorithm. *Cities, 151*, Article 105133. https://doi.org/10.1016/j.cities.2024.105133

Raya-Tapia, A. Y., Sánchez-Zarco, X. G., Cansino-Loeza, B., Ramírez-Márquez, C., & Ponce-Ortega, J. M. (2024b). A typology country framework to evaluate the SDG progress and food waste reduction based on clustering analysis. *Trends in Food Science & Technology, 143*, Article 104304. https://doi.org/10.1016/j.tifs.2023.104304

Reis, M. S., & Saraiva, P. M. (2021). Data-centric process systems engineering: A push towards PSE 4.0. *Computer Chemical Engineering,* 155, 107529. https://doi.org/10.1016/j.compchemeng.2021.107529

Retiti Diop Emane, C., Song, S., Lee, H., Choi, D., Lim, J., Bok, K., & Yoo, J. (2024). Anomaly detection based on GCNs and DBSCAN in a large-scale graph. *Electronics,* 13, 2625. https://doi.org/10.3390/electronics13132625

Rovira, M., Engvall, K., & Duwig, C. (2022). Identifying key features in reactive flows: A tutorial on combining dimensionality reduction, unsupervised clustering, and feature correlation. *Chemical Engineering Journal, 438*, Article 135250. https://doi.org/10.1016/j.cej.2022.135250

Ruiz, L. G. B., Pegalajar, M. C., Arcucci, R., & Molina-Solana, M. (2020). A time-series clustering methodology for knowledge extraction in energy consumption data. *Expert Systems with Applications, 160*, Article 113731. https://doi.org/10.1016/j.eswa.2020.113731

Sajid, S., Haleem, A., Bahl, S., Javaid, M., Goyal, T., & Mittal, M. (2021). Data science applications for predictive maintenance and materials science in context to Industry 4.0. *Materials Today Proceedings, 45*, 4898–4905. https://doi.org/10.1016/j.matpr.2021.01.357

Saxena, A., Prasad, M., Gupta, A., Bharill, N., Patel, O. P., Tiwari, A., Er, M. J., Ding, W., & Lin, C.-T. (2017). A review of clustering techniques and developments. *Neurocomputing, 267*, 664–681. https://doi.org/10.1016/j.neucom.2017.06.053

Seghers, E. E., Briceno-Mena, L. A., & Romagnoli, J. A. (2023). Unsupervised learning: Local and global structure preservation in industrial data. *Computers & Chemical Engineering, 178*, Article 108378. https://doi.org/10.1016/j.compchemeng.2023.108378

Sendín, J.-O.H., Banga, J. R., & Csendes, T. (2009). Extensions of a multistart clustering algorithm for constrained global optimization problems. *Industrial and Engineering Chemistry Research, 48*, 3014–3023. https://doi.org/10.1021/ie800319m

Seng Ng, Y., & Srinivasan, R. (2010). Multi-agent based collaborative fault detection and identification in chemical processes. *Engineering Applications of Artificial Intelligence, 23*, 934–949. https://doi.org/10.1016/j.engappai.2010.01.026

Shanker, R., Singh, R., & Bhattacharya, M. (2017). Segmentation of tumor and edema based on K-mean clustering and hierarchical centroid shape descriptor. In *2017 IEEE International*

Conference on Bioinformatics and Biomedicine (BIBM). Presented at the 2017 IEEE International Conference on Bioinformatics and Biomedicine (BIBM), IEEE, Kansas City, MO, pp. 1105–1109. https://doi.org/10.1109/BIBM.2017.8217810

Shi, H., Ge, X., & Liu, B. (2024). Adaptive fault diagnosis for high-purity carbonate process based on unsupervised and transfer learning. *Chemical Engineering Science, 300*, Article 120631. https://doi.org/10.1016/j.ces.2024.120631

Song, J., Wang, S., & Zhao, H. (2020). Traffic flow detection at road intersections based on K-means and NURBS trajectory clustering. *Mathematical Problems in Engineering, 2020*, 1–6. https://doi.org/10.1155/2020/1383198

Srinivasan, R., Wang, C., Ho, W. K., & Lim, K. W. (2004). Dynamic principal component analysis based methodology for clustering process states in agile chemical plants. *Industrial and Engineering Chemistry Research, 43*, 2123–2139. https://doi.org/10.1021/ie034051r

Tang, M., An, B., Qiu, J., Wang, S., Li, Z., Zhang, Y., Yan, Y., & Wang, W. (2024). Multi-model predictive control of SCR flue gas denitrification system in coal-fired power plant based on kernel fuzzy c-means clustering and integrated model. *Canadian Journal of Chemical Engineering, 102*, 748–764. https://doi.org/10.1002/cjce.25082

Tang, Y., & Zhang, S. (2021). Fault detection of FWTPs in coal-fired power plants using K-WD-KPCA in consideration of multiple operation conditions. *Asia-Pacific Journal of Chemical Engineering, 16*, Article e2599. https://doi.org/10.1002/apj.2599

Teh, S. Y., Chua, K. B., Hong, B. H., Ling, A. J. W., Andiappan, V., Foo, D. C. Y., Hassim, M. H., & Ng, D. K. S. (2019). A hybrid multi-objective optimization framework for preliminary process design based on health, safety and environmental impact. *Processes, 7*, 200. https://doi.org/10.3390/pr7040200

Thibault, É., Chioua, M., McKay, M., Korbel, M., Patience, G. S., & Stuart, P. R. (2023). Experimental methods in chemical engineering: Data processing and data usage in decision-making. *Canadian Journal of Chemical Engineering, 101*, 6055–6078. https://doi.org/10.1002/cjce.25014

Thomas, M. C., Zhu, W., & Romagnoli, J. A. (2018). Data mining and clustering in chemical process databases for monitoring and knowledge discovery. *Journal of Process Control, 67*, 160–175. https://doi.org/10.1016/j.jprocont.2017.02.006

Valencia-Marquez, D., Ortiz-Munguia, J. A., Maldonado-López, E., Quintana-Hernández, P. A., Louvier-Hernández, J. F., & Fuentes-Cortés, L. F. (2023). Multi-objective and machine learning strategies for addressing the Water–Energy–Waste nexus in the design of energy systems. *Sustainable Energy Technology Assessments, 60*, Article 103445. https://doi.org/10.1016/j.seta.2023.103445

Vassiliadis, V. S., Mappas, V., Espaas, T. A., Dorneanu, B., Isafiade, A., Möller, K., & Arellano-Garcia, H. (2024). Reloading process systems engineering within chemical engineering. *Chemical Engineering Research and Design, 209*, 380–398. https://doi.org/10.1016/j.cherd.2024.07.066

Vooradi, R., Bertran, M.-O., Frauzem, R., Anne, S. B., & Gani, R. (2018). Sustainable chemical processing and energy-carbon dioxide management: Review of challenges and opportunities. *Chemical Engineering Research and Design, 131*, 440–464. https://doi.org/10.1016/j.cherd.2017.12.019

Wang, J., Li, Y., Gao, R. X., & Zhang, F. (2022). Hybrid physics-based and data-driven models for smart manufacturing: Modelling, simulation, and explainability. *Journal of Manufacturing Systems, 63*, 381–391. https://doi.org/10.1016/j.jmsy.2022.04.004

Wang, S., Liu, H., Pu, H., & Yang, H. (2020). Spatial disparity and hierarchical cluster analysis of final energy consumption in China. *Energy, 197*, Article 117195. https://doi.org/10.1016/j.energy.2020.117195

Wang, Y., & Li, H. (2020). Industrial process time-series modeling based on adapted receptive field temporal convolution networks concerning multi-region operations. *Computers & Chemical Engineering, 139*, Article 106877. https://doi.org/10.1016/j.compchemeng.2020.106877

Warsito, B., Sumiyati, S., Yasin, H., & Faridah, H. (2021). Evaluation of river water quality by using hierarchical clustering analysis. *IOP Conference Series Earth Environment Science, 896*, Article 012072. https://doi.org/10.1088/1755-1315/896/1/012072

Xie, W.-B., Lee, Y.-L., Wang, C., Chen, D.-B., & Zhou, T. (2020). Hierarchical clustering supported by reciprocal nearest neighbors. *Information Sciences, 527*, 279–292. https://doi.org/10.1016/j.ins.2020.04.016

Xu, R., & Wunsch, D. C. (2010). Clustering algorithms in biomedical research: A review. *IEEE Reviews in Biomedical Engineering, 3*, 120–154. https://doi.org/10.1109/RBME.2010.2083647

Xu, S., Lu, B., Baldea, M., Edgar, T. F., Wojsznis, W., Blevins, T., & Nixon, M. (2015). Data cleaning in the process industries. *Reviews in Chemical Engineering, 31*,. https://doi.org/10.1515/revce-2015-0022

Zheng, S., & Zhao, J. (2020). A new unsupervised data mining method based on the stacked autoencoder for chemical process fault diagnosis. *Computers & Chemical Engineering, 135*, Article 106755. https://doi.org/10.1016/j.compchemeng.2020.106755

Ziaei-Halimejani, H., Zarghami, R., Mansouri, S. S., & Mostoufi, N. (2021). Data-driven fault diagnosis of chemical processes based on recurrence plots. *Industrial and Engineering Chemistry Research, 60*, 3038–3055. https://doi.org/10.1021/acs.iecr.0c06307

Chapter 5
Greenhouse Gas Emissions Clustering for Net Zero Goals

5.1 Introduction

The ongoing climate issue represents one of the most significant concerns confronting humanity (Ripple et al., 2023). Governments and international organizations have enacted measures and policies to facilitate the transition to a low-carbon economy, especially to reduce greenhouse gas (GHG) emissions from the energy sector. In 2023, worldwide carbon dioxide (CO_2) emissions attained a record 36.8 gigatons, reflecting a 0.9% rise from the prior year, with the energy sector accountable for 75% of these emissions (Glavič et al., 2023).

The Paris Agreement, ratified at COP21 in 2015, aimed to restrict global temperature rise to below 2 °C relative to pre-industrial levels, with an objective to cap warming at 1.5 °C (Rajamani & Werksman, 2018; Sadai et al., 2022). Without decisive and immediate actions to meet this objective, the Earth confronts catastrophic and permanent repercussions. This encompasses the exhaustion of natural resources, decline in biodiversity, and risks to human survival and world health (Muluneh, 2021). The World Health Organization has recognized climate change as the paramount threat to global public health, since its impacts intensify pre-existing vulnerabilities and generate new health hazards (www.who.int/publications/i/item/9789240095403).

The primary contributor to global warming is the persistent dependence on fossil fuels, such as coal, oil, and natural gas, which represent over 75% of global greenhouse gas emissions and almost 90% of all carbon dioxide emissions (Rahman et al., 2024). Despite substantial advancements in renewable energy use, fossil fuels continue to account for over 80% of global energy supply, highlighting the pressing necessity for a shift towards cleaner and more sustainable energy systems (Abbasi et al., 2022; Khaleel et al., 2024).

This energy transition entails significant alterations in energy production and use to diminish reliance on fossil fuels (Adelekan et al., 2024). It underscores the

A. Y. Raya-Tapia et al., *Machine Learning and Clustering for a Sustainable Future*, Studies in Computational Intelligence 1233,
https://doi.org/10.1007/978-3-032-03876-0_5

use of renewable energy sources, enhancement of energy efficiency, and electrification of critical sectors, including transportation and industry (Hassan et al., 2024). Attaining "Net Zero" emissions, where greenhouse gas emissions created are equal to those extracted from the atmosphere, has emerged as a critical goal for worldwide sustainability initiatives. Confronting this challenge necessitates revolutionary action of an unparalleled magnitude. Recent assessments suggest that meeting international climate objectives will need a threefold increase in installed renewable energy capacity and a twofold enhancement in energy efficiency by 2030 (Dadashi et al., 2022; Darby et al., 2024). These actions are essential not just to reduce GHG emissions but also to guarantee sustainable and secure energy supply for future generations.

Carbon dioxide (CO_2), methane (CH_4), and nitrous oxide (N_2O) are the predominant greenhouse gases contributing to climate change (Jones et al., 2023; Wei et al., 2016). The Global Warming Potential (GWP) meter quantifies the heat-retaining capacity of these gases in comparison to CO_2, which has a GWP of 1 (Rabbi & Kovács, 2024). CH_4 is around 25 times more efficient in heat retention over a century, but N_2O is 298 times more potent (Nisbet et al., 2021). Nonetheless, CO_2 is the predominant greenhouse gas owing to its substantial emission volumes and its capacity to last in the atmosphere for centuries or even millennia (Berry, 2021). Mitigating emissions of these gases is crucial for combating global warming, especially by targeting the origins of CH_4 and N_2O in agriculture, energy generation, and waste management.

Contemporary instruments, including artificial intelligence (AI), have arisen as formidable friends in the analysis and resolution of the intricate dynamics of greenhouse gas (GHG) emissions (Qerimi & Sergi, 2022). Advanced methodologies, including clustering algorithms like k-means, facilitate the detection of patterns and trends in emissions across many locations and industries (Govender & Sivarkumar, 2020). These technologies offer essential insights crucial for formulating efficient and targeted mitigation plans, facilitating expedited advancement towards net zero emissions.

Numerous studies have effectively utilized clustering approaches to examine GHG emissions, illustrating the efficacy of these methods in recognizing trends and informing targeted mitigation initiatives. Gallo et al. (2018) employed clustering analysis to assess the influence of the Kyoto Protocol on greenhouse gas emissions in the agriculture and forestry sectors. Their research underscored the geographical shift of emissions following the Kyoto Protocol, accentuating the significance of policy formulation and timing in surmounting innovation obstacles and enhancing resource efficiency.

Zhang et al. (2021) investigated urban rivers as significant sources of CO_2, CH_4, and N_2O emissions throughout the Chaohu Lake basin in eastern China. Their investigation revealed considerable regional diversity in greenhouse gas emissions, with urban rivers displaying markedly elevated fluxes due to nutrient-dense, hypoxic conditions. Likewise, Leng et al. (2023) examined geographical patterns and determinants of dissolved greenhouse gas concentrations throughout the Yangtze River, revealing regional clustering affected by wetlands and oxygen levels. Their findings emphasized the necessity of including wetland contributions and nutrient management in GHG mitigation methods for extensive river systems.

In agricultural contexts, Kolasa-Więcek (2013) utilized k-means clustering to categorize OECD nations according to characteristics associated with agricultural greenhouse gas emissions. The comparison of k-means results with hierarchical clustering methods demonstrated consistent groups that highlighted geographical and industrial disparities, such as the distinction between Nordic nations and highly industrialized regions.

These works together demonstrate how clustering analysis, including non-hierarchical techniques like k-means, may be tailored to various fields to discern emission trends and prioritize climate initiatives. This method allows policymakers to customize tactics to regional settings, targeting particular environmental and socio-economic aspects that affect emissions.

This chapter enhances existing approaches by doing an extensive investigation of worldwide emissions of CO_2, CH_4, and N_2O during a 30-year span (1990–2020) for 108 nations. K-means clustering categorizes nations into low, medium, and high emission categories based on total and per capita emissions. The study's methodology identifies critical emission patterns and provides actionable insights for policymakers, enabling the formulation of specific mitigation policies for each group. This chapter presents a comprehensive classification of various GHGs and worldwide areas, serving as a distinctive instrument for promoting international collaboration and prioritizing measures according to common emission profiles. Its contributions seek to enhance worldwide initiatives aimed at attaining net zero emissions and a sustainable future.

5.2 Methodology Implemented

The approach employed in this study comprises numerous interconnected processes (refer to Fig. 5.1). Initially, emissions data for CO_2, CH_4, and N_2O were gathered and structured into datasets encompassing both total and per capita emissions. Subsequently, data pretreatment was performed, including cleaning to address outliers or inconsistencies and normalizing via the z-score approach to assure comparability. The k-means method for time series was utilized to analyze the temporal structure of the data, employing Dynamic Time Warping (DTW) as a metric to assess similarities across emission trajectories over time, therefore capturing intricate dynamic patterns. Countries were categorized as low, medium, and high emission clusters. The found clusters were defined by total emissions, per capita emissions, and temporal trends, and were associated with geographic and socioeconomic characteristics, including Gross Domestic Product (GDP) and population. The data were examined to discern significant trends and their progression, correlating them with sustainability goals and offering clear visual representations, including temporal graphs and maps. Ultimately, distinct findings and suggestions were developed to inform climate mitigation plans according to the attributes of each cluster.

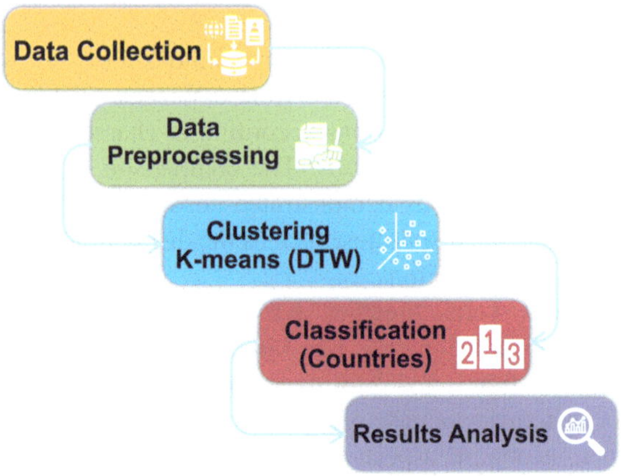

Fig. 5.1 Outline of the applied methodology

5.2.1 Dataset

This study's dataset offers an extensive examination of worldwide emissions of the three principal greenhouse gases (CO_2, CH_4, and N_2O) during a 30-year span (1990–2020), encompassing 108 nations across five continents. The data was obtained from the World Bank DataBank (2024), a widely esteemed library of economic and environmental indicators that aggregates vetted, high-quality information from international organizations and national statistical agencies. The gathered information comprises annual records categorized into total emissions (in metric tons) and per capita emissions (metric tons per individual) for each country, facilitating the analysis of both absolute and population-adjusted trends. This comprehensive dataset incorporates chronological, geographic, and comparative aspects, offering a dependable and thorough basis for assessing emission trends and facilitating significant conclusions.

5.2.2 Data Preprocessing

Data preparation was essential for successful clustering and thorough analysis of GHG emissions. Standardizing the numerical variables reduced errors caused by variations in scale and magnitude, enhancing the model's performance and comparability across datasets. This study employed the Z-score standardization approach, adjusting the data to achieve a mean of zero and a standard deviation of one, so improving the accuracy and reliability of the clustering process.

$$z_i = \frac{x_i - \mu}{\sigma} \tag{5.1}$$

$$\mu = \frac{1}{N}\sum_{i=1}^{N} x_i \tag{5.2}$$

$$\sigma = \sqrt{\frac{1}{N}\sum_{i=1}^{N} (x_i - \mu)^2} \tag{5.3}$$

Here z_i corresponds to the normalized score of x_i, μ represents the mean score of variable x_i, σ is the standard deviation of variable x_i, and N represents the overall number of samples.

5.2.3 K-means Algorithm with DTW Metric

The k-means method is fundamental to unsupervised machine learning, esteemed for its simplicity and efficacy in grouping data according to similarity (Bock, 2007). Nonetheless, its efficacy declines when utilized with time series data due to the constraints of the normal Euclidean distance metric, which includes susceptibility to slight temporal variations and the necessity for series of equal length. The Dynamic Time Warping (DTW) measure alleviates these issues by effectively aligning time series to reduce differences in their temporal structure (Cai et al., 2021).

In time series research, such as the GHG emissions examined in this chapter, DTW facilitates the recognition of complex dynamic patterns that conventional metrics frequently neglect. Dynamic Time Warping (DTW) facilitates the ideal temporal synchronization of two sequences, accommodating discrepancies in the velocity or timing of occurrences. This is achieved by a dynamic programming method to compute the cumulative deformation distance, guaranteeing that temporal differences do not skew the grouping outcomes.

DTW determines the distance among two time series, $X = [x_1, x_2, ..., x_n]$ and $Y = [y_1, y_2, ..., y_m]$, where the distance is determined as follows:

$$D(i,j) = d(x_i, y_j) + min\{D(x_{i-1}, y_{j-1}), D(x_{i-1}, y_j), D(x_i, y_{j-1})\} \tag{5.4}$$

Here $D(i,j)$ represents the cumulative distance to the point (i,j) and $d(x_i, y_j) = (x_i - y_j)^2$ corresponds to the Euclidean distance among the points x_i and y_j. The typical steps of the K-means approach with DTW are the following:

Step 1: Initially, k time series are randomly selected as the initial "centroids," denoting the temporary centers of each cluster.

Step 2: For each time series, the distance to each centroid is computed using the DTW metric (see to Eq. 5.4).

Step 3: Each time series is allocated to the cluster whose centroid exhibits the minimal DTW distance, so guaranteeing that it is categorized with analogous series.

$$\text{Cluster}(X) = min\sum\nolimits_{i} DTW(X, C_j) \tag{5.5}$$

Step 4: For each cluster, compute a new centroid that most accurately captures the series inside that cluster. The new centroid is the series inside the cluster that minimizes the aggregate of the DTW distances relative to all other series in the cluster:

$$C_j = min\sum\nolimits_{p}\sum\nolimits_{l} DTW(X_p, X_l) \tag{5.6}$$

This way, C_j corresponds to the "most representative" series of that cluster, and X_l is the individual time series inside the allocated cluster.

Step 5: Steps 2–4 are reiterated until the clusters settle, exhibiting little variation between iterations, or a stopping requirement is met, such as lowering the overall distance between the series and their centroids.

Despite DTW's greater computing demands compared to the Euclidean metric, its capacity to capture temporal alignments warrants its application in investigations where temporal patterns are essential for understanding emission dynamics.

5.3 Results and Discussion

The approach utilized categorizes the 108 nations into three groups according to their emission levels, employing a scale that differentiates between low, medium, and high emissions. This categorization is utilized for each of the three primary greenhouse gases, regarding both total emissions and per capita emissions.

5.3.1 Total and Per Capita N_2O Emissions

5.3.1.1 Global Per Capita N_2O Emissions

Figure 5.2 illustrates the outcomes of the clustering analysis categorizing nations into three groups according to worldwide N_2O emissions per capita, quantified in metric tons (mt), throughout the research period.

Figure 5.2a specifically illustrates the cohort of nations with the largest emissions, spanning from 0.45 million tons released by Cameroon to 5.6 million tons emitted by Australia. This group consists of 10 nations that together produce the greatest per capita N_2O emissions throughout the research period. Mongolia heads this group, accounting for 14.16% of the group's per capita emissions. This elevated proportion is mostly attributable to its mining operations and a modest population of 3.5 million, which exacerbates its per capita emissions, despite its position of 46th in overall

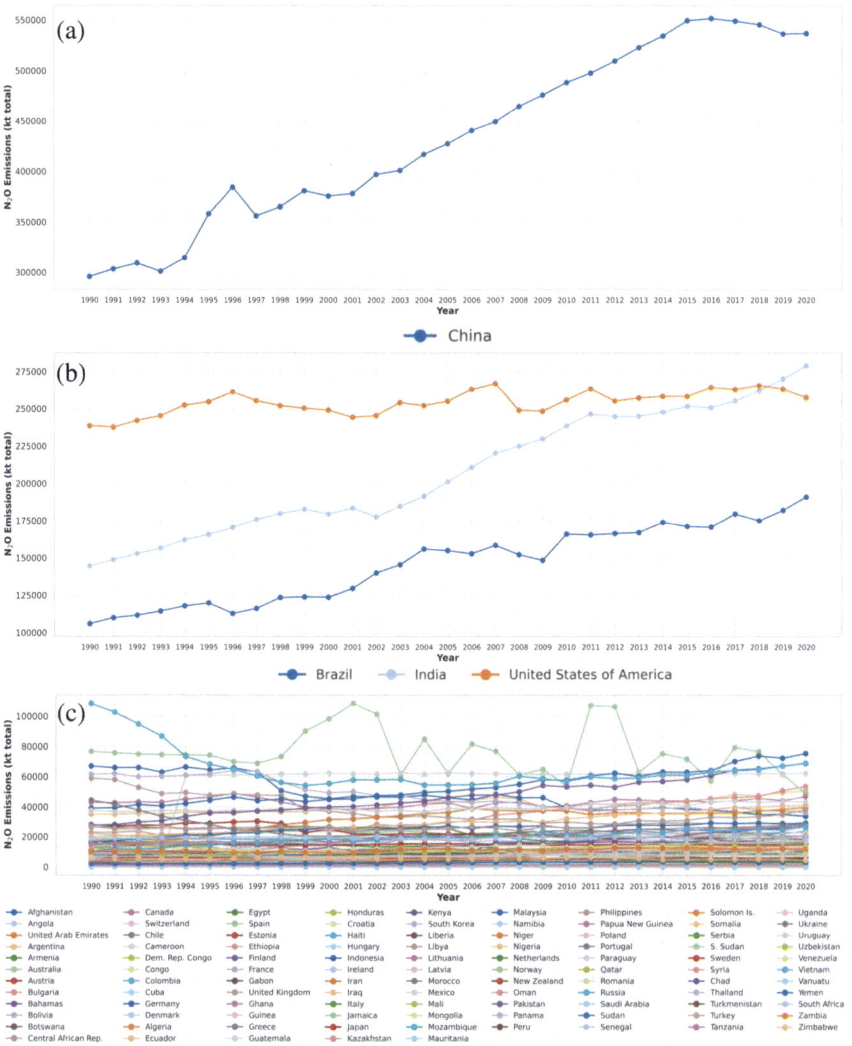

Fig. 5.2 Global N_2O emissions in mt per capita; **a** High emissions group, **b** Medium emissions group, **c** Low emissions group

N_2O emissions. Australia accounts for 13.14% of the group's emissions, mostly attributable to its agricultural and cattle operations, in addition to mining activities. New Zealand ranks next at 12.8%, exhibiting a comparable economic profile to Australia, where the agriculture industry is crucial.

Cameroon exhibits the most significant increase (968.73%) in emissions over the research period, mostly attributed to a sharp escalation between 1993 and 1994, coinciding with its conflict with Nigeria. The war indirectly elevated emissions by deforestation, damage of agricultural land, fires, and the deployment of explosives during the battle.

Figure 5.2b illustrates the cohort of nations exhibiting moderate N_2O emissions, consisting of 25 countries with per capita emissions varying from 4.49 to 3.12 metric tons. Lithuania heads this category, accounting for 6.66% of emissions, with a marginal rise of 2.08% throughout the research period. Paraguay ranks next with 6.56%, despite a 75.03% decrease in emissions from 1990 to 1995. This decline corresponded with Paraguay's shift from a 30-year dictatorship to a democratic one.

Figure 5.2c encompasses the cohort of nations exhibiting the lowest N_2O emissions per capita, varying from 0.05 to 1.03 metric tons. Somalia heads this category with a contribution of 2.68%, principally propelled by its agriculture and livestock industries, despite a population of about 1.8 million. Conversely, densely populated nations like China and India exhibit very reduced per capita emissions within this cohort. Nevertheless, these nations are significant worldwide contributors owing to extensive agricultural operations to satisfy their food requirements. During the research period, China had a 38% rise in per capita emissions, whilst India saw an 18% increase, attributed to population and economic expansion.

Additional details can be found in Appendix Sect. 5.6.1.1.

The average trajectory trends, expressed in mt, for each per capita N_2O emissions group are displayed in Fig. 5.9 (Appendix). Cluster 0 (low emissions) is shown in green, Cluster 2 (medium emissions) is shown in blue, and Cluster 1 (high emissions) is shown in red. The average trend of the high-emissions group, which includes 9.25% of the nations taken into consideration in this study, is shown in the Cluster 1 trajectory. The trend varied during the course of the analysis, with the largest emissions occurring between 1994 and 2020 and reaching a peak of 33.4 mt in 2000. Between 2013 and 2020, emissions were at their lowest, with 2019 recording at least 21.83 mt. The company demonstrated a 21.74% drop in overall emissions over the research period, as well as a 24.61% decrease between 1990 and 2020. These declines are consistent with slower population growth in 1994–2000 and faster growth in 2013–2020 in nations like Australia, Uruguay, and Mongolia. With an average of 88.29 mt per person and an annual average of 2.94 mt per person, Cluster 1 produced 882.9 mt in total throughout the 30-year research period, which is comparable to 36.86% of worldwide per capita N_2O emissions. This category is distinguished by industries including agriculture, cattle, and mining. The typical behavior of medium-level emissions is represented by the Cluster 2 trajectory, which includes 23.14% of the nations taken into consideration in this study. During the 1990–2020 study period, the trend was almost linear, with a few notable deviations. With total emissions of 29.17 mt and an average of 1.16 mt per person year, 1990 marked the highest. Between 1990 and 1995, there was a decrease, and until 2008, it stabilized at about 0.98 mt yearly per person. Between 2009 and 2020, there was another little decrease, with total emissions of 22.23 mt and yearly emissions per person of 0.88 mt in 2009. Between 1990 and 2020, emissions decreased by 21.84%, for a total reduction of 23.69% over the course of 30 years. With an average of 29.82 mt per person and an annual average of 0.99 mt per person, the group produced 745.70 mt of emissions, accounting for 31.13% of world per capita N_2O emissions. Agriculture and cattle are the primary economic activity in these nations, and they have a big impact on N_2O emissions.

A more detailed analysis is available in the Appendix Sect. 5.6.1.2.

The average trend of the low-emissions group, which includes 67.61% of the nations taken into consideration in this study, is depicted by the Cluster 0 trajectory. From the start of the era until its conclusion, this group shows a consistent decline with no discernible oscillations. With 24.44 mt and an average of 0.40 mt per person year, 1990 saw the highest emissions. 2018 saw the lowest amount—22.52 mt and 0.30 mt per person annually—before stabilizing until 2020. The group's conduct can be somewhat explained by the fact that this drop occurs at the same time as the world's population continues to expand. Its limited individual contributions and lack of variation, as it is the largest cluster, support the group's almost linear downward tendency. With an average of 10.50 mt per person and an annual average of 0.35 mt per person, Cluster 0 produced 766.57 mt throughout the research period, accounting for 32.00% of all worldwide N_2O emissions. These nations often have sizable populations, and the strong demand for food necessitates the intense management of the agricultural and agri-industrial sectors, which has a major impact on N_2O emissions. Further details are provided in Appendix Sect. 5.6.1.2.

The three categories' respective environmental impacts varied significantly, according to the study of total per capita emissions. With an annual average of 2.94 mt per person, Cluster 1, which makes up just 9.25% of the countries under study, is responsible for 36.86% of worldwide per capita N_2O emissions. With emissions 2.97 times greater than Cluster 2 and 8.40 times higher than Cluster 0, this indicates a disproportionate impact per capita. With an annual average of 0.99 mt per person and a contribution of 31.13% of total emissions, Cluster 2 is positioned as an intermediate category in terms of both total emissions and per capita effect. Cluster 0 produces just 32.00% of total emissions, with an average of 0.35 mt per person year, even though it includes 67% of the nations under study. These findings demonstrate how a small number of nations in Cluster 1 contribute disproportionately to global emissions, whereas Clusters 2 and 0 have less per capita impacts but, because of their greater populations, make considerable total contributions.

The world's nations are divided into groups based on their per capita N_2O emissions in Fig. 5.3. Three categories of nations are shown visually on the globe as colored groups: low emissions (Cluster 0) are shown in green, medium emissions (Cluster 2) are shown in blue, and high emissions (Cluster 1) are shown in red. This graphic illustrates how emissions of this GHG are distributed throughout several areas.

With four and nine countries, respectively, the majority of Latin American nations are classified as medium or low emitters. Because of its small population and the importance of agriculture to its economy, Uruguay is the only country with high emissions. Latin America is more prevalent among groups with moderate global contributions, according to this distribution. Nonetheless, there are noticeable variations in the region's economic activity, population, and consumer habits.

The United States and Canada are the two nations with the highest per capita N_2O emissions in North America. This pattern illustrates the impact of variables including

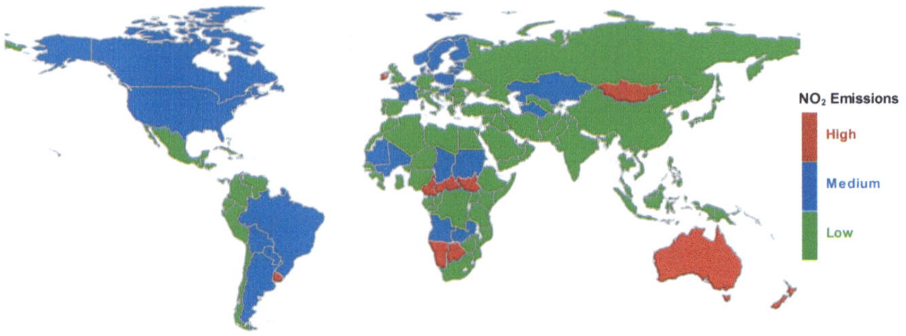

Fig. 5.3 Map of N_2O emissions (mt) per capita by country

economic status, population size, and technological advancement. Both nations rank among the top emitters of this greenhouse gas worldwide.

For additional details on segmentation by region, refer to Appendix Sect. 5.6.1.3.

5.3.1.2 Global Total N_2O Emissions

The findings of the clustering analysis of nations into three groups according to the worldwide total N_2O emissions in kilotons (kt) throughout the study period are shown in Fig. 5.4.

The high-emissions group, which includes just China, is shown in Fig. 5.4a. Total emissions in this group range from 295,990.80 kt to 551,682.72 kt. With 61.86% of the overall rise throughout the research period, compared to its population growth of 21.82%, China is now the world's highest emitter of N_2O in absolute terms. Its role as a global leader in chemical manufacturing, the widespread use of nitrogen-based fertilizers to fulfill food demand, and the country's increasing industrialization and urbanization are all blamed for this disproportionate increase. China ranks 72nd in terms of emissions per capita, while being the highest emitter in absolute terms, indicating the importance of its population size.

The medium-emissions group, which includes the United States, India, and Brazil, is shown in Fig. 5.4b. Emissions in this group range from 106,201.45 kt to 279,003.79 kt. 41% of this group's emissions come from the US, primarily from intensive agriculture and livestock. However, because of limited population growth, environmental laws, and agricultural technological advancements, emissions growth has remained relatively steady at 8.47%. India comes in second with 34.14% of emissions, a 66.9% increase over the research period, mostly due to its intensifying agriculture and expanding population (47.65%). Deforestation, agricultural growth, and the production of biofuels are the main causes of Brazil's 24.02% contribution and 61.34% increase in emissions.

For quantitative comparisons and further analysis of emission trends by country, see Appendix Sect. 5.6.1.4.

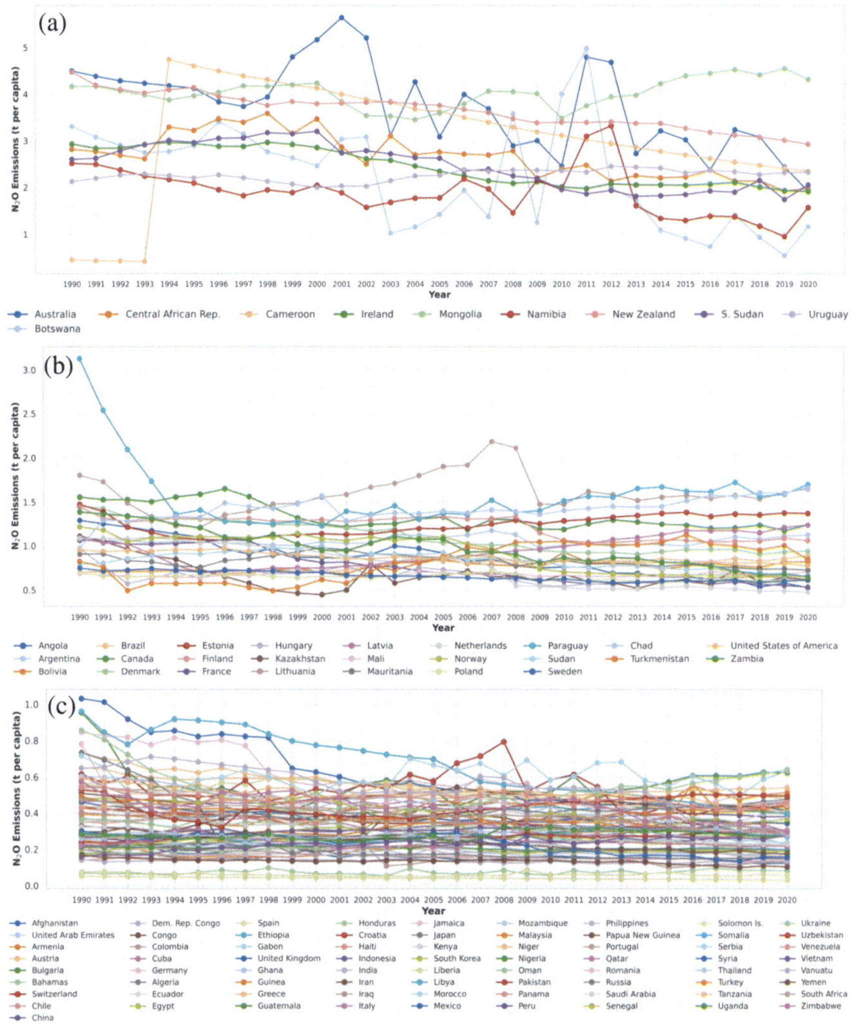

Fig. 5.4 Global N_2O emissions in total kilotons; **a** High emissions group, **b** Medium emissions group, **c** Low emissions group

104 of the 108 nations under study are included in the low-emissions category, shown in Fig. 5.4c. Their total emissions range from 17.58 kt (Bahamas in 1994) to 108,556.14 kt (Russia in 1990). Even though this group produces the fewest emissions overall, there are notable variations among its members. Australia (5.43%), Russia (4.63%), and Cameroon (3.90%) are the leading contributors in this category, demonstrating the impact of their mining and agricultural endeavors.

Over the course of the research, the combined emissions of all three groups' nations total 2,395.18 kt of N_2O. Although N_2O only makes up 9.5% of all combined

GHG emissions (N_2O, CH_4, and CO_2), its overall emissions rose dramatically due to industrialization, population growth, and agricultural development.

The behavior of the average trajectory of total N_2O emissions in kilotons (kt) for each group is shown in Fig. 5.10 (Appendix). The colors of the three groups are used to distinguish them: a red line indicates high emissions, a blue line indicates medium emissions, and a green line indicates low emissions.

In Cluster 2, N_2O total: China is the only nation in this cluster that is responsible for all of its emissions, hence it determines the trajectory. Because of agricultural reforms and economic changes, China's emissions varied between 1990 and 2003, reaching their lowest level in 1990 (295,990.80 kt). Emissions surged between 2004 and 2016, reaching a high of 555,682.72 kt in 2016. From 2017 to 2019, they decreased somewhat, and in 2020, they increased slightly once again. China's total emissions of N_2O over the 30 years under analysis amounted to 13,419,151.74 kt, or 17.73% of the world's total emissions.

N_2O total in Cluster 1: This group, which comprises the US, Brazil, and India, accounts for medium emissions and contributes 24.92% of world N_2O emissions. The group's emissions are almost linearly increasing, peaking in 2020 at 728,109.39 kt. The group's total emissions throughout the research period were 18,854,175.57 kt.

For a detailed breakdown of emissions trends, including individual contributions of countries in Clusters 1 and 0, refer to Appendix Sect. 5.6.1.5.

The global breakdown of nations by total N_2O emissions is shown in Fig. 5.11 (Appendix). nations with high emissions are shown in red (Cluster 1), those with medium emissions are shown in blue (Cluster 2), and those with low emissions are shown in green (Cluster 0). The map shows the distribution of nations by continent or area in a schematic manner. A worldwide view of N_2O emission patterns is provided by this categorization.

Only two nations in the Americas are classified as high-emissions: Brazil, which contributes 5.98% of world emissions, and the United States, which accounts for 10.42%. The effect of elements like population size, economic and technical advancement, and the prevalence of industries like agriculture and livestock is evident in these nations. Argentina (13th, 1.69%), Mexico (14th, 1.61%), and Canada (12th, 1.76%) are some of the major contributors to the low-emissions group, which includes the other nations in the Americas.

Since all of Africa's nations are categorized as low-emissions, the continent appears completely green on maps. However, the importance of agriculture and the lack of strong regulatory frameworks make nations like Cameroon (7th, 2.23%) stand out.

See Appendix Sect. 5.6.1.6 for a comprehensive overview of emissions rankings and contributions by nations in other areas.

5.3.2 *Total and Per Capita CH_4 Emissions*

5.3.2.1 Per Capita CH_4 Emissions

The findings of the clustering analysis of nations into three groups according to per capita CH_4 emissions, expressed in metric tons (mt), over the 1990–2020 study period are displayed in Fig. 5.5.

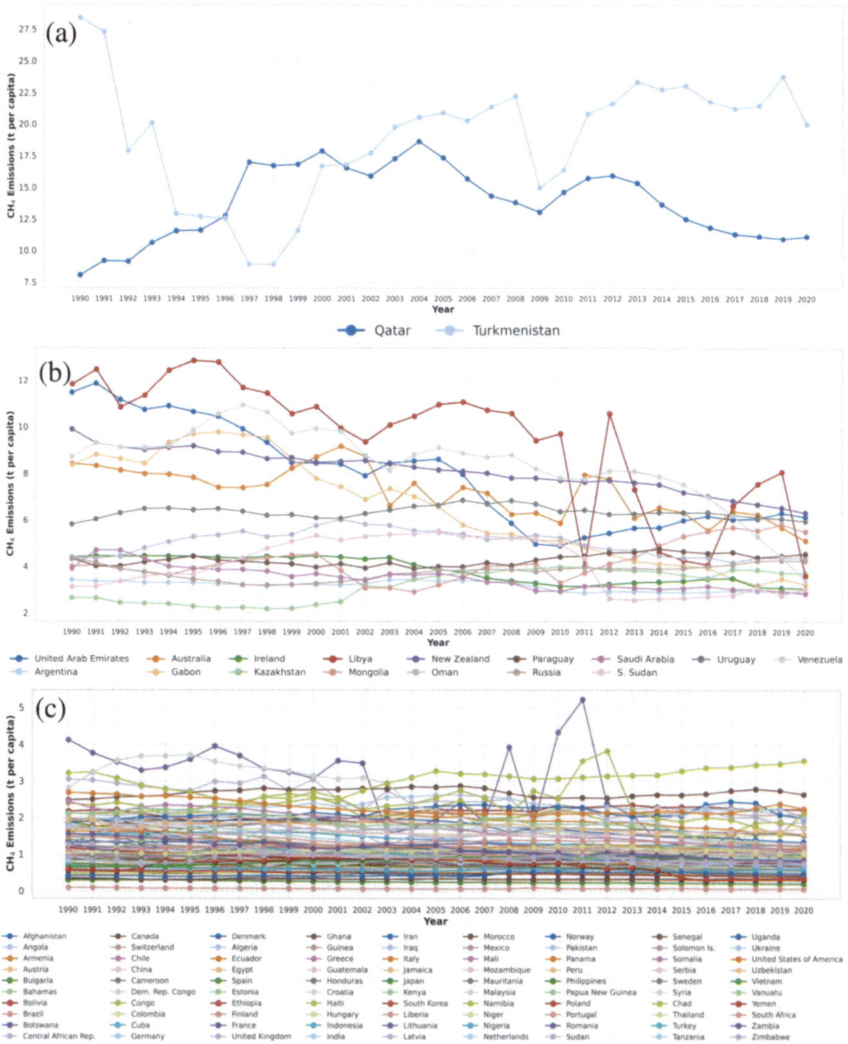

Fig. 5.5 Global CH_4 emissions in mt per capita; **a** High emissions group, **b** Medium emissions group, **c** Low emissions group

Only Turkmenistan and Qatar make up the tiny group of nations with the greatest per capita CH_4 emissions, as shown in Fig. 5.5a. With 57.85% of the total emissions, Turkmenistan tops the group, followed by Qatar with 42.11%. Both nations' economies, which mostly depend on the extraction and export of natural gas, are to blame for the excessive emissions. Even though Turkmenistan dominated this category, there were notable drops in emissions during times of economic upheaval, such 1997–1998 and 2009. Qatar's tiny population and economic dependence on LNG exports have an impact on its emissions.

The 16 nations in the medium-emissions category are depicted in Fig. 5.5b. New Zealand (9.13%), Venezuela (9.42%), and Libya (10.60%) are the top donors. While Venezuela's erratic pattern is a reflection of its economic turmoil and deteriorating oil infrastructure, Libya's emissions are linked to its production of gas and oil. Despite developments in clean energy, New Zealand's agriculture sector—particularly animal farming—remains a major emitter.

Ninety nations in the low-emissions category, where individual contributions are negligible, are included in Fig. 5.5c. With only 0.07% of the group's total emissions, Liberia has the lowest emissions per person. Because of their vast populations, notable nations like China, the US, and India continue to rank among the world's top polluters while having low emissions per person.

For a comprehensive breakdown of individual country contributions, trends, and factors driving emissions within each group, refer to Appendix Sect. 5.6.2.1.

The average trajectories of per capita CH_4 emissions in mt for each group are shown in Fig. 5.12. Colors are used to distinguish the three clusters: Cluster 0 corresponds to low emissions (green line), Cluster 2 to medium emissions (blue line), and Cluster 1 to high emissions (red line).

Only Qatar and Turkmenistan are included in Cluster 1, which makes up less than 2% of the nations examined. Over the course of the research, this group's trajectory exhibits noticeable peaks and valleys. Between 2000 and 2008, emissions peaked in 2004 at 39.18 mt per capita, mostly owing to Turkmenistan's volatility brought on by economic problems and gas pipeline disruptions. Despite being impacted by variations in its population growth, Qatar's emissions were comparatively constant. With yearly averages of 16.95 mt per person, Cluster 1 produced 1,017.04 mt of CH_4 per capita throughout the 30-year period, or 14.49% of world CH_4 emissions per capita.

Over the course of the 30 years, Cluster 2, which represents 14.81% of the nations examined, shows a falling tendency with some small peaks and oscillations. The biggest decrease in emissions was between 2013 and 2020, when overall emissions reached their lowest point at 68.07 mt and per capita at 4.25 mt. In total, this group produced 2,757.85 mt of CH_4 per person, which accounted for 39.31% of the total worldwide. Due to population growth, legislative changes, and economic considerations, the group's fall is ascribed to declines in important nations including the UAE, New Zealand, and Venezuela.

The biggest group, Cluster 0, which includes 83.33% of the nations under analysis, exhibits a consistent, linear decrease in CH_4 emissions per person over the research period. Between 2012 and 2020, emissions dropped from a peak of 118.15 mt in 1990 to 94.23 mt yearly. This group's total emissions came to 3,240.57 mt, or 46.19% of the world's CH_4 emissions per person. The main cause of the 46.19% drop over the 30 years under study is the world's fast population increase, which reduced emissions per person.

For further details on the specific trajectories, individual contributions, and economic influences of each country within these clusters, refer to Appendix Sect. 5.6.2.2.

Three categories are visually distinguished by color in Fig. 5.13, which shows the global segmentation of nations according to their per capita CH_4 emissions: low emissions are shown in green (Cluster 0), medium emissions are shown in blue (Cluster 2), and high emissions are shown in red (Cluster 1).

According to the worldwide per capita ranking, four American nations are in the medium emissions group (Cluster 2). Venezuela's economy is reliant on hydrocarbons, putting it in fourth place globally and accounting for 3.70% of global emissions. Livestock production is the main source of emissions in Uruguay, which comes in tenth place with a 2.80% contribution. Driven by cattle activity, Paraguay and Argentina come in eleventh and eighteenth place, respectively, with contributions of 1.89% and 1.39%. With nations like Mexico in 50th rank (1.14%), Brazil in 30th place (0.89%), and Colombia in 39th place (0.70%), the remainder of Latin America leads the low emissions group (Cluster 0).

The majority of African nations are classified as low emissions (Cluster 0). One notable exception is South Africa, which contributes 0.6% and is ranked 44th in the world in terms of emissions per capita. However, because of its substantial coal output, South Africa ranks 23rd in the world in terms of overall emissions. Despite having a tiny population of 2.6 million, Botswana ranks 95th in terms of overall emissions but 20th in terms of per capita CH_4 emissions (1.19%).

Most European nations fall into the low emissions category and score lower in the worldwide ranking of per capita CH_4 emissions. France is 60th (0.47%), Italy is 79th (0.36%), Spain is 74th (0.38%), and Germany is 75th (0.37%). These nations, however, rank far higher when it comes to worldwide total emissions, with Germany coming in at number 21, France in at number 22, Italy in at number 32, and Spain in at number 36. This demonstrates how population size affects emissions per capita.

Turkmenistan and Qatar, which rank first and second globally with contributions of 8.39% and 6.10%, respectively, are members of the high emissions category in Asia. Natural gas production and exports from these nations are a contributing factor to their high emissions.

For further details on regional trends and country-specific contributions, refer to Appendix Sect. 5.6.2.3.

5.3.2.2 Global Total Emissions of CH_4

Figure 5.6 displays the clustering analysis for the total CH_4 emissions worldwide throughout the study period, expressed in kt.

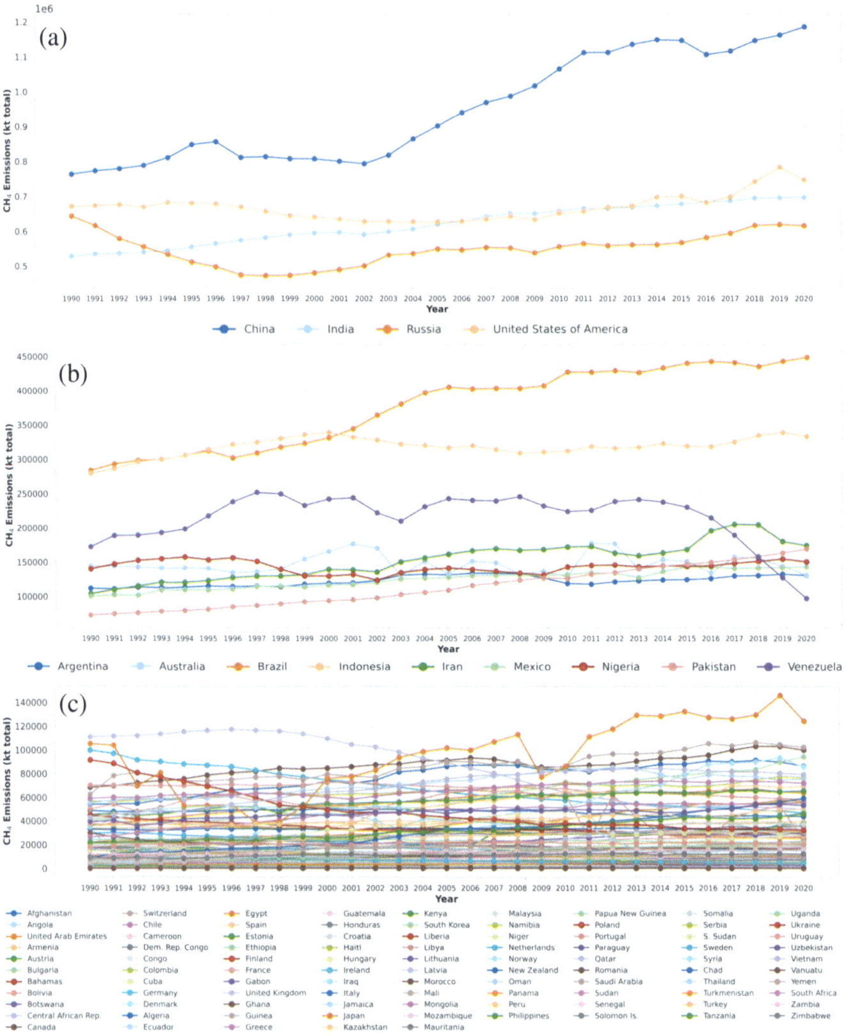

Fig. 5.6 Global CH4 emissions in total kt; **a** High-emission group, **b** Medium-emission group, **c** Low-emission group

The four nations that dominate global CH_4 emissions—China, the United States, India, and Russia—make up the high-emissions group (Fig. 5.6a). Emissions from this category vary from 1,186,285.18 kt (China, 2020) to 472,999.32 kt (Russia, 1998).

China tops the category with 34.00% of the total emissions, primarily from its large-scale agricultural and animal husbandry, as well as its energy sector, which includes coal mining, which emits a lot of CH_4. Between 1990 and 2020, China's emissions increased by 45.20%.

The hydraulic fracturing boom in natural gas extraction is the main driver of the United States' 24.02% share, with waste management and animal husbandry also making a contribution. Emissions rose more moderately by 11.36% in spite of these sources.

With emissions mostly from rice farming and its sizable animal population, which produces CH_4 through digestive processes, India accounts for 22.21% of the total. During the research period, India had a 28% growth.

As a result of a drop in natural gas production and shipments to Europe brought on by geopolitical considerations, Russia is the only nation in this group to have reduced its emissions by 3.17%.

Nine nations make up the medium-emissions group (Fig. 5.6b), with Brazil accounting for 21.96% of the group's total emissions. The main source of Brazil's emissions is its cattle ranching industry, which is among the biggest in the world and has grown dramatically in recent years.

18.55% comes from Indonesia, mostly from rice farming. In contrast, economic troubles and a decline in oil and gas production caused Venezuela, a major emitter, to cut its emissions by 45.43%. Iran, Australia, Nigeria, Mexico, and Pakistan are other major sources, with energy sectors, animal farming, and agriculture being the main drivers of emissions in these countries.

The biggest cluster of 95 nations, the low-emissions group (Fig. 5.6c), with emissions ranging from 144.81 kt (Bahamas in 1990) to 146,450.68 kt (Turkmenistan in 2019). The natural gas industry in Turkmenistan puts it at the top of this category, although industrial activity in Canada, Germany, and France also makes these nations significant emitters.

For a more detailed breakdown of emissions within each cluster, including individual country contributions and trends, refer to Appendix Sect. 5.6.2.4.

Figure 5.14 displays the average trend of total CH_4 emissions by group. The colors of the three groups are used to distinguish them: a red line indicates high emissions, a blue line indicates medium emissions, and a green line indicates low emissions.

Including 3.84% of the nations under analysis, Cluster 1, which stands for the high-emission group, has an overall rising trend interspersed with periods of stabilization. Four nations make up the group: Russia, China, India, and the United States. Russia has lowered its emissions, while the United States has kept its emissions comparatively flat, stabilizing the trend while China and India continue to increase their emissions. With 3,265,126.89 kt of CH_4, 2019 had the greatest emissions, while 2002 saw the lowest at 2,514,991.75 kt. Emissions increased steadily between 2002 and 2020, with a little decline of 1.24% in 2016. Cluster 1 saw a 22.29% rise during the 30-year period, producing 86,644,298.21 kt of CH_4, or 40.11% of world emissions. During the research period, the average emissions for each nation in this group were 21,610,574.55 kt, with an average of 720,352.48 kt each year. The energy industries of China, the US, and Russia are important suppliers, as is the Indian agriculture industry.

Cluster 0, which includes 8.33% of the nations in the analysis and represents the medium-emission category, shows modest swings in growth. From 1990 to 2001, there was constant expansion, and from 2002 to 2009, there was stability. 2017 had a high in emissions, which averaged 210,547.04 kt per nation year. The group suffered a drop of 6.08% from 2018 to 2020. Over the 30-year period, the total rise was 23.79%, with notable gains coming from nations like Iran (55.58%), Brazil (46.56%), and Pakistan (84.87%). With 53,249,993.19 kt of CH_4 produced overall, Cluster 0 accounted for 24.71% of worldwide CH_4 emissions. Over time, the average emissions per nation were 5,916,665.91 kt, with an average of 197,222.19 kt each year.

With a virtually linear increase trend, Cluster 2, which represents the low-emission category, comprises 87.96% of the nations under analysis. The group's peak weight was 2,758,551.96 kt in 2019, while its lowest weight was 2,145,200.62 kt in 1992. The group's growth during the 30-year period was 22.19%, and its contributions to world CH_4 emissions were 75,807,047.60 kt, or 35.17%. During that time, the average emissions per nation were 797,968.92 kt, with an annual average of 26,598.96 kt.

For a detailed breakdown of trends, specific country contributions, and comparative analyses within each cluster, refer to Appendix Sect. 5.6.2.5.

Figure 5.15 shows an analysis of the worldwide distribution of total CH_4 emissions across three groups. Colors are used to distinguish the clusters: low-emission nations are shown in green (Cluster 2), medium-emission countries are shown in blue (Cluster 0), and high-emission countries are shown in red (Cluster 1). A vivid visual depiction of CH_4 emissions across continents is given by this categorization.

With 20,771,297 kt of emissions throughout the research period, the United States ranks second globally and is the only nation in the Americas in the high-emission category. With minor contributions from waste management and livestock, the United States' massive energy sector—particularly the hydraulic fracturing method of extracting and distributing natural gas—is the main source of emissions.

With 11,694,947.05 kt (7.91%), Brazil is in fifth position globally in the medium-emission group. Its agriculture industry, particularly the production of livestock, is largely responsible for its emissions. Other nations in this group include Venezuela (7th, 6,674,910.81 kt, 3.09%), Argentina (12th, 3,815,357 kt, 1.77%), and Mexico (11th in the world, 3,876,629.91 kt, 1.79%). Argentina's emissions are mostly from cattle, Mexico mixes energy and livestock industries, while Venezuela's emissions are mostly from hydrocarbons.

Canada (15th internationally, 2,703,274.54 kt, 1.22%), Colombia (24th, 2,064,210.68 kt, 0.95%), and Chile (71st, 359,833.85 kt, 0.16%) are among the nations in the Americas' low-emission category. Despite falling into the low-emission category, Canada's rating emphasizes the importance of its energy industry.

Due to low levels of industrial and energy activity, almost all African nations fall into the low-emission category. Only Nigeria, which ranks 10th in the world with 4,453,910.26 kt (2.06%), is a member of the medium-emission category. Nigeria is a major producer of natural gas and crude oil, which accounts for the majority of its emissions.

With major contributions like Germany (ranked 21st internationally, 2,121,849.09 kt, 0.98%), France (ranked 22nd, 2,097,529.12 kt, 0.97%), Italy (ranked 32nd, 1,474,816.83 kt, 0.68%), and Spain (ranked 36th, 1,175,987.17 kt, 0.54%), the European Union is fully represented in the low-emission category. Though substantial industrial activity maintains these nations' relatively high rankings within the group, strict environmental controls in Europe have reduced emissions.

Countries in Asia are split evenly amongst the three groupings. With 29,398,255.03 kt (13.64%) of CH_4 emissions, China leads the globe in this category. Its energy industry, especially the production of coal, as well as its agricultural and animal industries, are the main drivers of these emissions. With 19,206,752.68 kt (8.91%), India comes in third place globally. Agriculture, particularly the raising of rice and animals, is the main source of emissions in this country. Russia comes in fourth place with 17,065,992.96 kt (7.91%), mostly from its oil industry. These three nations collectively generate 30.46% of the world's CH_4 emissions.

Notable contributions to the medium-emission category are Australia (9th, 4,576,671.96 kt, 2.12%), Pakistan (13th, 3,539,198.02 kt, 1.64%), and Iran (8th worldwide, 4,737,938.50 kt, 2.19%). Iran's emissions are linked to energy production, whereas Pakistan's emissions are mostly from agricultural, and Australia's emissions are mostly from cattle.

Turkmenistan stands out among low-emission nations because of its substantial natural gas output, ranking 14th in the world. Afghanistan (70th), Mongolia (60th), and Japan (29th) are some noteworthy nations. Turkmenistan has some of the highest per capita emissions in the world, even though it is in the low-emission category.

For additional data and detailed discussion, refer to the Appendix Sect. 5.6.2.6.

5.3.3 *Total and Per Capita CO_2 Emissions*

5.3.3.1 Global CO_2 Emissions Per Capita

Based on their global per capita CO_2 emissions, expressed in mt, for the 1990–2020 study period, nations are grouped into three clusters in Fig. 5.7.

Seven nations make up the high-emissions category (Fig. 5.7a), with emissions ranging from 6.6 mt in 1990 in Oman to 47.6 mt in 2004 in Qatar. With 27.35% of the total emissions, Qatar is the group leader. Its emissions show a tendency of oscillation, with substantial decreases in 2001 and 2009 and notable rises from 1996 to 2007. The 2001 decline is consistent with a population growth of 5.09% and steady total emissions (0.95% growth). In a similar vein, overall emissions increased by 1.93% in 2009, while the population increased by 11.49%, resulting in a more drastic decrease. Emissions per capita decreased after 2012 as population growth exceeded overall emissions. In addition to contributions from its petrochemical industry, construction, transportation, and high consumption rates, Qatar's emissions are mostly caused by energy-intensive activities including gas liquefaction, transportation, and power generation.

With 14.74% of emissions, the United Arab Emirates (UAE) comes in second. Due to its fast population expansion (168.87%), the UAE has a downward trend, with per capita emissions falling by 31.70% during the research period. The primary

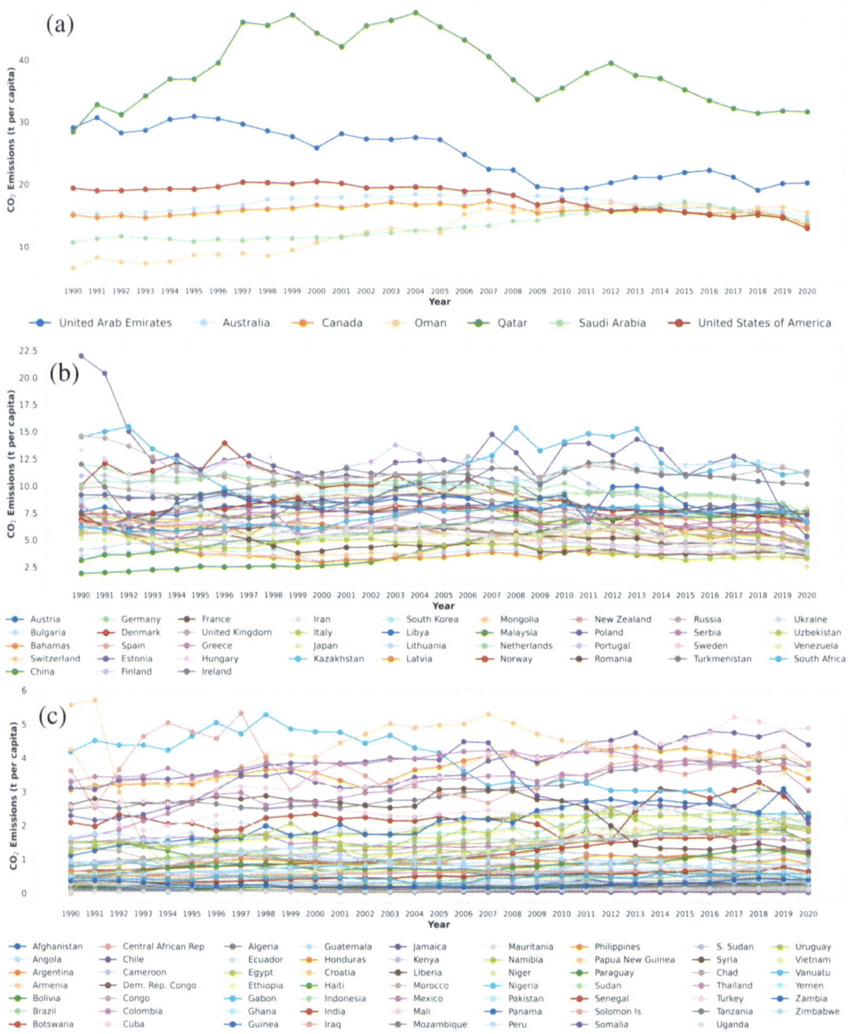

Fig. 5.7 Global CO_2 emissions in mt per capita; **a** High-emissions group, **b** Medium-emissions group, **c** Low-emissions group

source of emissions is the extraction and processing of hydrocarbons, and the energy matrix is dominated by gas-fired electricity production.

With a contribution of 12.82%, the US comes in third place and has reduced its emissions per capita the most (37.92%) of any country in the group. Even so, there are still a lot of emissions from burning fossil fuels for transportation, industry, and electricity.

Australia's 12.06% contribution shows high emissions per person because of its tiny population and energy dependence on gas and coal. Other significant sources are Saudi Arabia (9.52%), Canada (11.29%), and Oman (9.17%), whose emissions are mostly related to the production and processing of hydrocarbons.

39 nations make up the medium-emissions category (Fig. 5.7b), with 1990 emissions ranging from 1.91 mt (China) to 22.01 mt (Estonia). With 4.41% of emissions, Estonia is at the top of this group. Although emissions decreased by 105.06% throughout the research period as Estonia embraced renewable energy sources, its position is explained by its reliance on oil shale, a fuel with a higher CO_2 intensity, for electricity generation.

Kazakhstan comes in second with 4.14%, fueled by industrial processes and energy production from fossil fuels. Russia's energy sector is the main reason it ranks third at 4.01%. Russia cut emissions by 24.90% despite being a large producer of coal, gas, and oil, in part because of economic factors and a move toward the use of natural gas.

Germany (3.45%), Japan (3.22%), Iran (2.10%), and Venezuela (1.81%) are some noteworthy nations. The primary source of emissions from these countries is the burning of fossil fuels for energy and industrial purposes.

The 62 nations that make up the low-emissions group (Fig. 5.7c) have emissions that range from 5.71 mt (Armenia in 1991) to 0.03 mt (Uganda in 1994). With 5.20% of contributions, Croatia is in the lead, followed by Mexico (4.76%), Argentina (4.66%), Gabon (4.75%), and Iraq (4.62%).

Transportation, industry, and power generation based on fossil fuels are the main sources of Mexico's emissions. Mexico's enormous population dilutes its per capita emissions, despite the country's high total emissions. In contrast, Argentina's emissions are linked to the country's use of fossil fuels and agriculture.

Brazil (2.29%) makes a significant contribution through deforestation and changes in land use, whereas India (1.46%) shows the impact of its enormous population, even if per capita emissions have increased by 92.28%.

For a detailed breakdown of the trends, country-specific contributions, and additional analyses, refer to Appendix Sect. 5.6.3.1.

Significant trends are shown in the study of worldwide CO_2 per capita emissions across the three emission categories, which are distinguished in Fig. 5.16. Cluster 0 (high emissions, red), Cluster 2 (mid emissions, blue), and Cluster 1 (low emissions, green) are the three distinct color groups.

The countries in Cluster 0, which makes up 6.73% of all countries, have the greatest CO_2 emissions per person. Emissions were very stable over the course of the 30-year period, with just minor variations. The group went through two separate phases: a continuous drop until 2020, after a rise from 1990 to 2004 that peaked at 153.59 t in 2004 (an annual average of 21.94 mt per person). Emissions peaked in 2020 at 123.28 mt, with an average of 17.61 mt per person per year. Notwithstanding variations, the group produced a total of 4,347.32 mt per capita, or 27.74% of the world's total, representing a minor overall decline of 0.12%. The extraction, processing, and energy production of fossil fuels are the main causes of emissions in these nations. The United States, Saudi Arabia, Canada, Qatar, and Turkmenistan—all significant producers of fossil fuels worldwide—are important contributors.

36.11% of the nations are in Cluster 2, which has a typically falling trend with sporadic stability and minor oscillations. Emissions showed a gradual decrease between 1990 and 2009, then a more noticeable decrease of 22.0% between 2010 and 2020. By 2020, emissions have decreased to their lowest level ever, averaging 6.03 mt per person per year. The group accumulated 8,840.81 mt of CO_2 per person over the research period, accounting for 56.42% of the world total. Even with this decrease, nations like China, Russia, and Kazakhstan continue to be major emitters, mostly as a result of industrial activity and the generation of energy from fossil fuels.

Emissions in Cluster 1, which includes 57.40% of all nations, are on the rise, mostly due to higher CO_2 growth than population growth. Over the course of the 30 years, emissions increased 19.95%, peaking at 1.43 mt per person in 2019. Despite having smaller individual contributions, this group produced 2,479 mt of CO_2 per person or 15.82% of the total worldwide. Countries like Mali and India, whose population increase accelerates the total emissions trajectory, are notable contributors.

The substantial differences between the groups with high, medium, and low CO_2 emissions per capita are highlighted by these findings. The yearly CO_2 production of high-emission nations is 15.56 times that of low-emission nations and 2.74 times that of medium-emission nations.

Between 2019 and 2020, there was a notable decrease in all clusters, mostly as a result of the COVID-19 pandemic, which caused short-term drops in transportation and industrial activities. This emphasizes the necessity of sustainable strategies as opposed to depending on transient, outside disturbances in order to achieve long-term emission reductions.

For additional details on cluster-specific trends, historical breakdowns, and country contributions, refer to Appendix Sect. 5.6.3.2.

In order to visually identify the levels of emissions, nations are mapped into three clusters in Fig. 5.17, which presents the study of worldwide CO_2 emissions per capita: Cluster 0 (High Emissions, red), Cluster 2 (Medium Emissions, blue), and Cluster 1 (Low Emissions, green). A thorough comprehension of regional behaviors with regard to CO_2 emissions is made easier by this depiction.

All three groupings of nations in the Americas are included. The United States ranks third internationally and tops the area in per capita emissions. The United States contributed 3.55% of global emissions throughout the research period with an average of 557.57 mt per person, or 18.58 mt per person year. With 490 mt per capita for the time, or 16.36 mt yearly per person, Canada comes in fifth place globally and contributes 3.13% of global emissions. Both nations are major producers of natural gas and oil worldwide, and their energy and industrial sectors primarily create emissions from the extraction, processing, and burning of fossil fuels. Even if its use has decreased, coal is still a significant source of emissions.

With 160.64 mt per capita (or 5.35 mt yearly), Venezuela is the only Latin American nation in the medium emissions group and ranks 39th overall, accounting for 1.02% of global emissions. Despite its reliance on hydroelectric electricity, thermoelectric facilities that use natural gas and petroleum derivatives are the main source of Venezuela's emissions. Transportation that depends on fossil fuels and the oil and gas sector both make substantial contributions.

The low emissions category includes nations like Argentina, Brazil, and Mexico. Mexico provides 0.75% worldwide and ranks 48th with an annual per capita contribution of 3.39 mt. Argentina comes in at number 50 with 3.83 mt yearly per person and a contribution of 0.73%, while Brazil comes in at number 63 with 1.89 mt per person and a contribution of 0.36%. The energy sector, which is dominated by fossil fuels like petroleum and natural gas, is primarily responsible for Mexico's and Argentina's emissions. On the other hand, land-use changes and deforestation—two major environmental issues—are the main causes of Brazil's emissions.

The majority of African nations are classified as low emissions, meaning they make up a very small portion of world emissions. Reduced dependence on costly transportation networks, low energy consumption per person, and a lack of industrial infrastructure are the causes of this. South Africa and Libya, which are in the category of countries with modest emissions, are notable exceptions. With yearly emissions of 8.6 mt per person, Libya ranks 19th in the world and contributes 1.64% of global per capita emissions. Emissions from extraction, transportation, and refining operations as well as thermal power plants contribute to this major producer of oil and gas. With an annual contribution of 1.40% and emissions of 7.34 mt per person, South Africa comes in at number 26. About 80% of its energy is produced by coal-based electrical generation. Another important factor is the mining sector, namely for coal, gold, and platinum.

All of the nations that make up the European Union are classified as having medium emissions. At an average of 10.18 mt per person per year, Germany ranks 14th in the world in terms of emissions, accounting for 1.94% of total emissions. Despite continuous efforts to decrease its use, lignite coal continues to be a major component of Germany's industrial sector. With a ranking of 36th, France emits 5.74 mt per year (1.1%), followed by Italy in 27th place with 7.09 mt per year (1.35%) and Spain in 30th place with 6.64 mt per year (1.23%). Given its reliance on nuclear power for energy, industry and transportation account for the majority of France's emissions. Despite the fact that both Spain and Italy have heavily included renewable energy into their energy grids, they still rely on natural gas to generate electricity.

All three groups have nations in Asia. Russia is rated 10th (11.82 mt yearly per person) and China 42nd (4.81 mt annually per person) in the group of countries with medium emissions. Both countries top the world in overall emissions, especially China, where coal-fired energy predominates, while having moderate emissions per person. Along with industry and transportation, the natural gas sector is the main source of Russia's emissions.

Countries in the high emissions group include the United Arab Emirates (2nd, 25.75 mt yearly per person) and Qatar (1st internationally, 39.63 mt annually per person), whose emissions are caused by oil processing and natural gas liquefaction in addition to high energy use for transportation and infrastructure. Oman (7th, 12.8 mt yearly per person) and Saudi Arabia (6th, 13.79 mt annually per person) are also significant producers of hydrocarbons. Fourth-ranked Australia emits 17.48

mt per person yearly, primarily from mining, transportation, and coal-based energy generation.

With the largest population in the world, India comes in at number 74 with an annual emissions per capita of 1.17 mt, which indicates a dilution of overall emissions because of its size. However, India's energy sector's use of coal keeps it as the world's fourth-largest emitter overall.

For a detailed breakdown of regional country contributions, ranking positions, and specific trends within each emissions group, refer to Appendix Sect. 5.6.3.3.

5.3.3.2 Global Total CO_2 Emissions

The findings of grouping nations into three categories according to their total CO_2 emissions throughout the 30-year study period (1990–2020) in metric tons (mt) are shown in Fig. 5.8. The research emphasizes the substantial differences between high-, medium-, and low-emission groupings as well as the dominance of a small number of nations in global emissions.

China and the United States make up the high-emission group (Fig. 5.8a), which collectively contributed to the bulk of CO_2 emissions worldwide. China, which makes up 54.10% of this group, had a sharp rise of 169.37% between 1990 and 2020, going from 2,173,364.2 mt to 10,944,686.2 mt. Its significant reliance on coal-fired power production, urbanization, and industrial growth are the main causes of this increase. Notwithstanding ongoing emissions from industry, transportation, and the burning of fossil fuels, the United States, which contributed 45.89%, saw a 9.71% decline during the same time period, indicating a move toward natural gas and renewable energy sources.

The countries of Japan, Germany, Russia, and India are in the medium-emission category (Fig. 5.8b). With a 141.47% rise in emissions, India stands out due to its fast industrialization and reliance on coal for energy. In contrast, economic changes and the switch to cleaner energy sources caused net decreases of 26.79% and 44.13% in Germany and Russia, respectively. Japan showed stability with a slight decline of 5.87%.

There are 102 countries in the low-emission category (Fig. 5.8c), which includes countries in Africa, Asia, Europe, and Latin America. Despite having lower absolute contributions, some of these nations—like Ghana, Mozambique, and Nigeria—showed notable increases in emissions as a result of their fast population expansion and growing energy needs. On the other hand, nations like France and the United Kingdom saw notable decreases, demonstrating the success of environmental regulations and the switch to renewable energy.

For a detailed breakdown of annual trends, country-specific contributions, and variations within the medium- and low-emission groups, refer to Appendix Sect. 5.6.3.4.

The average trajectory trends of worldwide total CO_2 emissions in kilotons (kt) for each group are shown in Fig. 5.18. Color is used to identify the three clusters: low

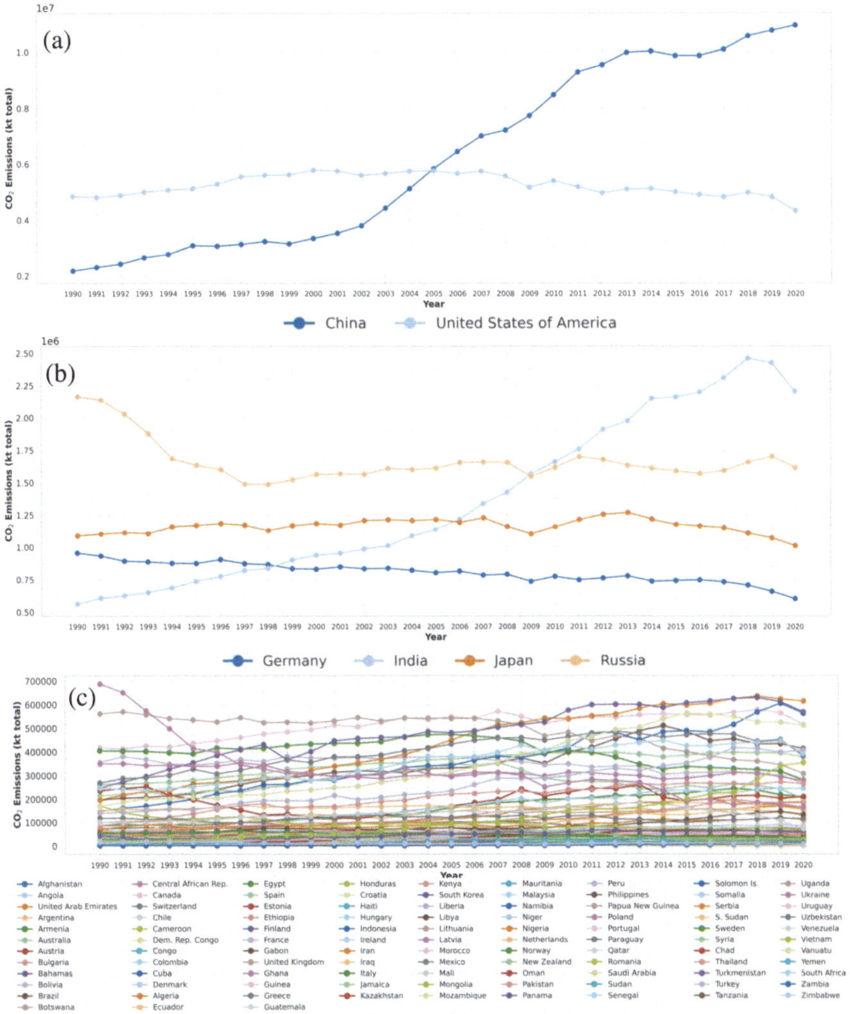

Fig. 5.8 Global total CO_2 emissions in kt; **a** High-emission group, **b** Medium-emission group, **c** Low-emission group

emissions (green), medium emissions (blue), and high emissions (red) are represented by Cluster 1.

China and the United States are the leading contributors to global CO_2 emissions in Cluster 1, which only comprises 1.85% of the nations under study. China's fast industrialization and coal-dependent energy production are driving its exponential expansion, whereas the US is showing a downward trend as a result of efficiency improvements and the use of renewable energy. Even still, their combined CO_2 emissions make up 44.00% of the world's total emissions, with China's ongoing coal dependency countering US decreases.

Emissions in Cluster 0, which includes Germany, Japan, Russia, and India, are growing more slowly but steadily. India propels growth through the use of coal, whereas Germany and Russia show decreases in emissions, especially as a result of energy system changes. 19.21% of worldwide emissions come from this category.

Cluster 2, which includes 94.44% of the countries under investigation, shows a linear rise in emissions during the research period, mostly due to population and economic expansion in emerging nations. Emissions peaked in 2018 with 11,296,852 kt, but the COVID-19 epidemic caused a significant drop in 2020. 36.77% of the world's total CO_2 emissions come from this category.

For detailed trends, specific emission data by year, and country contributions within each cluster, refer to Appendix Sect. 5.6.3.5.

The worldwide division of nations according to their total CO_2 emissions is shown in Fig. 5.19, where they are divided into three color-coded clusters: low emissions (green, Cluster 1), medium emissions (blue, Cluster 0), and high emissions (red, Cluster 2). The global differences in CO_2 emissions are clearly seen by this categorization, which also emphasizes the substantial contributions of a small number of nations to the total.

China and the US, which together produce about half of the world's CO_2 emissions, lead the high-emissions category. With a total of 191,755,756.2 mt over the research period, or 23.81% of the worldwide total, China is the world's greatest emitter. Rapid urbanization and infrastructural growth, as well as the extensive use of coal for industrial processes, especially the manufacturing of steel and cement, and the generation of electricity, are the main causes of its emissions. In second place with 162,638,318.1 mt, or 20.19% of worldwide emissions, is the United States. Despite a 9.71% decrease over the previous 30 years, the United States is still a significant emitter because of its reliance on fossil fuels for industrial production, transportation, and electricity.

Germany, Russia, Japan, and India are among the nations in the medium-emissions category. Their combined contributions result in a slower but consistent rise in emissions, which is impacted by both individual growth and reduction trends. On the other hand, because of growing energy consumption and population expansion, the low-emissions group, which includes most nations, has seen notable relative increases, especially in Africa, Latin America, and portions of Asia.

The cumulative CO_2, CH_4, and N_2O emissions from the 108 nations under analysis equal 805,263,389.3 mt, or 73.44% of the total emissions. Although emissions temporarily decreased globally as a result of the COVID-19 epidemic, the long-term trend shows a continuous growth, especially in poorer nations.

For a comprehensive regional breakdown, including country-specific contributions, temporal trends, and a detailed analysis of emission clusters, refer to Appendix Sect. 5.6.3.6.

5.3.4 Greenhouse Gas Emissions Overview

The increases and differences in emissions of the three main greenhouse gases (CH_4, CO_2, and N_2O) from 1990 to 2020 are assessed in this study, highlighting the combined contribution of these gases to global warming and the pressing need for focused mitigation measures. These gases together make up more than 90% of all greenhouse gas emissions, with CO_2 accounting for 73%, CH_4 for 16%, and N_2O for around 10%. 28 CO_2 emissions, which are mostly caused by burning fossil fuels for energy generation, transportation, and industrial processes, continue to be the principal cause of climate change (Friedlingstein et al., 2022). The report draws attention to the United States and China's disproportionate contributions, which combined make up about 44% of the world's CO_2 emissions. Since 1990, China's emissions have increased by 169% as a result of its reliance on coal, fast industrialization, and urbanization (Zheng et al., 2020). On the other hand, the shift to natural gas and renewable energy sources helped the United States cut emissions by 9.7% during the same time period (Le Quéré et al., 2021). Significant decreases are shown by European nations like Germany, the UK, and France, which are fueled by developments in renewable technology and climate regulations like the Paris Agreement (Climate Change, 2014). The second-largest emissions are CH_4, which makes a substantial contribution because of their potential for short-term warming, which is 28 times greater than that of CO_2 over a 100-year period. The main sources of CH_4 are coal mining, natural gas production, and agriculture (raising rice and cattle) (Saunois et al., 2020). Large-scale agricultural activity in China, Brazil, and India are major emitters. Due in large part to their sale and production of natural gas, nations like Qatar and Turkmenistan have the highest per capita CH_4 emissions (Jackson et al., 2020). Although energy systems have advanced, CH_4 emissions have not decreased as much as CO_2.

Despite having a lower percentage, N_2O is a serious hazard since it has a 298-fold higher potential for global warming than CO_2 (Suddick et al., 2013). Deforestation, soil degradation, and nitrogen-based fertilizers in agriculture are the main causes of N_2O emissions. While developing countries in tropical regions confront growing N_2O levels due to expanding agricultural frontiers and unsustainable land-use practices, Australia, New Zealand, and parts of Africa show considerable emissions per capita (Reay et al., 2012).

Lastly, although though CO_2 accounts for the majority of world emissions, CH_4 and N_2O also require equal attention because of their strong warming effects and rising tendencies in agricultural economies. A multimodal strategy is needed to address these emissions, including decarbonizing electricity systems, lowering reliance on coal and natural gas, and putting sustainable farming practices into place to reduce CH_4 and N_2O emissions worldwide.

5.4 Conclusions

108 nations' GHG emissions (N_2O, CH_4, and CO_2) during a 30-year period (1990–2020) are thoroughly evaluated in this study, which offers important insights into regional and temporal differences in emission patterns.

- The research emphasizes the disproportionate warming potential of CH_4 and N_2O while highlighting the worldwide dominance of CO_2 emissions, which are caused by the burning of fossil fuels and industrial activity. Despite making up a lesser portion, these gases pose serious problems because of their increased radiative forcing and tenacity in agricultural sectors.
- The study highlights geographic disparities, pointing out that while developing regions, especially in Africa and Latin America, exhibit rising trajectories driven by population growth and economic development, a small number of countries (such as China, the United States, and India) dominate global emissions.
- Per capita emissions in high-emission countries are diluted by population size, which obscures their absolute impact, according to the integration of total emissions and per capita emissions study.
- By highlighting similar emission patterns, clustering algorithms make it easier to classify nations into high, medium, and low emission categories. This information may then be used to guide mitigation measures that are particular to a certain region.
- The findings support the necessity of focused strategies to address certain GHGs. While energy efficiency and the switch to renewable energy can reduce CO_2, agricultural innovation, better land management, and a decrease in fugitive emissions from energy systems are needed to solve CH_4 and N_2O.
- Since big polluters have a disproportionate impact on global warming while emerging economies struggle to balance growth and sustainability, the study highlights the significance of international collaboration.
- More accurate policy suggestions might be possible if the analysis's scope was expanded to include sector-specific contributions (such as those from the energy, agricultural, and transportation sectors).
- Future emission inventories may be more accurate if real-time emissions monitoring is incorporated utilizing satellite data and cutting-edge machine learning algorithms.
- To create fair policies that strike a balance between environmental objectives and economic growth, it is crucial to investigate the socioeconomic effects of mitigation techniques, especially in developing nations.
- Future research might also assess how climate policies (such the Paris Agreement and carbon price) affect the paths taken by emissions reduction, providing information on how effective international accords are.
- The results emphasize how urgent it is to accelerate global decarbonization through investments in low-carbon energy systems, sustainable farming practices, and technology innovation.

- Financial and technological assistance is needed to address growing emissions in emerging nations, promoting a fair switch to clean energy while maintaining economic stability.
- This chapter provides a roadmap to identify important goals for lowering GHG emissions and mitigating climate change, which is a crucial starting point for regional and global climate action.

5.5 Appendix

5.5.1 Total and Per Capita N_2O Emissions

5.5.1.1 Global Per Capita N_2O Emissions

The top contributors to the group of nations with the greatest N_2O emissions per capita (Fig. 5.2a) are Australia (13.14%), New Zealand (12.8%), and Mongolia (14.16%). While Australia and New Zealand's emissions are mostly driven by agriculture and livestock, Mongolia's low population and reliance on mining make its impact disproportionate. With the largest rise in emissions (968.73%) throughout the research period, Cameroon accounts for 10.94% of the total. This rise is attributed to indirect emissions from land loss, deforestation, and fire-related activities during the 1993–1994 conflict with Nigeria. Other nations in this group include Namibia (7.86%), Uruguay (7.94%), and Botswana (7.86%), whose small populations increase the influence of their economic activity on emissions per capita. The main cause of Botswana's 303.11% increase in emissions was the country's 2008–2010 diamond mining crisis, which prompted a shift to cattle husbandry.

There are 25 nations in the medium emissions group (Fig. 5.2b), with Lithuania and Paraguay contributing the most. While Paraguay had a large decrease (75.03%) in emissions between 1990 and 1995, which was a result of its political change, Lithuania's emissions grew by just 2.08% over the research period. Despite being some of the biggest N_2O polluters in the world, Canada, Brazil, and the US are also included in this group, demonstrating decreases in per capita emissions as a result of population expansion.

Somalia leads the 73 nations in the low emissions category (Fig. 5.2c) with a contribution of 2.68%. Other nations in this group include Mexico, Chile, Venezuela, and Colombia in Latin America, and Germany, Greece, and Spain in Europe. Despite having high overall emissions, Asian nations like China and India have low emissions per person because of their sizable populations. The agriculture and livestock industries in Somalia account for the majority of its emissions profile, but China and India have seen rises of 38% and 18%, respectively, which are indicative of their expanding economies and populations.

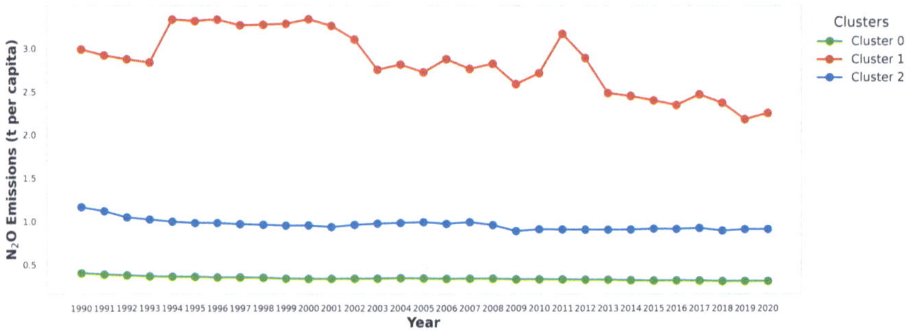

Fig. 5.9 Average trajectory of global N_2O emissions in mt per capita by group

5.5.1.2 Average Trends in Global Per Capita N_2O Emissions

The high-emissions group's erratic behavior is reflected in the Cluster 1 trajectory, which shows peaks and dips strongly correlated with demographic and economic factors. Countries like Australia, Uruguay, and Mongolia saw slower population growth between 1994 and 2000, then faster growth between 2013 and 2020 when overall emissions decreased. The primary economic activities that contributed to this cluster's emissions profile were mining, livestock, and agriculture (Fig. 5.9).

With distinct intervals of decrease and stabilization, the Cluster 2 trajectory shows almost linear emissions trends. Emissions decreased between 1990 and 1995 as a result of stabilization measures, and from 1995 to 2008, there were steady trends at about 0.98 mt per person year. Changes in agricultural techniques and population expansion were blamed for the reduction between 2009 and 2020. This group's nations mostly depend on agriculture and cattle, both of which continue to be major contributors to N_2O emissions.

The low-emissions group's decrease is shown to be linear and steady by the Cluster 0 trajectory. This group comprises nations that require extensive agricultural and agri-industrial management due to their huge populations and high food demand. The group's overall emissions were impacted by the cumulative impacts of several nations' modest contributions, and the stability of emissions after 2018 is consistent with patterns in population growth.

5.5.1.3 Global Per Capita N_2O Emissions Segmentation

With a comparable percentage in each category, Europe has a balanced distribution between medium and low emissions groups. Due to its small population (5.2 million) and the importance of its agricultural industry, especially the production of meat and dairy products, only Ireland is included in the high-emissions category.

With just two nations in the medium-emissions group (Kazakhstan and Turkmenistan) and two in the high-emissions group (Australia and Mongolia), the low-emissions group predominates throughout Asia. Population structure has a big influence on this area. Although nations like China, India, and Russia have low emissions, they nonetheless contribute considerably to global emissions because of their high food consumption as well as their transportation and industrial processes.

Of the 21 nations in Africa, the majority fall into the low-emissions category. Both the high emissions group (five nations) and the medium emissions group (six countries) have considerable presence, though. During the research period, countries in the high-emissions category had large increases in emissions, mostly as a result of economic crises, wars, and major industries like mining and agriculture.

The 108 nations in the three categories (high, medium, and low emissions) had combined emissions of 2,395.18 mt of N_2O per person. Nitrous oxide is the one with the lowest emissions per person among the three primary greenhouse gases (N_2O, CH_4, and CO_2). Just 9.55% of the total emissions of CO_2, CH_4, and N_2O are caused by this gas.

The 108 nations' per capita emissions decreased by a noteworthy 25.28% throughout the research period, which is consistent with a global trend toward lower emissions brought on by population growth. During this time, the average annual worldwide N_2O per person was 0.73 mt. Per capita N_2O emissions are declining, although this is not at the level of overall emissions because population increase is a significant factor. A perceived per capita drop is calculated by dividing total emissions by the population of each nation or area. This suggests that the apparent decline in per capita terms can be explained by population growth exceeding N_2O emissions.

5.5.1.4 Global Total N_2O Emissions

High-Emissions Group

China's emissions rose 61.86% between 1990 and 2020, which was over three times as much as the country's population growth (21.82%) during that time. Excessive fertilizer use, a major contributor to N_2O emissions, is connected to this rise. Furthermore, China's industrial emissions worsen its effects because it is a major manufacturer of chemicals and fertilizers worldwide. Its emissions profile is also influenced by burning biomass and urbanization.

Medium-Emissions Group

- Because of effective agricultural technology and regulations, the United States displayed a more consistent pattern, increasing by just 8.47% over a 30-year period.
- The demand for agricultural products to feed a growing population (47.65% growth) was the primary driver of India's notable increase (66.9%). Although the manufacture of fertilizer adds to emissions, agriculture has a greater impact.

- Brazil's emissions, which have increased 61.34% over the past 30 years due to deforestation and the production of biofuel, are out of proportion to the country's population growth (34.90%). The use of fertilizer in the manufacturing of ethanol from sugarcane is a major source of emissions.

Low-Emissions Group

104 nations with emissions ranging from 17.58 kt to 108,556.14 kt are included in this category. Contributions differ:

- The top 14 contributors, which include Australia (5.43%), Russia (4.63%), and Cameroon (3.90%), are responsible for 46.24% of the group's emissions, which are mostly caused by mining, deforestation, and agriculture.
- 25.13% come from mid-level donors (positions 15–31), including South Africa, Japan, and Turkey.
- Contributors in lower ranks (32–91), such as South Korea, Kenya, and Venezuela, add 28.03%.
- Countries like Haiti, Qatar, and the United Arab Emirates are among the minimal donors (positions 92–108), which provide only 0.58%.

Even though these nations emit very little, there are notable differences among them due to differences in economic activity, population size, and technological sophistication.

5.5.1.5 Average Trajectory of Global Total N_2O Emissions

China's trajectory in Cluster 2, N_2O total, reflects its position as the world's greatest emitter of N_2O. Emissions varied between 1990 and 2003 as a result of increasing agricultural efficiency and economic changes. In 1990, the lowest emissions (295,990.80 kt) were noted. Rapid economic growth, industrialization, and agricultural intensification—particularly the widespread use of nitrogen fertilizers—were the main drivers of the peak emissions, which reached 555,682.72 kt between 2004 and 2016. By 2020, emissions had risen 61.86% but per capita emissions had decreased 21.74%.

N_2O total in Cluster 1: This group, which includes the US, Brazil, and India, exhibits a consistent rise in emissions. Brazil and India are constantly growing as a result of population increase and agricultural expansion. Despite having the highest emissions in the group, the United States' total growth rate was lower (8.47%) as a result of technological developments and environmental laws. 2020 saw the greatest emissions (728,109.39 kt), with an average of 242,703.13 kt per nation.

96.29% of the nations under analysis are included in Cluster 0, N_2O total, which exhibits a virtually linear trend with just slight variations. 1993 had the lowest emissions (1,273,747.24 kt), which increased steadily until 2020 (1,548,654.47 kt). Over the course of 30 years, total emissions increased by 15.10%, accounting for 57% of

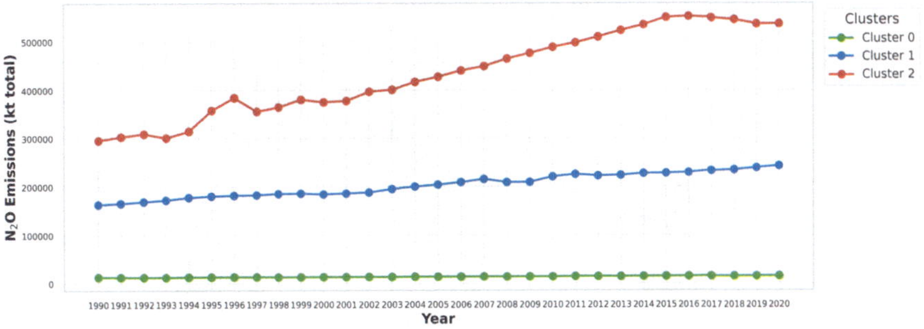

Fig. 5.10 Average trajectory of global total N_2O emissions in kilotons by group

worldwide N_2O emissions. Nevertheless, compared to the other clusters, the average yearly emissions per nation (13,901.17 kt) were much lower.

The clusters' research shows unique dynamics influenced by productive and economic forces. With 17.73% of global emissions, China is the only country in the high-emissions cluster. Due to the growth of industry and agriculture, the medium-emissions cluster—which includes the US, Brazil, and India—contributes 24.92%. Although the great majority of nations are part of the low-emissions cluster, which contributes 57% of global emissions, its contributions per nation are significantly smaller, suggesting that emissions per nation are growing very slowly (Fig. 5.10).

5.5.1.6 Global Total N_2O Emissions Segmentation

Canada, Argentina, and Mexico are significant contributors to the low-emissions group in the Americas, whereas the United States and Brazil lead the high-emissions cluster. Their emissions are largely caused by Mexico's burning of agricultural wastes and Canada's rich organic soils (Fig. 5.11).

Although all of Africa's nations are classified as having low emissions in absolute terms, the distribution of emissions per capita shows more variation because many of them have small populations. South Africa (30th), Zambia (39th), and Botswana (80th) are important contributors.

All of Europe's nations are categorized as low-emissions, with mid-range ranks worldwide. Germany (9th, 1.97%), France (10th, 1.95%), Spain (20th, 1.53%), and Italy (24th, 0.89%) are notable nations. The area gains from controlled environmental regulations and cutting-edge technologies.

With presence in all three groups, Asia exhibits the most varied patterns. With 17.73% of global emissions, China leads the high-emissions group, while Russia (2.66%) and India (8.51%) top the medium-emissions group. These three nations collectively are responsible for 28.9% of the world's N_2O emissions.

75,644,985.09 kt is the total quantity of N_2O emissions from the 108 nations in all clusters. Nitrous oxide contributes just 6.89% of total emissions, making it the least emittable of the three major greenhouse gases (N_2O, CH_4, and CO_2). Due to variables

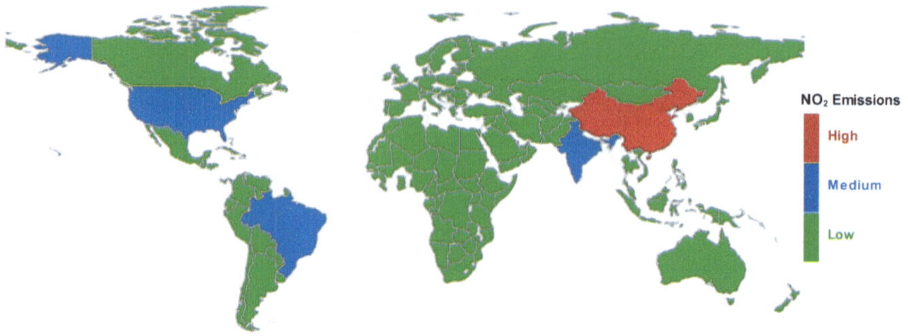

Fig. 5.11 Map of global total N_2O emissions in kt by country

including extensive fertilizer usage, livestock expansion, deforestation, burning of biomass, and shifting climatic patterns, worldwide total emissions rose by 28.53% throughout the research period, but per capita emissions decreased by 25.28%. As a result of these modifications, there is a feedback loop in which N_2O increases global warming, which speeds up agricultural and natural soil processes and increases N_2O production.

5.5.2 *Total and Per Capita CH_4 Emissions*

5.5.2.1 Global Per Capita CH_4 Emissions

Figure 5.5a shows the economy of Turkmenistan, which is in the high-emissions category, is largely dependent on the extraction of natural gas, which makes up over 80% of its exports. The decrease in 2009 came after a pipeline explosion that momentarily stopped output, whereas the decrease in 1997–1998 was linked to economic disagreements with Russia over gas pipelines. Despite a 192.39% rise in population throughout the research period, Qatar's per capita CH_4 emissions increased by 44.34%, mostly as a result of its tiny population and heavy reliance on LNG production.

Figure 5.5b shows the major contributors to the medium-emissions category are Australia, the United Arab Emirates, New Zealand, Venezuela, and Libya.

- Libya is the country with the highest emissions (10.60%), mostly from the exploitation of oil and gas and inadequate infrastructure brought on by protracted political and economic unrest.
- Due to its economic crisis and decreased oil output, Venezuela has seen a notable 83.71% drop in emissions per capita.
- New Zealand makes up 9.13% of this category, where enteric fermentation from animal production produces significant CH_4 emissions. Its agriculture industry

continues to be a significant source of emissions even after technical advancements and regulatory measures reduced its per capita emissions by 43.97%.
- Despite 90% of its power coming from gas, the UAE cut its per capita emissions by 56.49% as a result of its fast population increase.
- Australia makes up 8.11%, mostly from livestock-related operations.

Argentina (3.56%), Russia (4.25%), and Paraguay (4.81%) are among the nations that have made lower contributions. Russia's energy industry, especially the production and transportation of natural gas, is responsible for a substantial portion of its emissions, whereas Argentina and Paraguay mostly rely on the agricultural and animal industries.

Figure 5.5c shows the 90 nations in the low-emissions category show only modest individual contributions.

- Liberia contributes only 0.07% to the group and has the lowest per capita emissions (0.06 mt in 2005).
- Despite having low per capita emissions, China (ranked 68th), the US (7th), and India (79th) stand out for their high overall emissions. The United States and India cut their per capita emissions by 16.89% and 19.20%, respectively, while China raised them by 23%. India's decrease is consistent with its population increase of 47.65%, which reduced its emissions per person.

Germany (ranked 57th, 0.80%) and Japan (ranked 107th, 0.24% of the group's emissions) are two other major producers. The fact that both nations rank in the middle of the world in terms of overall emissions (Japan 42nd, Germany 21st) shows how population size affects emissions per person.

5.5.2.2 Average Trajectory of Global Per Capita CH_4 Emissions

Less than 2% of the nations in this analysis are in Cluster 1, which indicates the average trajectory of the group with the greatest emissions (CH_4 per capita). This group exhibits a pattern with notable peaks and dips throughout the course of the analysis period. The 30 years under study had the lowest emissions between 1994 and 1998. The group produced 101.22 mt throughout this period, with the lowest amount being in 1994 when they produced 24.34 mt collectively and an average of 12.24 mt per individual. Prior to this, emissions fell 32.49% between 1990 and 1994. On the other hand, two time periods were found to have the greatest emissions: the first was from 2000 to 2008, when the group created 323.77 mt, followed by a 32.53% increase from 1998 to 2000; the second was from 2011 to 2019, when 318.18 mt were generated. Emissions fell by 22.24% in 2009 compared to 2008, which came before this most recent era. Emissions did, however, rapidly rebound between 2010 and 2011, increasing by 28.40%, before the second phase of high emissions began. In 2004, group emissions hit their highest peak for the 30-year research period, totaling 39.18 mt per capita, with an average of 19.59 mt per person. When these variations were combined, the group's emissions decreased by 14.80% overall over the study period and by 14.80% in 2020 when compared to 1990.

Over the course of the 30 years under study, this group produced 1,017.04 mt of CH_4 per person, or 14.49% of all CH_4 emissions worldwide. Over the course of these 30 years, the average emissions per person were 508.52 mt, and the average emissions per person per year were 16.95 mt. This group is made up only of Turkmenistan and Qatar, two nations that are major producers and exporters of petroleum and liquefied natural gas worldwide. Their main economic activity are these two commodities. Turkmenistan is the country that has the most impact on the group's emissions swings. This is because of economic crises and gas pipeline accidents that disrupted gas shipments, temporarily lowering emissions. The trajectory of the group clearly reflects these decreases. In contrast, Qatar also had swings, but of a lesser size, mostly associated with changes in population growth.

Cluster 2, which includes 14.81% of the nations included in this study, shows the average emissions behavior for the group of countries with medium levels of CH_4 per capita. This group has a straight declining trend with minor peaks throughout the 30 years under analysis. The largest emissions were recorded between 1990 and 2003, when 1,370.70 mt of emissions were produced overall, with an average of 85.68 mt per person. Emissions decreased by 7.06% throughout this period, peaking in 1991 when the group recorded 101.15 mt of total emissions and 6.32 mt of emissions per person on average. An intermediate level of emissions was noted between 2003 and 2011; 702.51 mt of total emissions were reported during this period, with an average of 43.90 mt per person, representing a 15.97% decrease. Then, between 2011 and 2012, there was a 6.25% rise, and from 2013 to 2020, emissions were at their lowest. The group reported 600 total mt and an average of 37.55 mt per participant during this last phase, resulting in a 19.68% decrease. With total emissions of 68.07 mt and an average of 4.25 mt per person, 2020 was the lowest year.

Overall, the cluster's actions during the course of the research led to 2,757.85 mt of CH_4 per person, or 39.31% of all CH_4 emissions worldwide per person. For nations in this category, the yearly average emissions per person were 5.74 mt, whereas the average emissions per person were 172.36 mt. In comparison to 1990, the overall decrease in emissions over the study period and in 2020 was 31.39%. Significant individual declines in important group contributors like Venezuela, New Zealand, and the United Arab Emirates as a result of economic, regulatory, or population growth issues account for this decline.

Cluster 0, CH_4 per capita, which includes 83.33% of the nations examined in this study, represents the average trend of the group of low-emissions countries. With no notable peaks or valleys, this group's emissions exhibit a consistent, linear decline. There was a consistent decline in CH_4 per capita emissions from the start of the era to its end. The group's yearly average emissions were 1.32 mt per person, reaching a peak of 118.15 mt in 1990. Emissions decreased steadily between this year and 2003, during which time they decreased by 11.65%. With an average of 1.16 mt yearly per individual, emissions for the group stabilized around 104 mt annually after 2004, and they remained essentially constant until 2006.

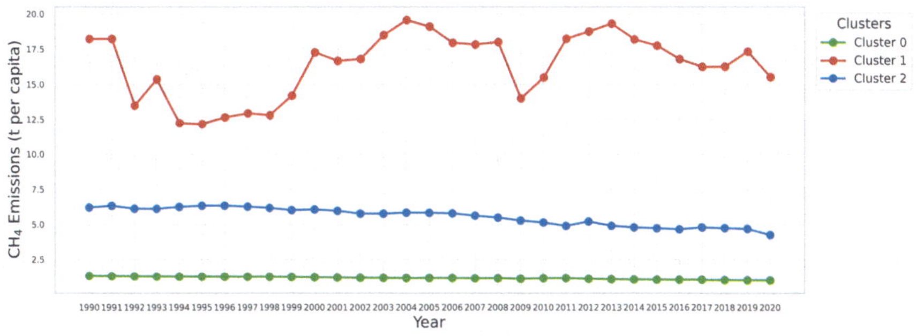

Fig. 5.12 Average trajectory of global CH_4 emissions in mt per capita by group

Beginning in 2007, there was a further little drop in emissions of 1.43% from 2006 to 2007. There were minor oscillations with minor peaks and valleys from 2007 to 2011. With average yearly emissions of 101.54 mt for the group and 1.12 mt per person, the balance indicates a 5.18% decrease throughout this time. Emissions started to decline steadily again in 2012 and continued to do so until 2020. The group's average yearly emissions during this time were 94.23 mt, with an average of 1.04 mt per person and an overall decrease of 8.19%.

Individual contributions are minimal because this group includes 90 of the 108 nations examined, which is the biggest number of countries examined (Fig. 5.12). Furthermore, when measured on a per capita basis, the involvement of densely populated nations like China and India as well as world population increase dilute overall CH_4 emissions. This dynamic explains the general downward trend in this group's per capita methane emissions, which showed a decline of 23.10% in 2020 compared to 1990 and a total decrease of 46.19% between 1990 and 2020. This group's total emissions came to 3,240.57 mt, accounting for 46.19% of all CH_4 emissions per person. Over the course of the time, the average emissions per person were 35.94 mt, with an average of 1.19 mt per person each year.

This indicates that this group produces the least amount of CH_4 emissions per person and shows the biggest decrease in emissions, which is mostly due to the world's population growing faster than expected.

5.5.2.3 Global Segmentation of Per Capita CH_4 Emissions

A thorough grasp of the global distribution of this greenhouse gas (GHG) is made possible by the division of nations by CH_4 per capita emissions, as shown in Fig. 5.13.

Four nations are part of the medium emissions category (Cluster 2) in America. Venezuela contributes 3.70% of world emissions, placing it in fourth position. Its economy is heavily reliant on hydrocarbons. Livestock production is the main source of emissions in Uruguay, which comes in tenth place with a 2.80% contribution. Livestock husbandry also drives Argentina, which is in 18th rank with 1.39%, and Paraguay, which is in 11th place with 1.89%. The other Latin American nations, which are in the middle of the world rankings, are in the low emissions category (Cluster 0). Mexico comes in at number 50 with 1.14%, mostly from the production of cattle and oil and natural gas. Despite being the fifth-largest worldwide emitter of CH_4, Brazil ranks 30th (0.89%) because of its substantial output of bovine cattle. Colombia, which ranks 39th, makes up 0.70% of the total, mostly from rice and cattle husbandry.

Liberia, which only contributes 0.03% of global emissions, is ranked 107th overall and 108th per capita in Africa. Oil and natural gas production are the main drivers of Libya's third-place worldwide contribution of 4.16% and Algeria's eighth-place contribution of 2.85% in the medium emissions group.

Ireland, which ranks 14th globally with a contribution of 1.70%, is the only European nation in the medium emissions group. However, Ireland is ranked 61st in the world for total emissions. One of the main sources of CH_4 emissions in Ireland is the country's agricultural industry, including dairy and cattle production.

All three emission categories include nations in Asia. With contributions of 8.39% and 6.10%, respectively, Turkmenistan and Qatar, both in the high emissions category (Cluster 1), are the top two countries in the world in terms of per capita CH_4 emissions. The large emissions per capita in both nations can be explained by their substantial natural gas production and exports.

Australia is in seventh place with 3.18%, Saudi Arabia in sixteenth place with 1.55%, and Russia in fifteenth place with 1.67% in the medium emissions category. Whereas Russia and Saudi Arabia's emissions come mostly from their energy sectors, specifically the production and processing of oil and gas, Australia's emissions are mostly associated with animal husbandry.

The 108 nations in the three categories (high, medium, and low emissions) had combined emissions of 7,015.47 mt of CH_4 per person. Methane contributes 27.97% of the total emissions, placing it in the middle of the three primary GHGs (N_2O, CH_4, and CO_2).

The 108 nations' per capita CH_4 emissions decreased by a considerable 25.15% throughout the research period, which was consistent with a general downward trend brought on by population growth worldwide. During this time, the average annual worldwide CH_4 emissions per person were 2.16 mt.

One of the main factors influencing per capita CH_4 emissions is the world's population growth. According to this dynamic, nations with large global emissions overall, like China and India, may rank lower in terms of emissions per capita, whereas those with lower or intermediate global emissions may rank higher. This finding emphasizes how crucial it is to take into account each person's contribution to emissions, irrespective of a nation or region's population size, main economic activity, or level of ecological and technical development.

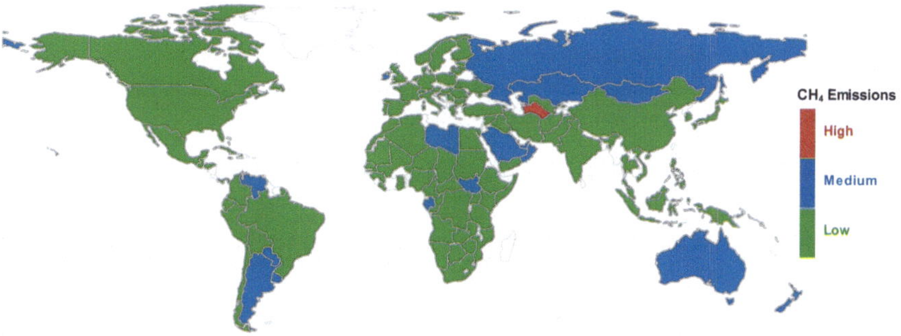

Fig. 5.13 Map of CH_4 emissions in mt per capita by country

5.5.2.4 Detailed Analysis of Global Total Emissions of CH_4

High-Emissions Group (Fig. 5.6a):

- China: In 2020, emissions reached 1,186,285.18 kt, a 45.20% increase. Rice agriculture, cattle husbandry, and coal mining are examples of primary sources.
- The United States: Made a 24.02% contribution, while emissions rose by 11.36%. Livestock farming and hydraulic fracturing for natural gas are the main sources.
- India: Agriculture (livestock and rice) was the main driver of the 28% increase in emissions. Contribution overall: 22.21%.
- Russia: Emissions dropped to 472,999.32 kt, a 3.17% decrease. Decreased natural gas shipments to Europe are the cause of the drop.

Medium-emissions group (Fig. 5.6b):

- Brazil: Made a 21.96% contribution, with emissions rising 46.56% as a result of cattle rearing.
- Indonesia: The production of rice was the main cause of the 18.06% rise in emissions.
- Venezuela: Due to a downturn in the economy and a decrease in energy output, emissions decreased 45.43% even though the country contributed 12.53%.
- Other sources of emissions include the energy industry, cattle husbandry, and rice agriculture in Iran (8.89%), Australia (8.59%), Nigeria (8.36%), Mexico (7.22%), and Pakistan (6.64%).

Low-emissions group (Fig. 5.6c):

- Turkey: Mainly from the production of natural gas, it contributed 3.88%.
- Canada: Energy-related activity accounted for 3.56% of the total.
- Industrial operations and coal extraction accounted for the majority of Germany's and France's contributions, which were 2.79% and 2.76%, respectively.
- Bahamas: Emissions were the lowest, at 144.81 kt (0.007%).

- With the exception of Nigeria, Africa continues to make up the majority of this group. Despite their industrialized economies, European nations enjoy the benefits of stringent environmental regulations that maintain comparatively low emissions.

5.5.2.5 Global Trends of Total CH_4 Emissions by Cluster

Significant differences in contributions to global CH_4 emissions are shown by the study of yearly emissions by country (Fig. 5.14). These differences are impacted by the number of nations in each category and the leading economic activity. Despite only consisting of four nations, the high-emission group accounts for 40.11% of worldwide CH_4 emissions, with an average of 720,352.48 kt per year. Agriculture and energy-intensive economies are the main drivers of this category.

Averaging 197,222.19 kt per year, the medium-emission category accounts for 24.71% of worldwide CH_4 emissions. According to this group, the primary sources of emissions are the energy and agriculture industries. Iran, Brazil, and Pakistan are just a few of the nations that have a big impact on general growth patterns.

With a far lower yearly average of 26,598.96 kt per country, the low-emission group, which comprises 87.96% of the nations, produces 35.17% of worldwide CH_4 emissions. Despite slight variations over the past 30 years, the group's size and very small individual contributions serve to maintain its trend. This group of nations typically has lower levels of industrialization and less reliance on activities that produce a lot of emissions.

The average yearly emissions of a nation in the high-emission group are 27 times higher than those of a country in the low-emission group (720,352.48 against 26,598.96 kt), according to a comparison analysis. Comparably, a medium-emission nation emits 7.41 times more CH_4 than a low-emission group country (197,222.19 kt compared 26,598.96 kt), while a high-emission group country emits 3.65 times more than a medium-emission country (720,352.48 kt versus 197,222.19 kt). Due to lower levels of industrialization, the low-emission group's contributions are comparatively small, whereas Cluster 1's dominance highlights the substantial influence of energy-intensive sectors in China, the US, and Russia.

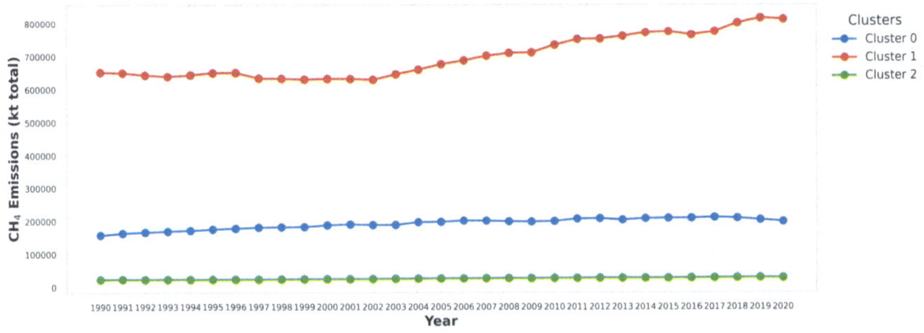

Fig. 5.14 Average trajectory of total global CH_4 emissions in kt by group

5.5.2.6 Total Global CH_4 Emissions Distribution

215,499,339.00 kt of methane were released globally from the 108 nations that were examined. Emissions of CH_4 make up 19.65% of total emissions of CO_2, N_2O, and CH_4. While per capita emissions decreased by 25.15% throughout the research period, overall emissions rose by 22.43%. The energy, agricultural, and animal industries are the main drivers of this overall increase in CH_4 emissions (Fig. 5.15).

The United States continues to be the largest emitter in the Americas, with Brazil, Mexico, Argentina, and Venezuela following in the medium-emission category. Canada, Colombia, and Chile are examples of low-emission nations, which represent a regional combination of agricultural and industrial activity.

Nigeria's medium-level emissions demonstrate its status as a significant producer of hydrocarbons, whereas the majority of African nations have low emissions because of their low levels of industrial activity.

Strict environmental regulations have made Europe a region with low emissions, yet industrial activity in Germany, France, Italy, and Spain continues to score highly.

The energy, agricultural, and livestock industries in Asia account for over one-third of worldwide CH_4 emissions, with China, India, and Russia sharing the lead. Due to its natural gas production, Turkmenistan leads among low-emission nations, whereas Iran, Pakistan, and Australia stand out as major medium-level contributors.

The differences within clusters show how agricultural, livestock, and energy production account for the majority of CH_4 emissions worldwide, with high-emission countries making disproportionate contributions despite only making up a small percentage of the total number of countries.

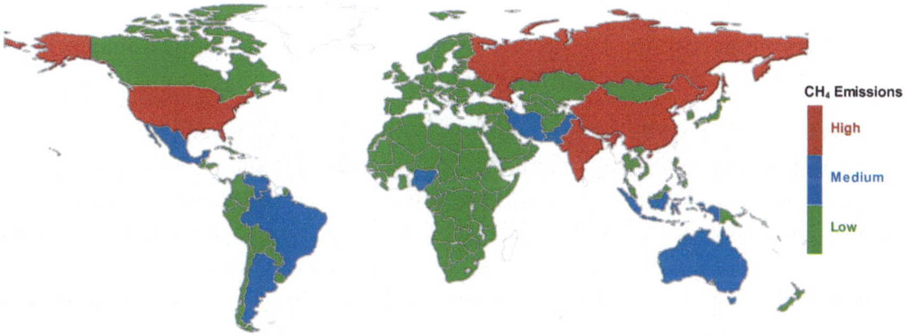

Fig. 5.15 Map of total global CH_4 emissions in kt by country

5.5.3 *Total and Per Capita CO_2 Emissions*

5.5.3.1 Global Per Capita CO_2 Emissions

Within each group, the clustering analysis reveals unique behaviors:

High-Emissions Group: This group is dominated by the United States, Australia, Canada, Saudi Arabia, Oman, Qatar, and the United Arab Emirates.

- Qatar: The country's overall emissions vary greatly, with infrastructural growth and gas liquefaction operations driving increases. 2001 and 2009 declines are consistent with population growth exceeding emissions increase.
- UAE: Efforts to diversify energy sources and embrace renewable technology resulted in a decrease in emissions per capita despite notable population increase.
- United States: Although industry and transportation continue to be significant sources, the country's steady switch from coal to natural gas and renewable energy helped to reduce emissions.
- Australia: Mining operations, agriculture, and reliance on coal all contribute to high emissions.

Countries in the medium-emissions group include Estonia, Kazakhstan, Russia, Germany, Japan, Iran, and Venezuela.

- Estonia: Adoption of renewable energy significantly reduced early high emissions caused by reliance on oil shale.
- Kazakhstan: Because of the production of fossil fuels and coal-based energy, emissions are still high.
- Russia: Emissions were reduced as a result of economic slowdowns and a move toward natural gas consumption.

Low-Emissions Group: This group, which consists of 62 nations, represents a variety of population trends and emission sources.

- Mexico: Because of its large population, its per capita CO_2 emissions are lower even though it ranks eleventh in terms of overall emissions.
- Brazil: Deforestation and changes in land use are the main causes of emissions.
- India: Despite being the world's largest emitter, the country has one of the lowest emissions per capita due to its rapid population growth, which balances rises in overall emissions.

5.5.3.2 Cluster Trends in CO_2 Per Capita Emissions

The trend of CO_2 per capita emissions for each cluster is shown in detail in Fig. 5.16.

Cluster 0 (High Emissions): There were two significant behavioral episodes in this group. Emissions rose by 8.16% between 1990 and 2004, reaching a peak of 153.59 mt in that year. After that, emissions steadily decreased, reaching 123.28 mt in 2020. Throughout the whole time, the average yearly emissions per person were 20.70

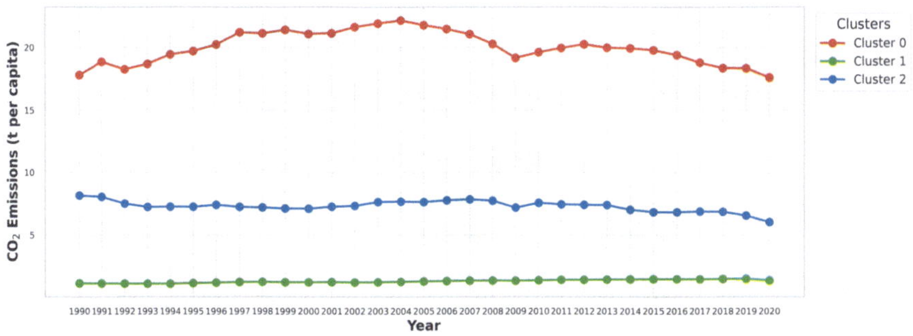

Fig. 5.16 Average trajectory of global CO_2 emissions in mt per capita by group

mt, indicating consistency in spite of variations. Emissions from energy production, the exploitation of fossil fuels, and industrial demand are the main causes in Saudi Arabia, the United States, and Qatar.

Cluster 2 (Medium Emissions): Emissions decreased gradually, especially between 2010 and 2020, when they decreased by 22.0%. The group's yearly average emissions were 7.55 mt per person, for a total of 8,840.81 mt. China, Russia, and Kazakhstan are the biggest producers, with emissions mostly coming from the burning of fossil fuels for industrial and energy production.

Cluster 1 (Low Emissions): This group showed an overall rise in emissions of 19.95%, and it had the biggest number of nations (57.40%). In nations like India and Mali, whose population rates increased by 141.47% and 238%, respectively, emissions increased more quickly. With an average yearly contribution of 1.33 mt per person, the group's total emissions came to 2,479 mt per capita.

The overall research shows how population increase, energy reliance, and economic structure affect CO_2 per capita emissions in various clusters. Although they are underrepresented, high-emission nations account for the majority of global contributions, while low-emission nations highlight the difficulties in striking a balance between sustainability and development.

5.5.3.3 Global Segmentation of CO_2 Emissions Per Capita

The global breakdown of nations by CO_2 emissions per person is shown in Fig. 5.17. As mentioned in the main text, the clustering reveals notable regional differences. Medium and low-emission nations point to variables like population growth, the integration of renewable energy, and decreased industrial activity as predictors of their emissions trajectories, whereas high-emission countries like the U.S., Australia, and Qatar dominate per capita outputs.

Notable contributors in the Americas include Mexico, Brazil, and Argentina in the low emissions category and Venezuela in the medium emissions group; Brazil's land-use changes and deforestation are particularly noteworthy. South Africa's and Libya's contributions from the coal and oil industries contrast with Africa's generally low

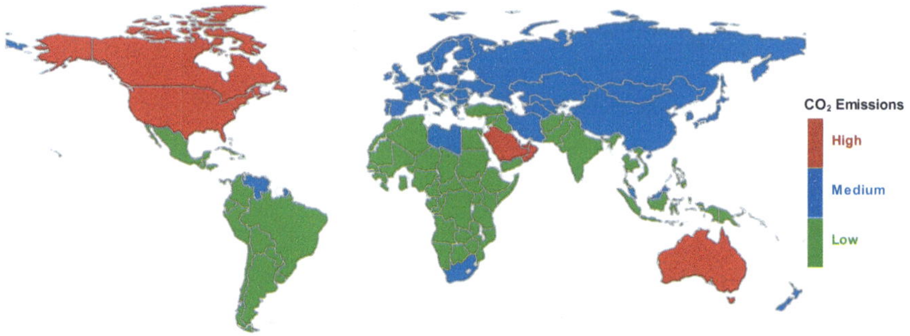

Fig. 5.17 Map of CO_2 emissions per capita in mt by country

emission status. Europe's medium emissions status is a reflection of its continued shift to renewable energy sources despite its continued reliance on fossil fuels. Asia has a wider range, from high-emission countries like the United Arab Emirates and Qatar to India's low emissions per person, despite the region's high overall contributions from its coal-dependent energy industry.

Over the research period, the total CO_2 emissions for all 108 nations under analysis amounted to 15,667.79 mt per capita, with CO_2 accounting for 62.47% of the total CO_2, CH_4, and N_2O emissions. This demonstrates how CO_2 accounts for the majority of greenhouse gas emissions worldwide.

5.5.3.4 Detailed Trends of Global Total CO_2 Emissions

The United States and China are in the high-emission category (Fig. 5.8a). The biggest contributor, China, saw isolated drops of 2.55% in 1999 and 1.61% in 2015, but overall it continued on a steady increasing track. The main causes of its emissions are urbanization, fossil fuel-powered transportation, and businesses that rely on coal, especially the manufacturing of steel and cement. Emissions rose 1.68% in 2020 compared to 2019. With a peak of 5,753,493.2 mt in 2005, the United States started a gradual drop that culminated in 4,320,532 mt in 2020, mostly as a result of less coal being used in favor of natural gas and renewable energy sources. The decline from 2019 to 2020 was 10.31%.

Germany, Japan, Russia, and India are all part of the medium-emission category (Fig. 5.8b).

- Russia: Due to economic difficulties, emissions fell 36.00% between 1990 and 1997 before stabilizing with very slight variations. Emissions reached 1,618,271 mt by 2020, a net decrease of 26.79%.
- India: Since 1990, emissions have increased by 152.07%, reaching a peak of 2,458,175.9 mt in 2018. Even with a brief decline of 10.59% in 2020, India's reliance on coal for electricity is still a major factor.

- Japan: Because of its reliance on coal, oil, and natural gas to generate energy, the country has recorded consistent emissions with a 5.87% decrease over the past 30 years.
- Germany: Although lignite coal is still a source of emissions, the country's switch to renewable energy has resulted in the biggest reduction in this group (44.13%).

There are 102 nations in the low-emission category (Fig. 5.8c), with Canada, the UK, and South Korea making major contributions.

- Canada: Despite having the highest emissions in this category (5.38%), they managed to reduce them by 22.37% throughout that time.
- United Kingdom: Contributed 5.10%, demonstrating the effectiveness of energy transitions and climate policy with a noteworthy 57.09% decrease.
- African countries: Due to population growth and energy needs, countries like Ghana (230.09%), Mozambique (207.10%), and Nigeria (152.47%) had the biggest rises.
- Latin America: Paraguay and Guatemala had even greater relative gains, while Mexico (38.82% rise), Brazil (78.32% increase), and Argentina (46.81% increase) showed noteworthy growth.
- Europe: Despite exhibiting the smallest growth, industrial activity allowed nations like Spain (0.10% rise) and Italy (33.96% increase) to maintain their leading positions.

All 108 of the nations that were examined had CO_2 emissions totaling 215,499,339 mt. Significant increases in emissions in Asia and Africa are contrasted with decreases in Europe and stable patterns in the Americas, highlighting regional differences.

5.5.3.5 Trends and Data of Global Total CO_2 Emissions by Clusters

The detailed average trajectory trends of each cluster's total worldwide CO_2 emissions in kilotons (kt) are shown in Fig. 5.18.

Emissions in Cluster 1 shows three different time periods:

- Emissions rose at a moderate pace of 1.28% per year between 1990 and 2002, with an average of 2,337,085 kt of emissions per nation.
- 2. The group grew at the fastest rate between 2003 and 2007, averaging 6.28% per year, reaching 57,333,078.9 kt in total emissions.
- 3. Emissions decreased by 1.83% in 2015 and 2.02% in 2020, but they grew more erratic between 2010 and 2020.

The United States' emissions peaked at 5,753,493.2 kt in 2005 and subsequently fell to 4,320,532 kt by 2020, representing a 9.71% overall decline, whereas China's emissions rose 169.37% over the same time, reaching 10,944,686.2 kt in 2020. Together, they released 354,394,074.3 kt of CO_2, with each nation emitting 5,906,567.9 kt on average each year.

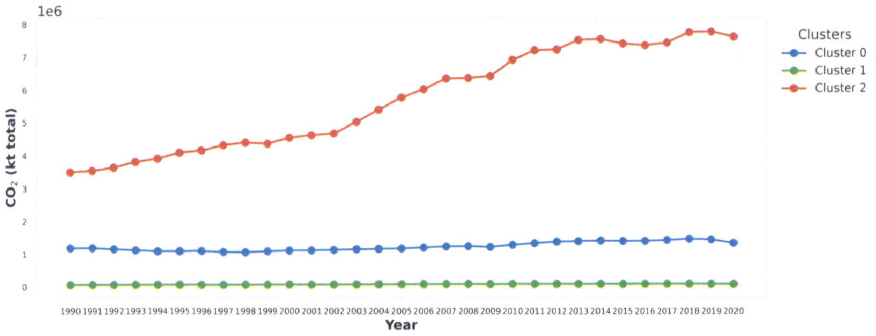

Fig. 5.18 Average trajectory of global total CO_2 emissions in kt by group

Over a 30-year period, Russia, India, Germany, and Japan have contributed 154,755,166.4 kt to Cluster 0. Because of its reliance on coal, India's emissions increased 141.47% while Russia's decreased 26.79%. Germany's energy transition was the primary factor in its highest drop of 44.13%.

With 102 nations, Cluster 2 is the biggest group and exhibits slow, linear expansion. The lowest emissions were 7,589,454.19 kt in 1990, while the highest emissions were 11,296,852 kt in 2018. The fast industrialization in Asia, Africa, and Latin America was a major factor in the group's 33.75% increase in emissions during that time. Global growth stayed steady despite notable declines in Europe.

5.5.3.6 Detailed Analysis of Global and Regional CO_2 Emission Trends

An extensive examination of CO_2 emission patterns is given in the Appendix, which covers national and regional contributions, cluster-specific behaviors, and temporal variations that were noted throughout the course of the 30-year research period.

Only China and the United States are in the high-emissions category; China's fast industrialization and heavy reliance on coal have propelled it to the top of the global rankings. Over the course of the research, China's emissions rose by 169.37%, peaking at 10,944,686.2 mt in 2020. The United States, on the other hand, reduced emissions by 9.71% between 1990 and 2020, while being the second-largest emitter. This was made possible in large part by transportation, industry, and energy generation based on fossil fuels.

With emissions ranging from 563,575.4 mt (India, 1990) to 2,458,175.9 mt (India, 2018), Germany, Russia, Japan, and India make up the medium-emissions group. India's emissions increased by 152.07% as a result of its significant reliance on coal, whereas Russia's emissions decreased by 26.79% as a result of economic downturns and changes in its energy sector. Germany led in emission reductions with a 44.13% loss, indicating its success in the shift to renewable energy, while Japan showed relative stability with a minor 5.87% decrease.

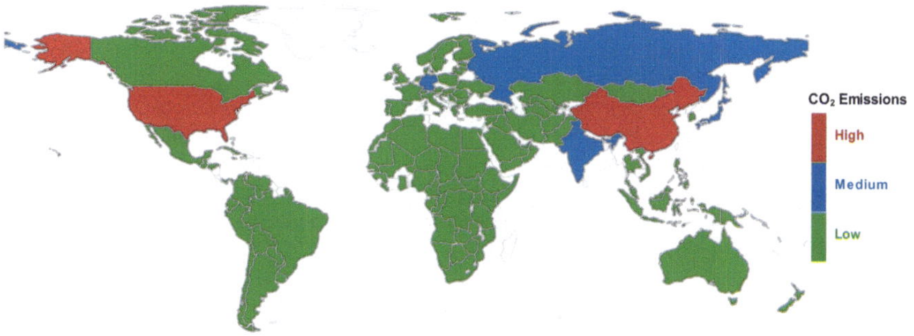

Fig. 5.19 Map of total CO_2 emissions in kt by country

With emissions ranging from 688,618.9 mt (Ukraine, 1990) to as low as 66.9 mt (Vanuatu, 1990), the 102 nations that make up the low-emissions category are primarily from Africa, Latin America, and Asia. This group had the largest relative increases, especially in Africa, where emissions have increased due to population growth and growing energy consumption. Nigeria (+152.47%), Mozambique (+207.10%), and Ghana (+230.09%) are noteworthy instances. In a similar vein, growing urbanization, reliance on fossil fuels, and industrial growth led to significant rises in several Asian nations, including Afghanistan, Iran, and Saudi Arabia.

Over the course of the 30-year period, worldwide CO_2 emissions rose by 48.78%, with the high-emissions category contributing the most. Nonetheless, the 2020 pandemic-related drop emphasizes how decreased economic activity and the use of fossil fuels may affect world emissions.

References

Abbasi, K. R., Shahbaz, M., Zhang, J., Irfan, M., & Alvarado, R. (2022). Analyze the environmental sustainability factors of China: The Role of fossil fuel energy and renewable energy. *Renewable Energy, 187*, 390–402. https://doi.org/10.1016/j.renene.2022.01.066

Adelekan, O. A., Ilugbusi, B. S., Adisa, O., Obi, O. C., Awonuga, K. F., Asuzu, O. F., & Ndubuisi, N. L. (2024). Energy transition policies: A global review of shifts towards renewable sources. *Engineering Science Technology of Journal*, *5*(2), 272–287. https://doi.org/10.51594/estj.v5i2.752

All for Health, Health for All: investment case 2025–2028. https://www.who.int/publications/i/item/9789240095403. Accessed 20 Dec 2024.

Berry, E. X. (2021). The impact of human CO_2 on atmospheric CO_2. *Science Climate Change.*

Bock, H.-H. (2007). Clustering methods: A history of k-means algorithms. In *Selected Contributions in Data Analysis and Classification*; Brito, P., Cucumel, G., Bertrand, P., De Carvalho, F., Eds.; Bock, H.-H., Gaul, W., Vichi, M., Arabie, Ph., Baier, D., Critchley, F., Decker, R., Diday, E., Greenacre, M., Lauro, C., Meulman, J., Monari, P., Nishisato, S., Ohsumi, N., Opitz, O., Ritter, G., Schader, M., Weihs, C., Series Eds.; Studies in Classification, Data Analysis, and Knowledge Organization; Springer Berlin Heidelberg: Berlin, Heidelberg, pp. 161–172. https://doi.org/10.1007/978-3-540-73560-1_15.

Cai, B., Huang, G., Samadiani, N., Li, G., & Chi, C.-H. (2021). Efficient time series clustering by minimizing dynamic time warping utilization. *IEEE Access, 9*, 46589–46599. https://doi.org/10.1109/ACCESS.2021.3067833

Climate Change 2014: Synthesis Report; Pachauri, R. K., Mayer, L. (2015). Intergovernmental Panel on Climate Change, Eds.; Intergovernmental Panel on Climate Change: Geneva, Switzerland.

Dadashi, Z., Mahmoudi, A., & Rashidi, S. (2022). Capacity and strategies of energy production from renewable sources in Arab countries until 2030: A review from renewable energy potentials to environmental issues. *Environmental Science and Pollution Research, 29*(32), 47837–47866. https://doi.org/10.1007/s11356-022-20544-z

Darby, S. J., Hammond, G. P., & Wu, J. (2024). Briefing: Stocktaking global warming: The outcomes of the 2023 Dubai climate summit (COP28). *Proceedings of the Institution of Civil Engineers—Energy,177* (5), 193–204. https://doi.org/10.1680/jener.24.00005.

DataBank | The World Bank. https://databank.worldbank.org/. Accessed 20 July 2024.

Friedlingstein, P., O'Sullivan, M., Jones, M. W., Andrew, R. M., Gregor, L., Hauck, J., Le Quéré, C., Luijkx, I. T., Olsen, A., Peters, G. P., Peters, W., Pongratz, J., Schwingshackl, C., Sitch, S., Canadell, J. G., Ciais, P., Jackson, R. B., Alin, S. R., Alkama, R., … Zheng, B. (2022). Global carbon budget 2022. *Earth System Science Data, 14*(11), 4811–4900. https://doi.org/10.5194/essd-14-4811-2022

Gallo, C., Faccilongo, N., & La Sala, P. (2018). Clustering analysis of environmental emissions: A study on Kyoto protocol's impact on member countries. *Journal of Cleaner Production, 172*, 3685–3703. https://doi.org/10.1016/j.jclepro.2017.07.194

Glavič, P., Pintarič, Z. N., Levičnik, H., Dragojlović, V., & Bogataj, M. (2023). Transitioning towards net-zero emissions in chemical and process industries: A holistic perspective. *Processes, 11*(9), 2647. https://doi.org/10.3390/pr11092647

Govender, P., & Sivakumar, V. (2020). Application of K-means and hierarchical clustering techniques for analysis of air pollution: A review (1980–2019). *Atmospheric Pollution Research, 11*(1), 40–56. https://doi.org/10.1016/j.apr.2019.09.009

Hassan, Q., Viktor, P., J. Al-Musawi, T., Mahmood Ali, B., Algburi, S., Alzoubi, H. M., Khudhair Al-Jiboory, A., Zuhair Sameen, A., Salman, H. M., & Jaszczur, M. (2024). The renewable energy role in the global energy transformations. *Renewable Energy Focu, 48*, 100545. https://doi.org/10.1016/j.ref.2024.100545.

Jackson, R. B., Saunois, M., Bousquet, P., Canadell, J. G., Poulter, B., Stavert, A. R., Bergamaschi, P., Niwa, Y., Segers, A., & Tsuruta, A. (2020). Increasing anthropogenic methane emissions arise equally from agricultural and fossil fuel sources. *Environmental Research Letters, 15*(7), Article 071002. https://doi.org/10.1088/1748-9326/ab9ed2

Jones, M. W., Peters, G. P., Gasser, T., Andrew, R. M., Schwingshackl, C., Gütschow, J., Houghton, R. A., Friedlingstein, P., Pongratz, J., & Le Quéré, C. (2023). National contributions to climate change due to historical emissions of carbon dioxide, methane, and nitrous oxide since 1850. *Science Data, 10*(1), 155. https://doi.org/10.1038/s41597-023-02041-1

Khaleel, M., Yusupov, Z., Alderoubi, N., Abdul_jabbar, R. L., Elmnifi, M., Nassar, Y., Majdi, H. S., Habeeb, L. J., & Abulifa, S. (2024). Evolution of emissions: The role of clean energy in sustainable development. *CiS, 12* (2), 122–135. https://doi.org/10.56578/cis120203.

Kolasa-Więcek, A. (2013). The use of cluster analysis in the classification of similarities in variables associated with agricultural greenhouse gases emissions in OECD countries. *Wieś i Rolnictwo, 1*(158), 59–66. https://doi.org/10.53098/wir.2013.1.158/03.

Le Quéré, C., Peters, G. P., Friedlingstein, P., Andrew, R. M., Canadell, J. G., Davis, S. J., Jackson, R. B., & Jones, M. W. (2021). Fossil CO_2 Emissions in the post-COVID-19 Era. *Nature Clinical Practice Endocrinology & Metabolism, 11*(3), 197–199. https://doi.org/10.1038/s41558-021-01001-0

Leng, P., Li, Z., Zhang, Q., Koschorreck, M., Li, F., Qiao, Y., & Xia, J. (2023). Deciphering large-scale spatial pattern and modulators of dissolved greenhouse gases (CO_2, CH_4, and N_2O) along the Yangtze River. *China. Journal Hydrology, 623*, Article 129710. https://doi.org/10.1016/j.jhydrol.2023.129710

Muluneh, M. G. (2021). Impact of climate change on biodiversity and food security: A global perspective—a review article. *Agriculture & Food Secure, 10*(1), 36. https://doi.org/10.1186/s40066-021-00318-5

Nisbet, E. G., Dlugokencky, E. J., Fisher, R. E., France, J. L., Lowry, D., Manning, M. R., Michel, S. E., & Warwick, N. J. (2021). Atmospheric methane and nitrous oxide: Challenges Alongthe Path to Net Zero. *Philosophical Transactions of the Royal Society A, 379*(2210), 20200457. https://doi.org/10.1098/rsta.2020.0457

Qerimi, Q., & Sergi, B. S. (2022). The case for global regulation of carbon capture and storage and artificial intelligence for climate change. *International Journal of Greenhouse Gas Control, 120*, Article 103757. https://doi.org/10.1016/j.ijggc.2022.103757

Rabbi, M. F., & Kovács, S. (2024). Quantifying global warming potential variations from greenhouse gas emission sources in forest ecosystems. *Carbon Research, 3*(1), 70. https://doi.org/10.1007/s44246-024-00156-7

Rahman, A., Murad, S. M. W., Mohsin, A. K. M., & Wang, X. (2024). Does renewable energy proactively contribute to mitigating carbon emissions in major fossil fuels consuming countries? *Journal of Cleaner Production, 452*, Article 142113. https://doi.org/10.1016/j.jclepro.2024.142113

Rajamani, L., & Werksman, J. (2018). The legal character and operational relevance of the Paris agreement's temperature goal. *Philosophical Transactions of the Royal Society A, 376*(2119), 20160458. https://doi.org/10.1098/rsta.2016.0458

Reay, D. S., Davidson, E. A., Smith, K. A., Smith, P., Melillo, J. M., Dentener, F., & Crutzen, P. J. (2012). Global agriculture and nitrous oxide emissions. *Nature Climate Change, 2*(6), 410–416. https://doi.org/10.1038/nclimate1458

Ripple, W. J., Wolf, C., Gregg, J. W., Rockström, J., Newsome, T. M., Law, B. E., Marques, L., Lenton, T. M., Xu, C., Huq, S., Simons, L., & King, S. D. A. (2023). The 2023 state of the climate report: Entering uncharted territory. *BioScience, 73*(12), 841–850. https://doi.org/10.1093/biosci/biad080

Sadai, S., Spector, R. A., DeConto, R., & Gomez, N. (2022). The Paris agreement and climate justice: inequitable impacts of sea level rise associated with temperature targets. *Earth's Future, 10*(12), e2022EF002940. https://doi.org/10.1029/2022EF002940.

Saunois, M., Stavert, A. R., Poulter, B., Bousquet, P., Canadell, J. G., Jackson, R. B., Raymond, P. A., Dlugokencky, E. J., Houweling, S., Patra, P. K., Ciais, P., Arora, V. K., Bastviken, D., Bergamaschi, P., Blake, D. R., Brailsford, G., Bruhwiler, L., Carlson, K. M., Carrol, M., … Zhuang, Q. (2020). The global methane budget 2000–2017. *Earth System Science Data, 12*(3), 1561–1623. https://doi.org/10.5194/essd-12-1561-2020

Suddick, E. C., Whitney, P., Townsend, A. R., & Davidson, E. A. (2013). The role of nitrogen in climate change and the impacts of nitrogen-climate interactions in the United States: Foreword to thematic issue. *Biogeochemistry, 114*(1–3), 1–10. https://doi.org/10.1007/s10533-012-9795-z

Summary for Policymakers. (2023). In *Climate Change 2021—The Physical Science Basis: Working Group I Contribution to the Sixth Assessment Report of the Intergovernmental Panel on Climate Change*; Intergovernmental Panel on Climate Change (IPCC), Ed.; Cambridge University Press: Cambridge, pp 3–32. https://doi.org/10.1017/9781009157896.001.

Wei, T., Dong, W., Yan, Q., Chou, J., Yang, Z., & Tian, D. (2016). Developed and developing world contributions to climate system change based on carbon dioxide, methane and nitrous oxide emissions. *Advances in Atmospheric Sciences, 33*(5), 632–643. https://doi.org/10.1007/s00376-015-5141-4

Zhang, W., Li, H., Xiao, Q., & Li, X. (2021). Urban rivers are hotspots of riverine greenhouse gas (N_2O, CH_4, CO_2) emissions in the mixed-landscape CHAOHU lake basin. *Water Research, 189*, Article 116624. https://doi.org/10.1016/j.watres.2020.116624

Zheng, Y., Zhang, Q., Tong, D., Davis, S. J., & Caldeira, K. (2020). Climate effects of China's efforts to improve its air quality. *Environmental Research Letters, 15*(10), Article 104052. https://doi.org/10.1088/1748-9326/ab9e21

Chapter 6
Optimizing Power Systems with Representative Week Selection

6.1 Introduction

Initiatives to transition energy systems toward decarbonization have prompted the development of policy frameworks focused on strengthening the role of renewables in power generation over the coming decades. Nevertheless, one significant drawback of renewable technologies lies in their non-continuous availability, as they are influenced by meteorological variability. This leads to mismatches between supply and demand, creating operational difficulties for the grid and compromising both reliability and cost-effectiveness (Li et al., 2024). Storage solutions are essential in enabling low-carbon electricity systems, as they support the assimilation of renewable generation technologies. They help smooth fluctuations arising from solar and wind energy sources, reduce the risk of supply shortfalls, and promote the effective use of excess electricity during times of decreased consumption or heightened system stress (de la Cruz & Castellanos, 2024).

Unit commitment (UC) formulations are used as short-term decision-making tools in power systems, operating with fine-grained time steps (e.g., hourly or even more frequent). These approaches schedule generation while accounting for both technical limitations and broader system-level constraints (Heuberger et al., 2017). On the other hand, generation expansion planning (GEP) focuses on long-term infrastructure decisions, typically disregarding small-scale temporal variations to perform evaluations over extended time horizons (Tejada-Arango et al., 2018).

Historically, flexible thermal power technologies have been predominant in power systems, enabling modeling frameworks to extend planning horizons and overlook short-duration operational limitations with negligible effects on output accuracy (Lara et al., 2018). Given the growing presence of wind and solar energy in electricity networks, it becomes necessary to evaluate hourly operational decisions to determine the grid's adaptability (Albadi & El-Saadany, 2010; Islam & Roy, 2023). This approach enables a systematic and rigorous evaluation of the balance between long-term investment decisions and short-term operational decisions. Researchers have

A. Y. Raya-Tapia et al., *Machine Learning and Clustering for a Sustainable Future*, Studies in Computational Intelligence 1233,
https://doi.org/10.1007/978-3-032-03876-0_6

integrated system operation constraints such as UC, reserve margins, startup limitations, and ramping thresholds into Generation Expansion Planning (GEP) models (Lara et al., 2018).

Modeling every hour across an extended planning timeframe leads to models containing billions of variables, rendering them impractical for current optimization solvers because of the substantial computational burden required (Li et al., 2022; Yeganefar et al., 2020). A feasible approach to simplify the optimization task involves choosing a limited number of representative time periods (days or weeks) extracted from time series datasets, ensuring that they preserve the fine temporal granularity and key trends of the original dataset (Helistö et al., 2020; Tahir et al., 2022).

There are three common approaches for selecting representative periods: clustering approach, optimization-based approach, and heuristic approach. Koltsaklis and Dagoumas (2018) selected a set of eight typical days that correspond to working weekdays and weekends across the four annual seasons in their heuristic methods. Yeganefar et al. (2020) selected twelve days at random, each matching a month from the year, while Mavrotas et al. (2008) employed monthly averages. To reflect both intra-annual and daily fluctuations in load and power supply, Short et al. (2011) selected sixteen periods that spanned various seasons, including one super-peak summer instance. Lozano et al. (2009) employed a 24-day period to delineate two typical intervals per month, using mean values for weekdays and weekends tied to yearly electricity demand. Nonetheless, this basic method of choosing representative time points does not ensure that the chosen set preserves the defining patterns of the source time series. Furthermore, the process used to select criteria is inconsistent, relying on empirical judgments, and there is no established procedure to validate the chosen periods (Gao et al., 2021).

De Sisternes and Webster (2013) validated their optimization-based methods by evaluating every possible weekly combination. They identified key representative weeks (RWs) that reduce the deviation between the NLDC estimated from the RWs and the original NLDC. As the number of representative weeks increases, both computational burden and processing time escalate.

Recent progress in machine learning methodologies suggests the implementation of clustering algorithms. In this context, typical periods featuring various standardized characteristics (demand, solar and wind production, weather conditions, etc.) are aggregated. One representative period, such as the centroid, medoid, or a selected instance, is chosen after performing the clustering process. Moreover, the importance factors given to each representative time span are implicitly established through the clustering-based approaches. Various clustering approaches, including k-means applied to historical datasets, have been utilized to identify typical periods and generate representative time points (Niu et al., 2021; Wang et al., 2021; Zhou et al., 2022). The k-medoids algorithm was utilized by Domínguez-Muñoz et al. (2011) to identify key representative days, which were assessed using two clustering quality metrics and a manual review of the calendar-based distribution. Moradi-Sepahvand and Amraee (2021) employed hierarchical clustering, utilizing historical records on electricity demand and renewable energy generation to derive 96 representative clustering-based hourly profiles, encapsulating the full year.

The most widely used approach identified is representative days; however, a standardized approach for determining the number of clusters in advance, specifically the amount of representative days, is lacking (Li et al., 2022). An iterative testing method has been employed to determine the optimal number of representative days by incrementally augmenting their quantity or utilizing partition quality metrics (Almaimouni et al., 2018). Consequently, there is contention on the appropriate duration of representative groups (Tejada-Arango et al., 2018). Utilizing representative groupings composed of days or weeks is stressed to support the preservation of temporal order; nevertheless, this approach also escalates the computing burden. Chronology can be ensured to cover extended durations of diminished injection from variable renewable sources and to better illustrate the flexibility needed between days (Nahmmacher et al., 2016).

6.2 Methods for Selecting Representative Periods

Several prior studies have contrasted various clustering techniques. Poncelet et al. (2017) conducted a comparison of methodologies utilizing heuristic methods, hierarchical approaches, random selection, and optimization techniques for choosing representative days, employing 15-min resolution time series for wind and solar generation. The optimization-driven strategy yielded the best results; nevertheless, it grew computationally intensive as the profile count increased. Teichgraeber and Brandt (2022) conducted a comparison of clustering methodologies embedded in the objective function framework of two basic operational optimization scenarios: battery charge or discharge control and gas turbine scheduling.

Pfenninger (2017) assessed k-means, k-medoids, hierarchical clustering, and atypical value detection techniques in the context of a generation expansion problem, choosing sample data from wind and solar hourly time series. Their analysis showed that outcomes could differ markedly depending on the methodologies, and the optimal way for selecting representative periods is contingent upon the input data and the configuration of model constraints. Kotzur et al. (2018) assessed different clustering techniques for selecting four representative days across five temporal datasets, employing 3 optimization tasks pertinent to energy system design applications: combined heat and power systems, residential energy networks, and isolated (islanding) operational modes. Performance evaluation of the modeled representative time periods is significantly contingent upon the optimization framework used. Moreover, they observed that no single clustering method is consistently better than the rest; nonetheless, k-medoids more effectively captures temporal variability, as it maintains data granularity more accurately than the centroid method.

In the selection process, continuity between sample periods is sometimes disregarded while aggregating representative days. A key advantage of using sample weeks is the preservation of chronology throughout each segment. Helistö et al. (2020) conducted a comparative analysis of k-means clustering, regular interval decomposition, random sampling strategies, as well as the approach proposed by

Sisternes and Webster for selecting representative weeks derived from three datasets: electric load, wind power, and solar photovoltaic data originating in Northern Europe. The techniques were analyzed based on how each technique affected the cost optimization framework in planning power systems across multiple regions. Findings indicated that regular decomposition is a suitable method for identifying planning weeks, particularly for system design. The effectiveness of these techniques is contingent upon both the data provided and the total number of weeks chosen. Understanding the importance of comparing multiple strategies for selecting representative weeks across systems and objectives is evident, particularly in energy networks where variable renewables comprise a significant portion of the generation mix.

Choosing the appropriate temporal resolution is crucial for power system planning, garnering the interest of scholars and industry professionals. This study selected four representative weeks because, according to the methodology of Sisternes and Webster's (2013) approach, the computational workload increases proportionally with the number of representative weeks included in the analysis, as the total number of potential combinations grows factorially with the selected number of weeks. Furthermore, for over four weeks, the mistakes have diminished, albeit not in a monotonic manner. The calculation time for selecting weeks using clustering techniques is unaffected by the number of weeks to be chosen. Research indicates that planning problems yield steady results after 4 to 5 weeks (Helistö et al., 2020). Consequently, the option of utilizing four sample weeks was evaluated. The determination of weeks was contingent upon the processing time required by the selection technique, as well as the magnitude and intricacy of the planning issue to be resolved. This is due to the potential inadequacy of computational resources when incorporating a substantial number of time periods into the optimization model.

- This chapter employs an approach designed to derive a set of four weekly representations of the NLDC, encompassing demand fluctuations and variations in wind and solar power generation, from multiple zones utilizing k-means clustering with both Euclidean and DTW distance measures, alongside the combinatorial technique proposed by De Sisternes et al. (2013). The error of the NLDC, derived based on the representative weeks selected by each method, is contrasted against the original NLDC. These weekly groupings are analyzed within the objective function framework of a mathematical formulation that integrates strategic long-term planning decisions alongside short-term system operation tactics for a multi-zonal power production network. Four weeks, each comprising 168 h, are selected to ensure a high degree of granularity and alignment within the optimization-based evaluation. Furthermore, the algorithm's computational demand is considered to guarantee both accuracy and feasibility. The selection of the optimal method must consider the data characteristics, needs, and analysis goals of power system modeling. Consequently, this study concentrates on the following targets: Examine how effective multi-regional weekly selection methods are regarding NLDC representation to more precisely reflect ramp changes across intervals, since this factor influences system flexibility needs.

- Examine how power generation technologies contribute to the outcomes of long-range planning scenarios, including operational-level decisions, alongside the cost-related metrics and emission outputs, produced during various representative weeks derived from k-means grouping using the Euclidean distance, DTW approach, and the Sisternes–Webster method.

6.3 Definition of the Research Issue

The primary obstacle in long-horizon planning frameworks that incorporate operational-level decisions is temporal complexity. Consequently, it becomes necessary to explore techniques to simplify the planning challenge without compromising adequate precision. Utilizing representative time periods is a viable alternative; nevertheless, the selection of these periods is crucial, since an improper choice might substantially skew the model's results. The selection is contingent upon the attributes and duration of the time series, as well as the optimization issue at hand. Consequently, it is imperative to evaluate many choices and select the most appropriate option. No clustering technique or distance measure is universally superior; thus, it is essential to emphasize the trade-off between precision and computational expense. Consequently, this study's approach relies on identifying four representative weeks from a yearly NLDC time series. The link between wind, solar, and demand variability is considered collectively. Three methodologies for picking weekly representatives are employed: k-means utilizing the Euclidean distance, k-means employing the DTW metric, and an optimization-driven approach that evaluates all feasible combinations. The Euclidean distance was selected as a simple yet consistent metric for comparison against a more complex one like DTW. For each technique, the group of four weeks exhibiting the smallest RMSE relative to the original NLDC are chosen for validating model fidelity. For every representative week set derived, a yearly increase of 2% was presumed across a 10-year horizon, which was then used as scenario input for long-term energy planning, incorporating short-term operational factors in power systems characterized by significant integration of variable renewable energy (VRE) across multiple regions; thereby facilitating the comparison and analysis of how operations are represented, and evaluating performance in the optimization objective function for each four-week selection strategy.

This challenge involves the management of a varied array of energy-related technologies, encompassing both traditional and green energy resources, within the regional electricity production. These sources must provide sufficient energy to satisfy demand at designated periods, while effectively controlling the direction of power delivery to address the energy needs within a defined area, or to retain it for later availability. Initially, green energy systems must effectively channel the generated energy either toward storage systems or to immediately meet energy requirements. Conventional technologies fall into two types: those utilizing natural gas and those employing alternative fuels. Renewable-based solutions are divided into two main groups depending on the availability of natural inputs: variable and non-variable.

The disjunctive optimization issue seeks to establish the most suitable arrangement of simultaneously operating power generation methods, and the requisite energy retention systems to facilitate effective process functionality. Additionally, the capabilities of these devices and the expenses related to their utilization are taken into account. This optimization challenge centers on planning over a multi-year time horizon while accounting for hourly operating conditions.

6.4 Implemented Methodology

Initially, information is collected to compute the Net Load Duration Curve (NLDC) in each zone of the analyzed electrical network. This information source is used to derive sample weeks through various clustering methods, specifically: K-means clustering employing the Euclidean distance, K-means clustering for time series utilizing the DTW technique, and the combinatorial procedure introduced by de Sisternes and Webster (2013). To evaluate the example weeks generated by each approach, they are implemented within a mathematical formulation for the operational design and strategic planning of a power grid over a ten-year period (see Fig. 6.1).

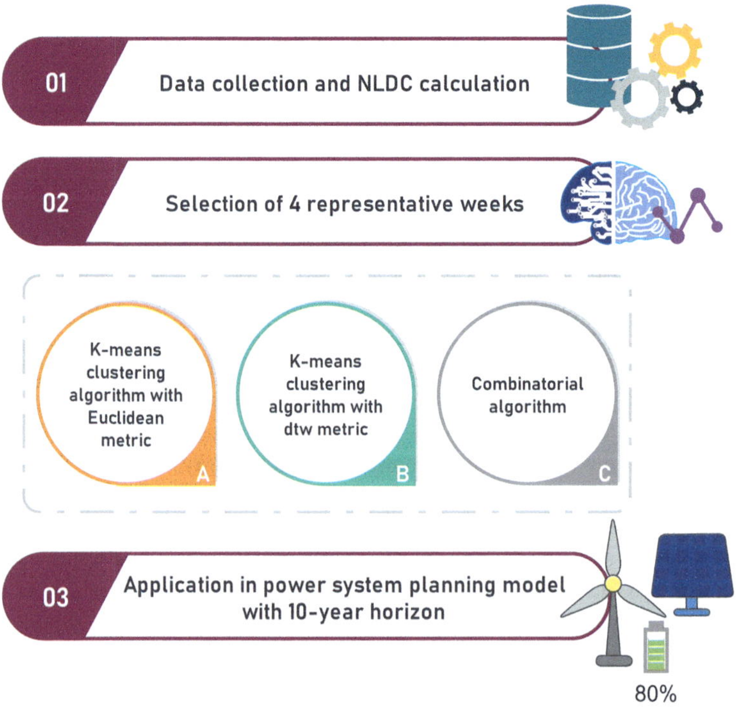

Fig. 6.1 Schematic depiction of the executed overarching technique

The accessibility and quality of the data required for model parameterization are significantly enhanced, facilitating more precise and credible model abstractions. The present study utilized actual hourly load and installed generation capacity information for the technologies in each examined location to develop the NLDC. Hourly meteorological data were utilized to determine the hourly performance factor for solar and wind sources. The volume of input variables has a direct impact on the magnitude of the problem addressed by the optimization model, hence affecting the necessary computational demands and the feasibility of obtaining a satisfactory solution within acceptable time limits (Ehrgott et al., 2014). This highlights vast quantities of information and expert-based evaluations that exceed what can be effectively governed by rudimentary mathematical procedures.

6.4.1 Computation of the NLDC for Each Region

A year has 52 weeks, from which 4 typical weeks must be chosen (each week consists of 168 h) (see Fig. 6.2) for utilization in the operational and energy system planning framework in the next part of this document.

The three techniques rely on identifying four representative weeks from a net load dataset covering an entire year. Net load is utilized to represent variability while balancing demand and variable power generating technologies with dispatchable generation. The disparity between the demand and variable energy output supplies, including solar PV and wind sources, is used to build the net load duration profile in each location. The hourly variable output is determined by multiplying the hourly capacity factor of each wind or solar energy source by its respective installed capacity.

$$\mathrm{NLDC}_{r,t} = \mathrm{D}_{r,t} - \mathrm{CF}_{r,t}^{wind}\overline{\mathrm{W}} - \mathrm{CF}_{r,t}^{solar}\overline{\mathrm{S}} \tag{6.1}$$

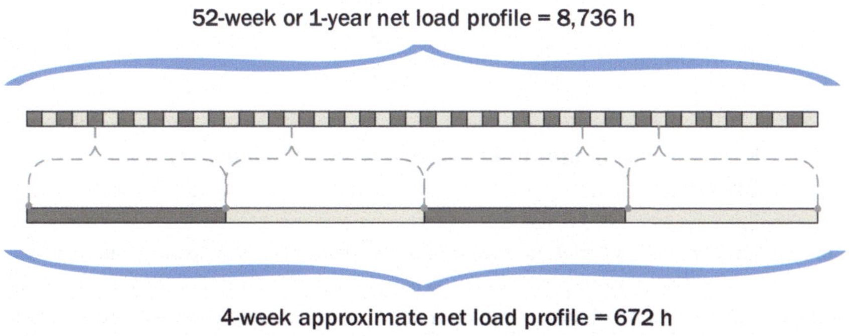

Fig. 6.2 The net load pattern over four selected weeks serves as a representative sample of the entire annual series of fifty-two weeks

$$\text{s.t.}: NLDC_{r,t-1} - NLDC_{r,t} \geq 0,\ \forall r \in R, t \in T$$

The net load duration profile is obtained by adjusting the hourly values from a set of four representative weeks to represent a full year (52 weeks) and thereafter sorting the data in descending fashion. To choose among all potential week approximations, the least squares error is minimized between the original value and its estimation. The solution to the subsequent optimization problem provides the optimal approximation (see Eq. 6.2).

$$min\|NLDC - N\tilde{LDC}\|^2 \tag{6.2}$$

Equation (6.3) applies RMSE as a quantitative indicator to assess the accuracy of the net load duration curve representation.

$$RMSE = \sqrt{\frac{\sum_{t=1}^{8,736}\left(NLDC_t - N\tilde{LDC}_t\right)^2}{8,736}} \tag{6.3}$$

The selection routine proposed by De Sisternes and Webster (2013) thoroughly explores all possible weekly groupings to identify the combination that minimizes error. The quantity of nested loops employed in this enumeration method corresponds to how many weeks are chosen, which is four in this instance. The weight assigned to each hour in the selected representative weeks to build a full-year synthetic profile is 52/4 = 13. The dataset comprises 52 weeks, with each week totaling 168 h; this collection represents the aggregate NLDC profile corresponding to each zone.

6.4.2 Clustering

During the preprocessing stage, the dataset is standardized using StandardScaler from the preprocessing module of sklearn (Scikit-learn, 2023). Mean values are removed, and the data are rescaled to ensure unitary variance.

6.4.2.1 K-Means Algorithm Implemented with a Euclidean Distance Criterion

K-means refers to a clustering method under the unsupervised learning category that forms partitions by assigning weeks into distinct subgroups depending on regional similarity. The steps involved in the k-means routine are outlined as follows: Step 1: Centroids are initially assigned at random according to a specified number of k groupings, with duration h = 0.

Step 2: Each weekly instance is defined by d dimensions, where these dimensions correspond to the 168 hourly values per week in every location. The proximity from each reference center is computed using the Euclidean distance metric, as defined in Eq. (6.4).

$$d(w_{nd}, m_{kd}) = \sqrt{\sum\nolimits_d (w_{nd} - m_{kd})^2}\, \forall n, \forall k \tag{6.4}$$

Step 3: Each entity is allocated to the cluster that exhibits the minimal distance relative to its centroid.

Step 4: Each cluster center is updated by computing the average of the associated data points within the group, as demonstrated in Eq. (6.5).

$$m_k^{h+1} = \frac{1}{\left|C_k^h\right|} \sum\nolimits_{w_n \in C_k^h} w_n \, \forall k \tag{6.5}$$

Steps 2 to 4 are reiterated until the termination criterion is satisfied, specifically the minimization of the total squared Euclidean distance, as shown in Eq. 6.6.

$$\sum\nolimits_{w_n \in C_k^t} \sum\nolimits_k d(w_n - m_k)^2 \tag{6.6}$$

6.4.2.2 Time Series Clustering via K-means with a DTW-Based Similarity Measure

Clustering of time series represents an approach used to investigate hidden structures shared by a collection of time series, specifically pertaining to each region. Dynamic Time Warping (DTW) seeks to temporally align sequences so as to reduce the Euclidean deviation across matched points. Overall distance at any index is detailed in Eq. (6.7).

$$D(x_p, y_q) = d(x_p, y_q) + min\{D(x_{p-1}, y_q), D(x_p, y_{q-1}), D(x_{p-1}, y_{q-1})\} \tag{6.7}$$

Here $d(x_p, y_q) = (x_p - y_q)^2$ where the Euclidean distance is the most frequently adopted measure, though other similarity or dissimilarity metrics can be considered as well.

Equation (6.8) illustrates the optimization problem, representing the DTW-based metric between two sequential data, X and Y, of identical duration, which determines the optimal alignment trajectory.

$$dtw(p, q) = min \sum \|D(x_p, y_q)\| \tag{6.8}$$

The Euclidean distance is a widely utilized metric owing to its straightforwardness and computational speed, rendering it a viable option. Nonetheless, this metric necessitates that time series exhibit the same duration. Moreover, it proves highly vulnerable to minor discrepancies across temporal sequences. For instance, if two time-ordered observations exhibit a minor offset while remaining otherwise identical, the resulting Euclidean measurement may become excessively large (Kate, 2016). DTW mitigates these problems.

Determining the DTW-based similarity between two time series requires identifying the optimal alignment from all possible configurations. Dynamic Time Warping (DTW) is computed through a recursive problem-solving approach to determine an effective alignment trajectory. Evaluating every possible route is "combinatorially exhaustive". Such computation is demanding and time-consuming. DTW has shown superior performance over Euclidean distance in time series-based classification and clustering. Nevertheless, most research efforts have adopted Euclidean distance due to its efficiency and the notable reduction in processing time demanded by DTW (Ratanamahatana & Keogh, 2004).

6.5 Implemented Mathematical Model

A mathematical formulation for optimal power generation planning in a large-scale network over a defined planning period is offered. Figure 6.3 presents a diagrammatic depiction of the proposed framework. A set of available technologies (j = 1, 2, …, J) fulfills energy needs in a specific geographical region (r = 1, 2, …, R) over hourly (t = 1, 2, …, T) and annual (s = 1, 2, …, S) intervals. Green energy sources produce both intermittent (jrt = 1, 2, …, J) and non-intermittent (jrh = 1, 2, …, J) energy, stored in energy storage systems (i = 1, 2, …, I) to meet power requirements as needed. Traditional technologies (jt = 1, 2, …, J) are categorized into those that use natural gas (jg = 1, 2, …, J) as fuel and those that rely on fuel oil (jc = 1, 2,…, J) as fossil fuel. The mathematical formulation underlying this work is thoroughly described in Raya-Tapia et al. (2025).

6.5.1 Objective Function (OF)

The *OF* seeks to minimize the *TOC*, which is calculated as follows:

$$O.F. = min\ \{TOC\} \tag{6.9}$$

The mathematical model represents a Mixed-Integer Linear Programming issue aimed at minimizing the Total Operating Cost (TOC) related to electricity generation. The created mathematical model is applicable in virtually all global regions, provided

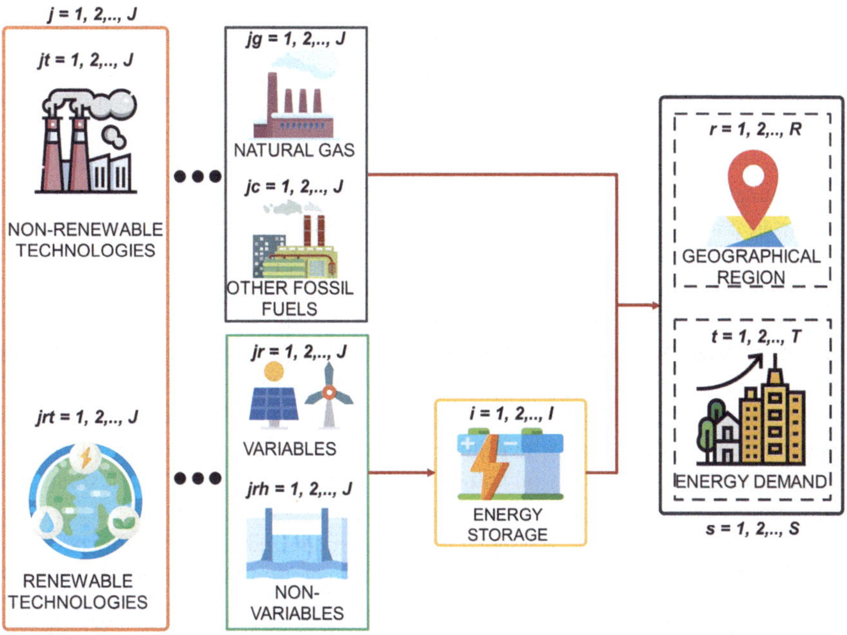

Fig. 6.3 Illustrative model of the superstructure applied to operational-level planning of electricity generation

that the specific limits and conditions of the power grid, together with the various renewable energy production alternatives, are included.

The optimization framework delineated in this study incorporates several simplified approaches to address the various decision-making levels that impact the energy system architecture. The ramping rate restrictions are represented by linear equations, employing network and technology grouping, while the temporal resolution challenges are addressed using a time-discretized structure. Numerous streamlining techniques aimed at progressively diminishing model dimensionality and complexity have been applied in different contexts to manage the intricacies involved in optimizing energy systems, as shown by Kotzur et al. (2021). Prior scientific studies have examined transformation of the energy landscape within this area. For instance, the establishment of hydrogen infrastructure (Serrano-Arévalo et al., 2023) and the evaluation of circular economy metrics (Serrano-Arévalo et al., 2024). While all these techniques are beneficial for strategizing the energy transition, the majority fail to guarantee the reliability of optimal operational outcomes. It is important to acknowledge a conceptual limitation in deploying these optimization schemes, particularly our ability to express them as formal mathematical formulations (Zhang et al., 2015).

6.5.2 Considered Case Study

The peninsular power system of Mexico is chosen as a case study for the application of the proposed framework, owing to the accessibility of data and the possibilities for renewable energy implementation arising from issues related to the availability of natural gas. The system encompasses power production and includes five key transmission areas: Lerma (RC), Chetumal (RD), Merida (RE), Cancun (RF), and Cozumel (RG). These regions are linked to the eastern power system by two additional regions, Grijalva (RA) and Tabasco (RB). Figure 6.4 illustrates how each area contains an electric transmission infrastructure with designated capacity. The production of electricity energy in the eastern and peninsular power systems utilizes various technologies; however, only the eight that significantly contribute to total production are studied. Electricity generation systems are categorized into two main types: conventional sources, which may use natural gas and oil-derived fuels, and renewable systems, which rely on naturally available inputs such as wind, solar, and water.

The modeling framework, grounded in a MILP model, adopts a planning perspective focused on operational decisions to ascertain generation using fluctuating renewable resources (e.g., wind and solar), stable renewable sources (such as hydro), and conventional technologies (e.g., combined cycle, internal combustion engines, gas turbines, thermoelectric units, and cogeneration systems) across seven regions, alongside the evaluation of energy storage system capacity and placement to meet electricity needs during each time step. Total costs are optimized, encompassing power generation and grid-related expenditures across regions, infrastructure investments

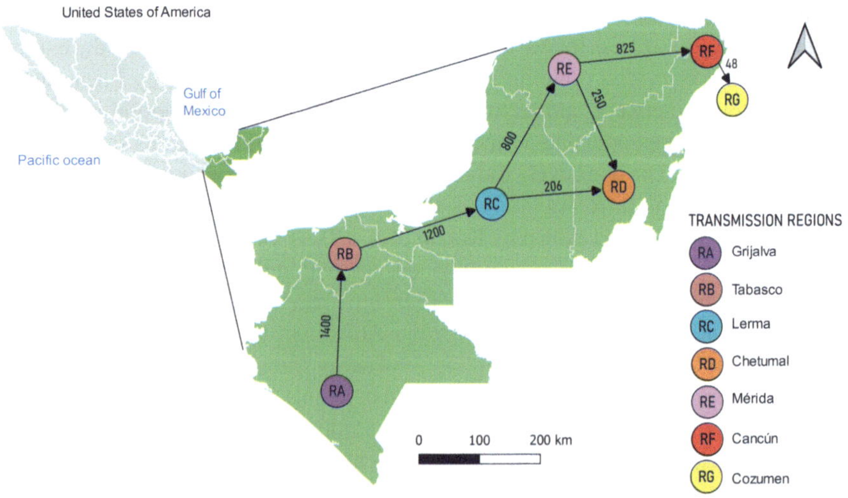

Fig. 6.4 Examined transmission areas of the model in Mexico

for energy storage units, their operational expenses, and associated environmental impacts (e.g., carbon taxation).

The contents of Table 6.1 detail the characteristics of the optimization framework, grounded in recent scientific publications (INECC, 2020; SENER, 2018). An electricity transmission efficiency of 95% is considered. The considered calorific values for natural gas and fuel oil were 50.03 GJ/ton and 40.10 GJ/ton, respectively. This formulation incorporates Vanadium redox flow batteries (VRB) and Lithium-ion technology (Li-ion) (INECC, 2020). The carbon tax was valued at $3 per ton of CO_2 equivalent (Renner, 2018), while the penalty for unmet energy demand was set at $2600 per MWh (SENER, 2018).

Four example weeks were chosen using three distinct techniques to define the NLDC, which integrates power demand and fluctuating renewable output across time, while minimizing temporal complexity and high computational cost. A selection of four representative weeks was made from a full-year horizon (8760 h), with each week representing 168 h. This suggested framework compares the accuracy of the

Table 6.1 Specifications of installed generation capacity and storage expenses

Generation technology	*Installed capacity (MW)*						
	Grijalva region (RA)	Tabasco region (RB)	Lerma region (RC)	Chetumal region (RD)	Mérida region (RE)	Cancún region (RF)	Cozumel region (RG)
Intermittent power							
Solar	–	–	–	–	–	–	–
Wind	17	–	40	8	145	18	5
Non-intermittent energy							
Hydro	4828	–	–	–	–	–	–
Traditional generators							
Turbogas	121	217	47	71	60	190	53
Combined cycle	–	–	252	–	1229	–	–
Efficient cogeneration	–	367	112	–	243	–	–
Internal combustion	1	47	–	–	13	–	–
Conventional thermoelectric	–	–	–	–	1	3	–
Energy storage							
Battery Type	Discharge efficiency	Charging efficiency	Variable costs ($/MWh)	Fixed costs ($)	Minimum capacity (MW)	Maximum capacity (MW)	Energy at time zero (MWh)
VRB	0.85	0.85	1000.00	944.00	6.65	100.00	0.0
Li-ion	0.98	0.98	2220.00	540.00	6.00	100.00	0.0

resultant representative weeks from each method in representing the NLDC profiles per region and their influence on the optimization objective space.

6.6 Results and Discussion for the Addressed Problem

6.6.1 Approach to Selecting Typical Weeks

Four representative weeks were selected using three methodological approaches, categorized as follows: (A) generated via the K-means algorithm combined with the Euclidean distance metric, (B) derived from time-series K-means clustering utilizing the DTW metric, and (C) obtained through a combinatorial strategy. As shown in Table 6.2, set C achieved the lowest root mean square error (RMSE), followed by sets B and A, respectively. It is evident that the weights assigned to each week in sets A and B are adjusted to reflect their contribution to an annual representation. For example, in set A, week 1 is assigned a weighting factor of 23, while week 48 appears 10 times. This allocation pattern is extended to the remaining weeks to complete the 52-week annual horizon.

n_init denotes the frequency with which the k-means routine is executed using distinct centroid initializations. The runtime for set A was about 30 s, set B required 35 min, and set C took 2 min. When using K-means along with DTW, no restrictions were imposed on the allowed deformation, which led to longer runtimes compared to the other two strategies. In contrast, employing K-means with Euclidean distance leads to a significant reduction in processing time compared to DTW-based calculations.

The NLDC results from calculating the gap between energy demand and the generation provided by fluctuating sources, sorted in descending magnitude. This is illustrated in Figs. 6.6, 6.7, 6.8, 6.9 and 6.10 within the Appendix, which also displays the estimated load profiles for each approach.

Table 6.2 Outcomes of the three methodologies

n_init	Representative weeks	Weights of each week for scaling	RMSE	Week Set ID
500	[31,48,13,4]	23,10,6,13	5.49	A
100	[35,47,44,36]	7,10,13,22	3.21	B
	[11,34,36,46]	52/4 = 13	2.80	C

6.6.2 Implemented Model in GAMS

The previously introduced optimization framework was implemented using GAMS (v43.4.1) and formulated as a mixed-integer linear programming (MILP) problem. The optimization task included approximately 2.95 million continuous variables, over 94,000 binary variables, and 2.4 million algebraic constraints. The computations were performed on a machine equipped with a 2.30 GHz Intel Core i7-12700H processor and 16 GB of RAM, with a processing time of around 1 min and 30 s. Table 6.3 provides a summary of environmental performance, cost implications, and energy indicators across the evaluated typical week groupings.

The environmental assessment reveals that Period A results in the lowest total emissions, amounting to 1.6516×10^6 $tonCO_{2eq}$. Nevertheless, it displays the peak emission intensity, reaching 0.49 $tonCO_{2eq}$ per MWh. This intensity is calculated as the ratio between emissions stemming from traditional power technologies and their corresponding total energy generation. On the other hand, Period C demonstrates greater emissions volumes but yields a lower emission intensity. Figure 6.11 from the Appendix section visualizes the yearly emissions attributed to each evaluated group.

From an economic standpoint, Group A is associated with the lowest total expenditure, amounting to $\$4.41 \times 10^{10}$. It also reflects the minimum spending in thermal power production, evaluated at $\$1.05 \times 10^7$, and the least investment in power transmission, reported at $\$4.64 \times 10^6$. Meanwhile, Group B records an aggregate cost of $\$4.42 \times 10^{10}$, including $\$1.08 \times 10^7$ for heat-based energy generation and a peak transmission expenditure reaching $\$4.73 \times 10^6$. In contrast, Group C demonstrates the most elevated total cost, equal to $\$4.42 \times 10^{10}$, alongside the highest thermal generation outlay of $\$1.12 \times 10^7$ and a transmission budget of $\$4.70 \times 10^6$.

Table 6.3 Acquired outcomes for weeks A, B, and C, while limiting expenses

Concept	A	B	C
Emissions ($tonCO_{2eq}$)	1.65×10^6	1.70×10^6	1.75×10^6
Emissions factor ($tonCO_{2eq}$/ MWh)	0.49	0.49	0.49
Total cost ($)	4.41×10^{10}	4.42×10^{10}	4.42×10^{10}
Energy production (MWh)	1.75×10^7	1.77×10^7	1.78×10^7
Production from conventional technologies	3.33×10^6	3.45×10^6	3.58×10^6
Production from variable renewable technologies	1.83×10^6	1.83×10^6	1.83×10^6
Production from non-variable renewable technologies	1.23×10^7	1.2508×10^7	1.23×10^7
Curtailment (MWh)	1.54×10^5	1.4645×10^5	1.42×10^5
Energy not supplied	163	163	163
Natural gas used (ton)	6.00×10^5	6.19×10^5	6.39×10^5
Fuel oil used (ton)	1.18×10^4	1.17×10^4	1.17×10^4

In comparison to the other representative week sets, Scenario A yields the lowest total energy within the evaluated decade-long horizon, totaling 1.75×10^7 MWh. Of this, 81% derived from clean energy sources and 19% from fossil-based generation. During the specified weeks, B generates 1.77×10^7 MWh, with 80.6% supplied by renewables and 19.4% from fossil-based generation. During the weeks in set C, the maximum energy output is 1.78×10^7 MWh, with 80% derived from clean energy sources and 20% from fossil-based generation.

Traditional power-generation methods rely on fossil-based inputs, the minimum consumption recorded in weeks A at 6.12×10^5 tons, including 98% natural gas and 2% fuel oil. Subsequently, set B comprises 6.30×10^5 tons, consisting of 98% natural gas with the remainder being fuel oil. The maximum consumption of fossil fuels occurred in set C, totaling 6.51×10^5 tons; this involved a marginally greater share of natural gas at 98% and 2% of fuel oil. Visual summaries of natural gas and fuel oil usage across all regions over ten years are provided in Figs. 6.12 and 6.13 of the Appendix.

The set of weeks C exhibits total costs that are 0.2% and 0.1% greater than those of sets A and B, respectively. Additionally, thermal energy production costs are 6.2% and 3.3% higher compared to sets A and B, while emissions generation is elevated by 6% and 3% relative to sets A and B. However, it is noteworthy that energy production is also 1.3% and 0.05% higher in comparison to sets A and B. This may be due to its more accurate portrayal of the fluctuation in renewable power generation and production, since it exhibits the smallest deviation in the NLDC representation. The analysis indicate that the weeks with elevated error in the NLDC often lead to underestimated system costs, potentially due to irregular fluctuations caused by high shares of variable renewable input or reduced variability. Consequently, when these dynamics are incorporated into the model, operational behaviors, including load-following responses and battery energy transfer cycles, may not be fully captured, leading to undervaluation of system expenses.

Table 6.4 presents the overall energy output by each technology across various regions throughout the 10-year planned period for the designated representative weeks A. The Grijalva region (RA) generates the highest proportion of electrical energy at 71.4%, with 98.9% derived from hydroelectric sources. The Merida region (RE) generated 18.9% of the total electrical energy over the past decade, comprising 55.5% from combined cycle and 34.7% from wind sources. The Lerma region (RC) generates 3.8% of its energy, with 47.4% derived from wind sources and the remainder from conventional energy sources, including combined cycle (34.7%), cogeneration (13.1%), and turbogas (4.7%). The Tabasco region (RB) accounts for 3.3% of generation, exclusively with conventional technologies, comprising 65.9% from cogeneration, 27.9% from turbogas, and 6.2% from internal combustion. The RF, RD, and RG regions produced 1.6%, 0.6%, and 0.4% from wind power technology (52.1%, 56.7%, and 52.3%) and Turbogas at 47.2%, 43.3%, and 47.7%, respectively.

Table 6.5 illustrates the aggregate energy output by technology across various regions across the 10-year planning interval for the designated representative weeks B. The Grijalva region (RA) accounts for 71.1% of total energy production over the

Table 6.4 Energy generated over a decade in each transmission zone within weeks A

Subset A of Representative Weeks							
Generation technology	*Energy output across a 10-year horizon*						
	Grijalva region (RA)	Tabasco region (RB)	Lerma region (RC)	Chetumal region (RD)	Merida region (RE)	Cancun Region (RF)	Cozumel region (RG)
Turbogas	86,286	159,277	31,982	48,751	40,973	130,548	36,506
Combined cycle	–	–	234,301.95	–	1,842,557.45	–	–
Efficient cogeneration	−0	376,883.68	88,520.51	–	210,431.21	–	–
Internal combustion	749.48	35,635.85	–	–	9,070.93	–	–
Conventional thermoelectric	−0	–	–	–	714.33	2,157.87	–
Wind	54,701.58	–	319,768.66	63,953.73	1,151,167.18	144,109.08	39,971.08
Solar	−0	–	–	–	62,453.95		–
Hydroelectric	12,395,662.08	–	–	–	–	–	–
Total energy (MWh)	12,537,400.12	571,796.80	674,573.75	112,704.85	3,317,368.74	276,815.68	76,477.55

past decade, generating 12,649,704.9 MWh, with 98.9% derived from hydroelectric power, being the highest electrical generation volume. The Merida region (RE) is the second largest energy-producing area, generating 19.3% of total energy, with 57.1% derived from combined cycle and 33.5% from wind sources. The power output percentage from the Lerma area (RC) is reduced to 3.8%, comprising 47.3% from wind energy and the remainder generated by conventional technologies: Combined Cycle at 35.3%, Cogeneration at 12.7%, and Turbogas at 4.7%. The Tabasco region (RB) produced 3.2% solely by conventional methods, comprising 66.0% from cogeneration, 27.8% from turbogas, and 6.2% from internal combustion. The Cancun area (RF) has a generation of 1.6%, predominantly utilizing renewable technology at 52.1% (wind) and 47.9% from conventional technologies (turbogas and thermoelectric). The RD and RG regions produce 0.6% and 0.4% from wind technology (56.8% and 52.4%, respectively), with the remainder derived from Turbogas.

Table 6.6 presents the overall energy output by each technology across various regions throughout the 10-year planned period for the designated representative weeks C. The Grijalva region (RA) generates 70.4% of the total electricity throughout the 10-year planning horizon, amounting to 12,528,854.20 MWh, with 98.9% derived from hydroelectric sources. The Merida area (RE) follows with 3,531,655.37 MWh, representing 19.8%, of which 58.1% is derived from combined cycle and 32.6% from wind energy. In the remaining regions, the proportion of electricity generation declines substantially. In the Lerma area (RC), electricity generation constitutes 3.9%, comprising 46.1% from wind, 36.2% from combined cycle, 13% from cogeneration, and 4.6% from Turbogas. The Tabasco area (RB) follows, generating 3.3% only using conventional technologies, comprising 6.0% from internal combustion, 26.8% from Turbogas, and 67.1% from Cogeneration. The Cancun area (RF) has a generation of 1.6%, comprising 51.9% from wind, 47.4% from Turbogas, and 0.8% from Thermoelectric sources. The regions with the lowest generation are RD and RG, producing 0.6% and 0.4% respectively, with 56.6% and 52.3% derived from wind technology and the remainder from Turbogas. Appendix Tables 6.8, 6.9, 6.10, 6.11, 6.12, 6.13, 6.14, 6.15 and 6.16 present the annual power generation by region for each set of weeks.

Figure 6.5 illustrates the energy generated by the areas for their internal use, the power transmitted to other regions, and the curtailment experienced by the regions for each set of typical weeks. In set A, it is indicated that in year 1, the energy generated for internal demand is 45.3%, the energy sent to other areas is 53.3%, and the curtailment is 1.4%. In year 10, the percentage of curtailment decreases to 0.6%, while production for internal demand rises to 56.2% and energy transmitted increases to 60.8%. During the weeks categorized as B, year 1 exhibited 45.1% of production dedicated to internal need, 53.7% of transmitted energy, and 1.2% of curtailment. In contrast, year 10 showed 56.1% of production for internal demand, 61.0% of transmitted energy, and merely 0.6% of curtailment. In the weeks of set C, there is a notable enhancement in curtailment, decreasing from 1.1% in year 1 to 0.6% in year 10, thereby augmenting production for internal demand from 45.4% to 56.5% and transmitted energy from 53.5% to 60.6%. Figure 6.14 in the Appendix

Table 6.5 Energy generated throughout the decade in each transmission location within the specified weeks of set B

Subset B of Representative Weeks							
Generation technology	*Energy output across a 10-year horizon*						
	Grijalva region (RA)	Tabasco region (RB)	Lerma region (RC)	Chetumal region (RD)	Merida region (RE)	Cancun Region (RF)	Cozumel region (RG)
Turbogas	85,882.89	157,914.22	31,936.56	48,694.89	40,917.65	130,281.57	36,269.29
Combined cycle	–	–	238,497.13	–	1,965,501.20	–	–
Efficient cogeneration	–	374,327.75	86,196.16	–	209,736.59	–	–
Internal combustion	745.42	35,338.34	–	–	9,032.46	–	–
Conventional thermoelectric	–	–	–	–	712.32	2,150.40	–
Wind	54,701.58	–	319,768.66	63,953.73	1,151,167.18	144,109.08	39,971.08
Solar	–	–	–	–	62,453.95	–	–
Hydroelectric	12,508,375.04	–	–	–	–	–	–
Total energy (MWh)	12,649,704.93	567,580.30	676,398.51	112,648.62	3,439,521.34	276,541.05	76,240.37

Table 6.6 Energy generated throughout a decade in each transmission zone within the specified weeks of set C

Subset C of Representative Weeks							
Generation technology	*Energy output across a 10-year horizon*						
	Grijalva region (RA)	Tabasco region (RB)	Lerma region (RC)	Chetumal region (RD)	Merida region (RE)	Cancun Region (RF)	Cozumel region (RG)
Turbogas	85,696.51	157,205.29	32,064.96	49,034.44	41,143.03	131,595.13	36,495.59
Combined cycle	–	–	251,537.80	–	2,052,454.35	–	–
Efficient cogeneration	–	393,311.33	90,537.88	–	214,558.29	–	–
Internal combustion	741.87	35,231.62	–	–	9,164.98	–	–
Conventional thermoelectric	–	–	–	–	713.59	2,154.56	–
Wind	54,701.58	–	319,768.66	63,953.73	1,151,167.18	144,109.08	39,971.08
Solar	–	–	–	–	62,453.95	–	–
Hydroelectric	12,387,714.25	–	–	–	–	–	–
Total energy (MWh)	12,528,854.20	585,748.24	693,909.30	112,988.17	3,531,655.37	277,858.77	76,466.67

section illustrates the generating structure of total electricity output and production for internal demand over the past decade.

The RA, RC, and RE regions are the sole areas that convey energy to other regions across the three sets of representative weeks. In set A, the Grijalva area (RA) transmits 7,944,528.00 MWh, accounting for 85.5% of the total electricity sent over the past decade; the Merida region (RE) transmits 1,262,335.98 MWh, representing 13.6%; and the Lerma region (RC) sends merely 1%. In set B, the Grijalva area (RA) transmits 8,056,836.04 MWh, accounting for 85.1% of the total transmitted energy; the Merida region (RE) transmits 1,327,829.82 MWh, representing 14.0%; and the Lerma region (RC) transmits 0.9%. In set C, the Grijalva area (RA) transmits 7,935,991.58 MWh, constituting 84.3% of the total transmitted energy; the Merida region (RE) transmits 1,393,788.30 MWh, representing 14.8%; and the Lerma region (RC) sends a mere 0.8%. Figure 6.15 in the Appendix illustrates the progression of energy transmitted across each location for every set of weeks.

The RC, RE, RD, and RG zones are the only locations subjected to curtailment across the three typical week groups. For set A, the Merida subregion (RE) presents an unutilized energy potential of 141,518 MWh, representing 92% of the total. The Cozumel subregion (RG) contributes 11,948 MWh (8%), while the RC and RD subregions contribute 0.2% and 0.4%, respectively. In set B, RE retains 134,396 MWh of unexploited energy, comprising 92%; RG provides 10,535 MWh (7%), with RC and RD contributing marginally (0.2% and 0.9%). In set C, RE again dominates with 129,812 MWh (91%), followed by RG at 10,689 MWh (8%), while RC and RD contribute 0.7% and 0.6%, respectively. Among all sets, A shows the greatest quantity of unused energy (1.54×10^5 MWh), followed by B (1.46×10^5 MWh), and C with the smallest value (1.42×10^5 MWh). Figure 6.16 in the Appendix displays how untapped energy evolves across zones in each representative week.

The Grijalva area (RA) was the sole area where energy storage systems were implemented; Table 6.7 presents the cumulative stored power throughout the ten-year planning horizon. Set A of the representative week possesses the most energy storage in ST-1 and ST-2 systems, whilst set C exhibits the least quantity stored. The ST-1 units have an installed capacity of 6 MWh in years 9 and 10, whereas the ST-2 units possess an installed capacity of 6.6 MWh in year 9 and 7.0 MWh in year 10 over all three sets of weeks. The procurement expenses of the storage equipment amount to $\$1.31 \times 10^7$, whereas the operation costs total $\$3.6112 \times 10^7$; these values remain unchanged for sets A, B, and C due to uniform installed capacities.

Today's energy landscape must respond to pressing global challenges, including rising environmental degradation, climate instability, and the depletion of non-renewable resources. This reality highlights the urgency of moving toward a resilient and low-emission energy paradigm. Adapting generation expansion strategies to be more responsive is fundamental, particularly under scenarios with substantial integration of variable renewables. This is especially relevant due to the fluctuating output of technologies like solar and wind. Therefore, addressing these dynamics is a priority in modern power system development and planning.

The selected representative weeks must accurately reflect the NLDC characteristics across the five analyzed zones of the power network. To simulate both common

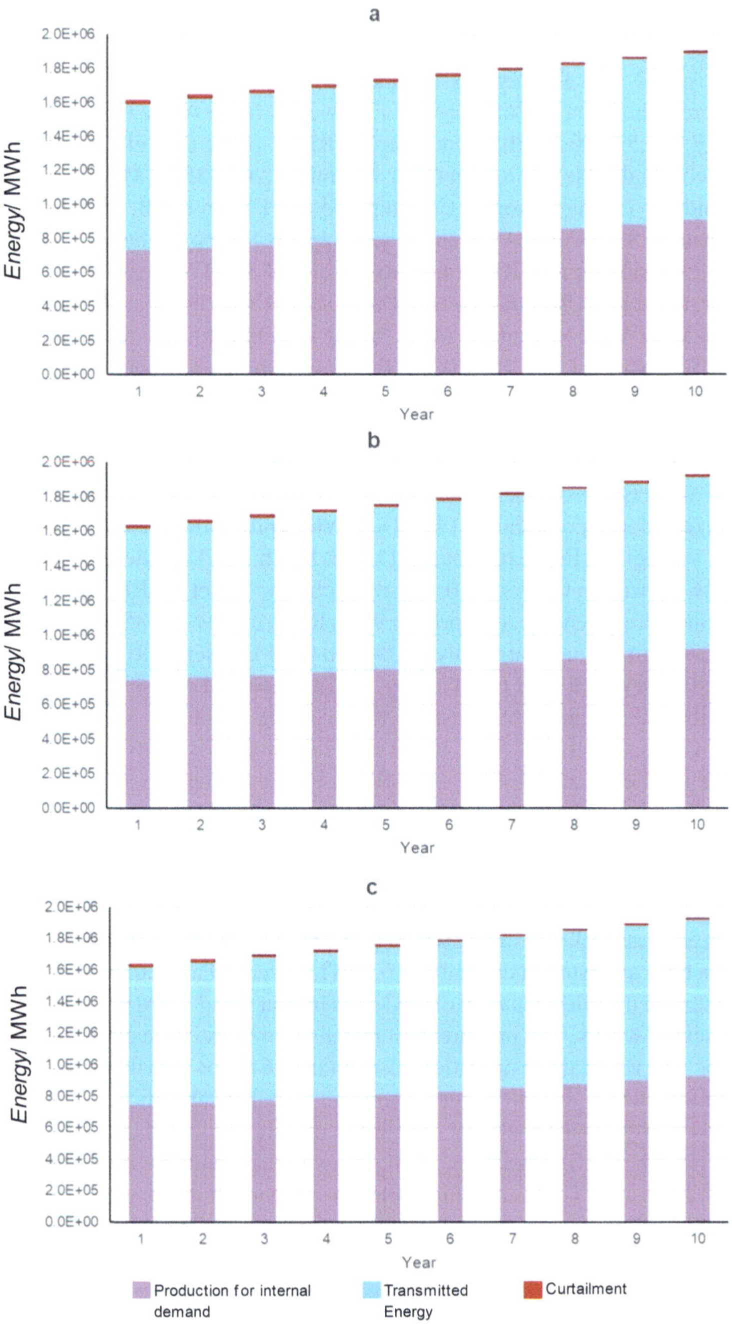

Fig. 6.5 Energy generated, transmitted, and curtailed throughout a decade for each representative week specified

Table 6.7 Total electricity storage for a decade in the RA

RA			
Storage units	Cumulative energy storage over a 10-year period (MWh)		
	Weeks A	Weeks B	Weeks C
ST-1	3,715	3,596	3,507
ST-2	5,053	4,974	4,335
Total	8,769	8,571	7,843

and uncommon load conditions, along with renewable supply patterns, it is essential to incorporate their occurrence probability. This temporal adjustment progressively streamlines the planning model by minimizing its dimensionality and computational burden. Notably, the values of the objective function are influenced by the sampling method, affecting both operational strategies and infrastructure investment outcomes.

6.7 Conclusions

Addressing temporally dynamic processes within complex energy system design requires substantial computational resources, which limits the practicality of applying mixed-integer linear programming models. To manage this, selected time blocks, termed representative intervals, are introduced to reduce computational burden through temporal discretization. This work explores three distinct methods for identifying representative weeks, using the fidelity of the NLDC across five regional subsystems as the evaluation metric. For each selected time set, a 2% annual increase was assumed over a decade and incorporated into a planning framework that integrates operational and design restrictions. This modeling approach facilitates the assessment of system performance, where the objective function reflects the total system cost. Three different strategies were applied to determine representative weeks: one based on k-means clustering with Euclidean distance, another using k-means with DTW, and a third involving a full combinatorial enumeration. Among these, the combinatorial approach produced the lowest RMSE (2.80), outperforming the DTW-based k-means (3.21) and the Euclidean-based k-means (5.49). Regarding the performance within the optimization framework, clustering-based techniques using Euclidean and DTW distances yielded more cost-effective outcomes when compared to the combinatorial strategy, which captures a wider range of temporal variability in the system.

Nevertheless, the use of k-means with the DTW distance metric requires roughly 17 times more processing effort to generate the representative time segments than the combinatorial strategy, and approximately 70 times more than when using k-means with the Euclidean distance. This increase is attributed to the lack of limitations on allowed temporal distortion. On the other hand, k-means based on Euclidean distance tends to form more robust groupings; in fact, each run with different initial centroid positions generally leads to consistent representative selections, unlike its DTW-based counterpart.

This study seeks to ensure an accurate assessment of the attributes inherent to a multi-area interconnected power network, considering factors like the aggregation of various geographical zones. The proposed methodology enables the identification of representative weeks, thereby decreasing the computational effort required for extended-horizon planning tasks associated with operational strategies in energy systems.

The selected representative weeks must align with the unique net load duration patterns specific to each of the five analyzed areas of the electrical grid. This ensures the accurate reproduction of both common and infrequent demand profiles, alongside the availability patterns of wind and solar resources. The main goal is to perform a thorough assessment of the structural traits of a geographically distributed energy network, considering the aggregation behavior of multiple zones. This method supports the identification of critical weeks, thereby reducing the computational burden associated with extended-horizon energy planning models and operational strategy analysis. It enhances forecasting precision and supports the development of flexible capacity deployment. Ultimately, it enables identifying the optimal balance between renewable and conventional energy technologies across each grid region evaluated.

6.8 Appendix

See Figs. 6.6, 6.7, 6.8, 6.9, 6.10, 6.11, 6.12, 6.13, 6.14, 6.15, 6.16 and Tables 6.8, 6.9, 6.10, 6.11, 6.12, 6.13, 6.14, 6.15, 6.16.

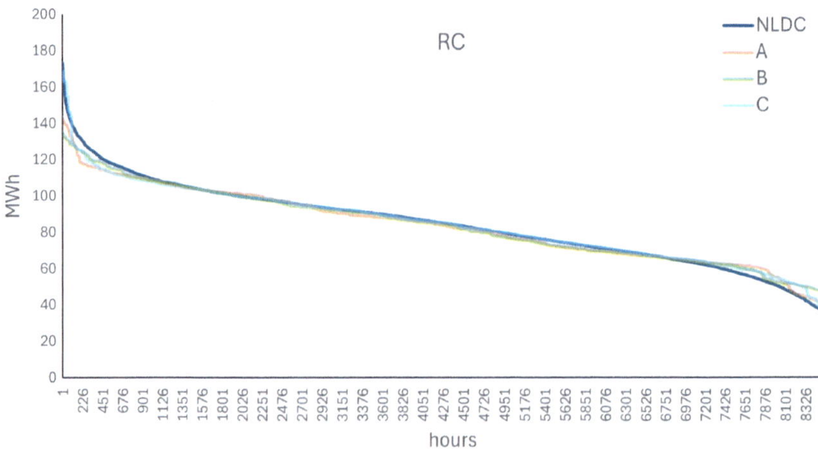

Fig. 6.6 The net load duration curve (NLDC) and its approximations using methods A, B, and C in the Lerma region (RC)

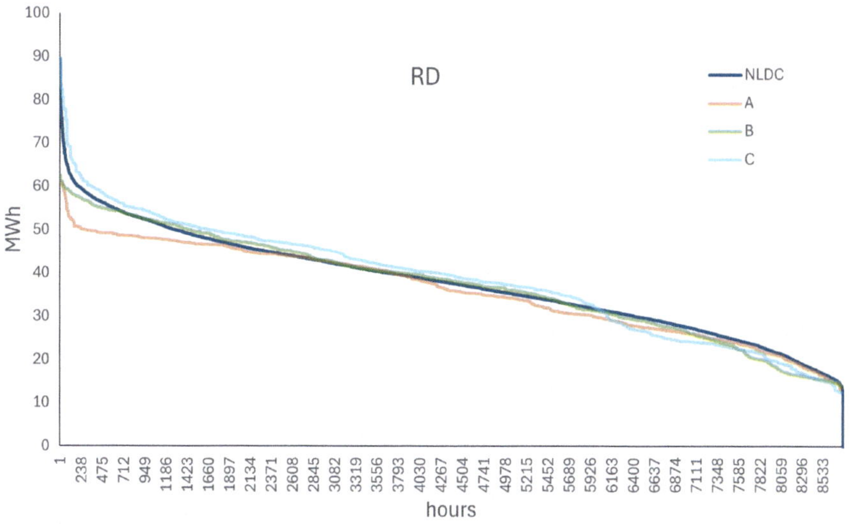

Fig. 6.7 The net load duration curve (NLDC) and its approximations using methods A, B, and C in the Chetumal area (RD)

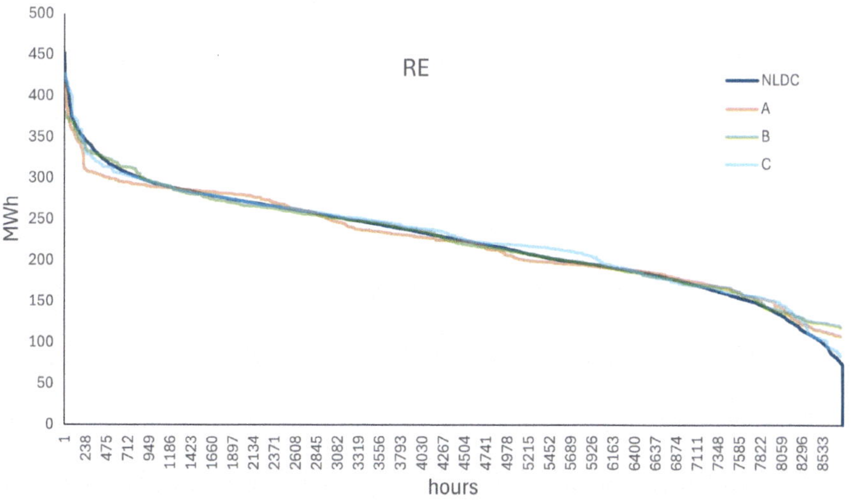

Fig. 6.8 The net load duration curve (NLDC) and its approximations using techniques A, B, and C in the Merida region (RE)

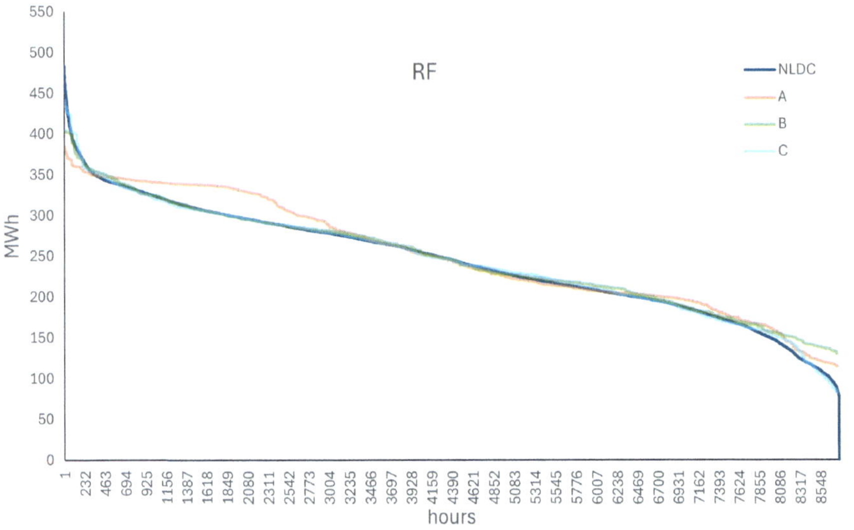

Fig. 6.9 The net load duration curve (NLDC) and its approximations using techniques A, B, and C in the Cancun region (RF)

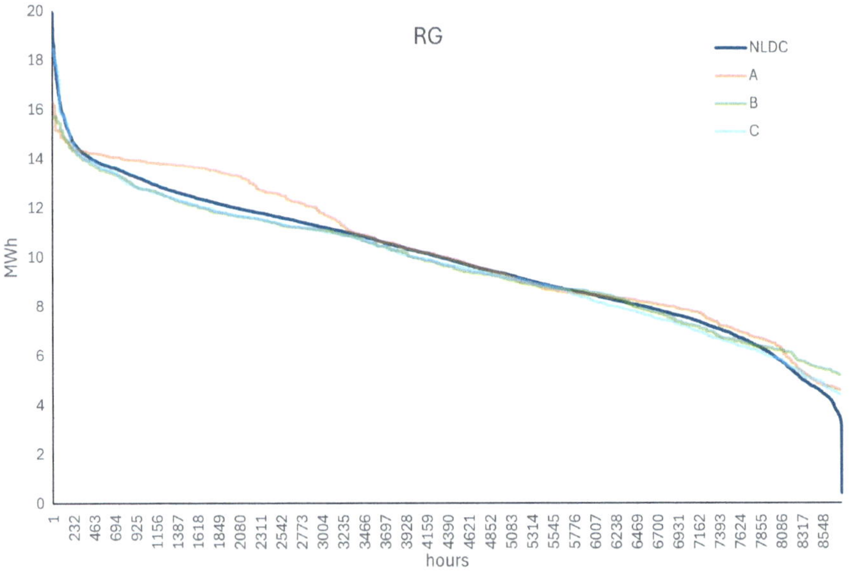

Fig. 6.10 The net load duration curve (NLDC) and its approximations using techniques A, B, and C in the Cozumel region (RG)

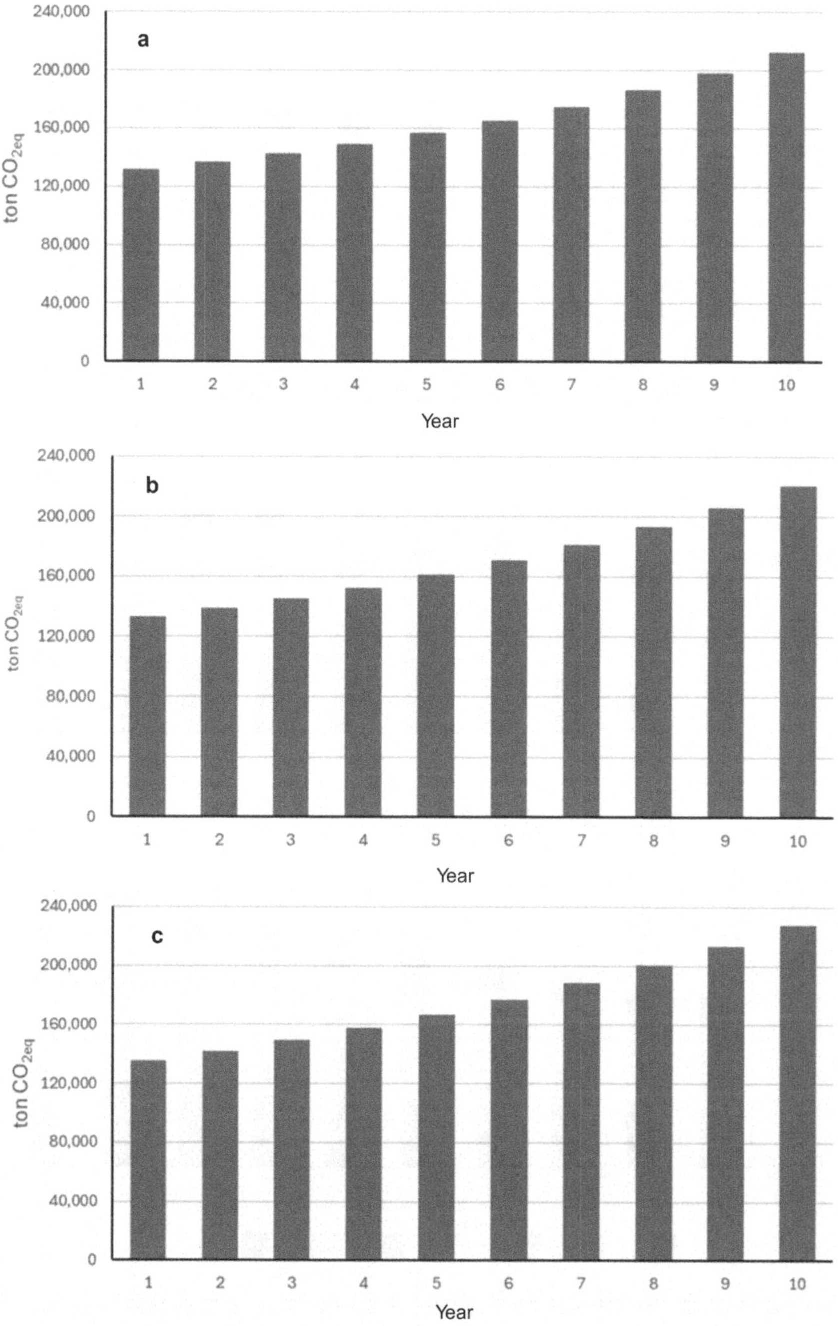

Fig. 6.11 Progression of emissions throughout each planning year in the designated weeks A, B, and C

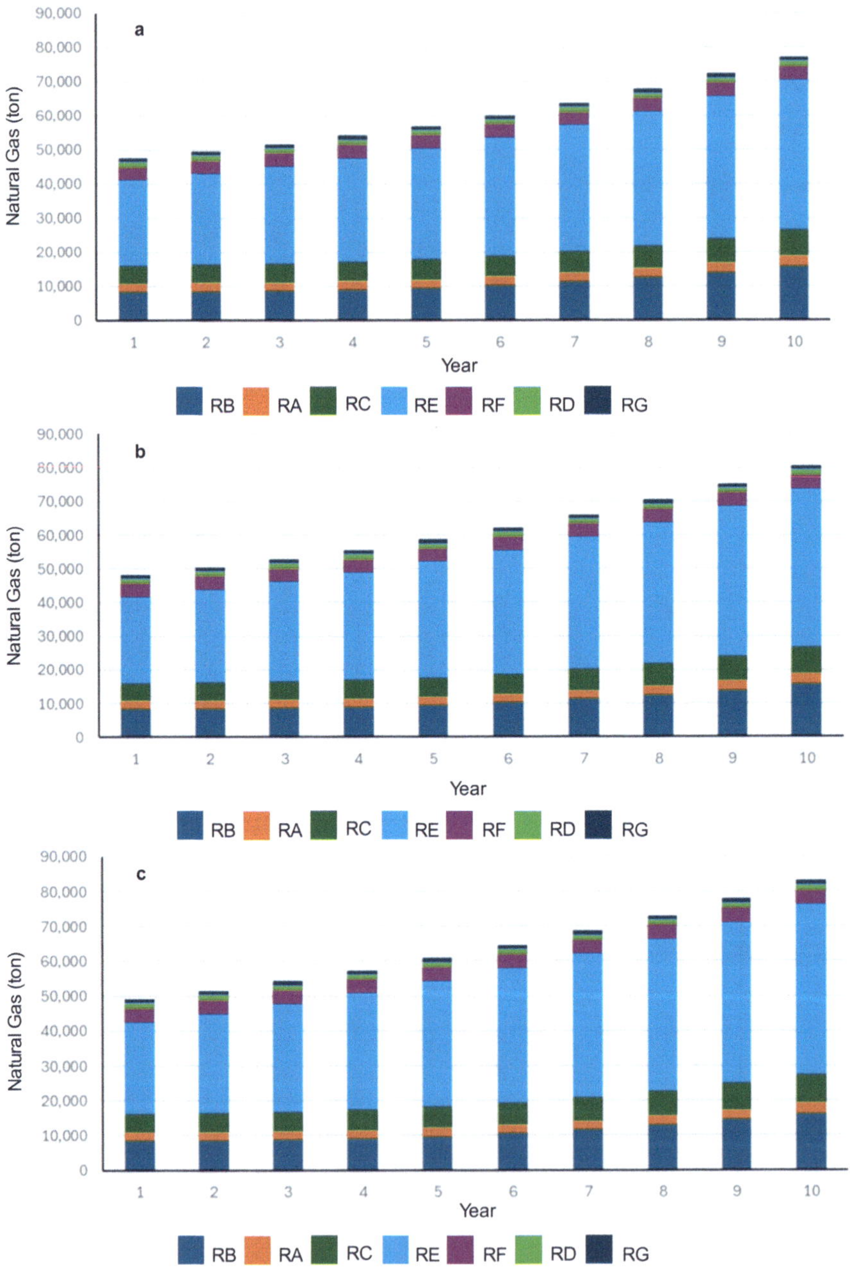

Fig. 6.12 Progression of natural gas utilization at each node during the weeks A, B, and C

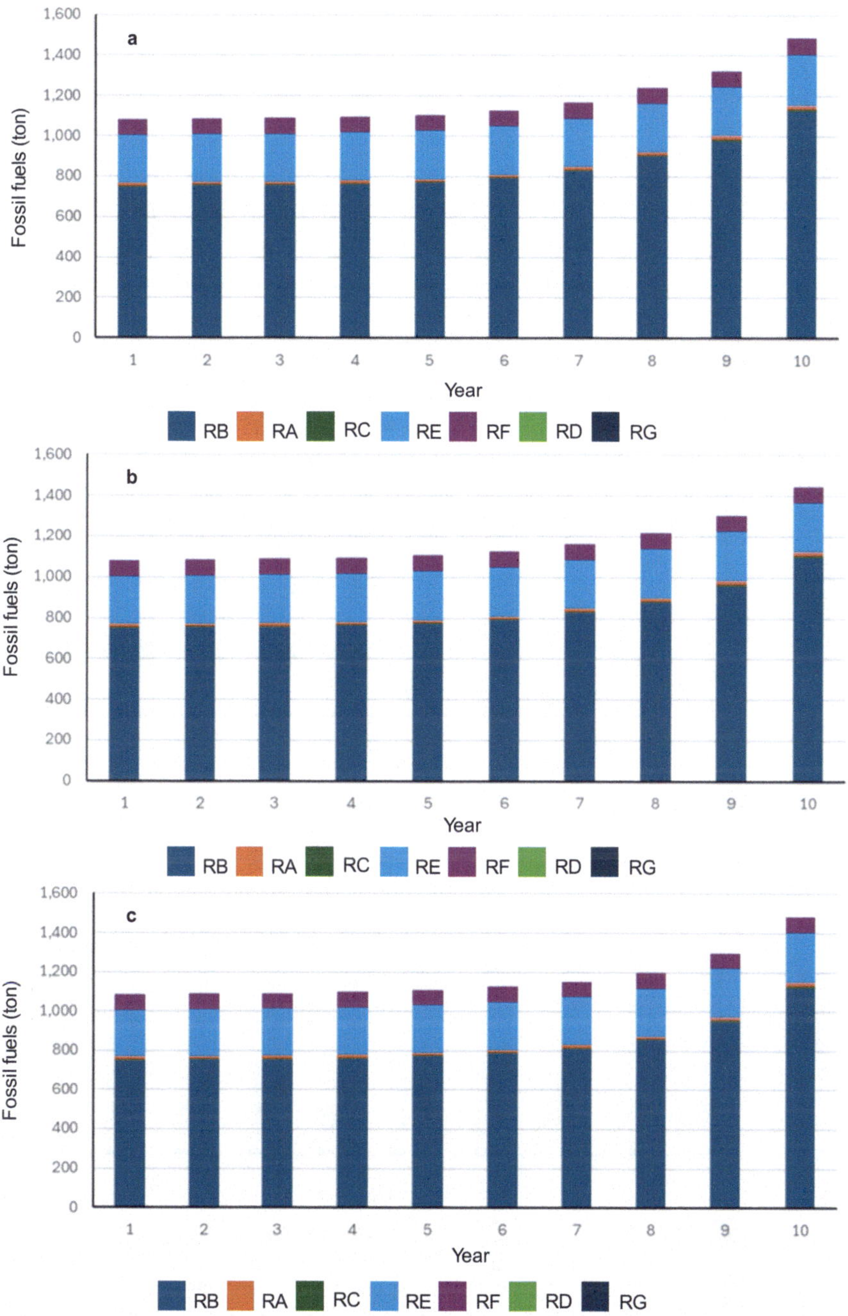

Fig. 6.13 Progression of fuel oil utilization at each node for the designated weeks A, B, and C

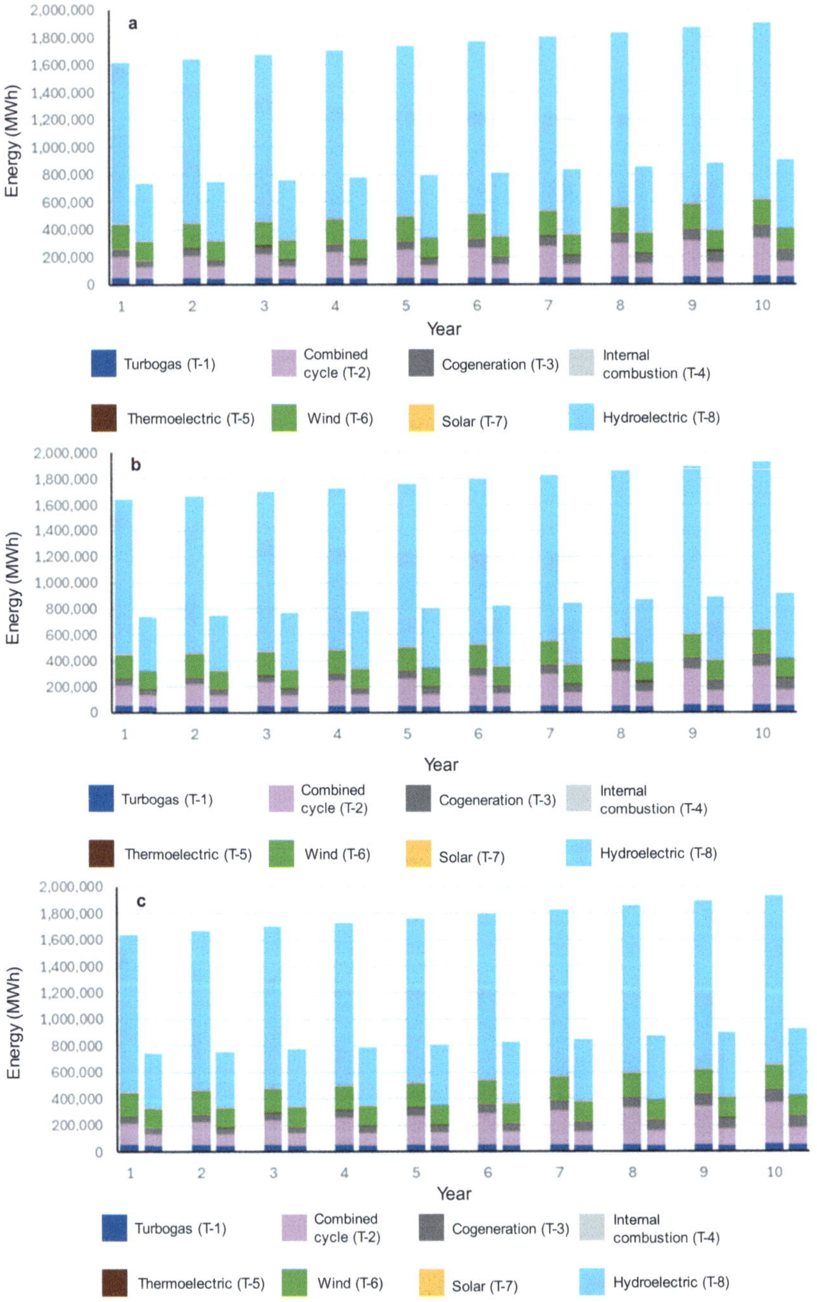

Fig. 6.14 The initial column illustrates the comprehensive generation framework of all regions. The second column illustrates the entire generation structure of the regions to meet their internal demand. The disparity between the columns represents the energy conveyed to other areas

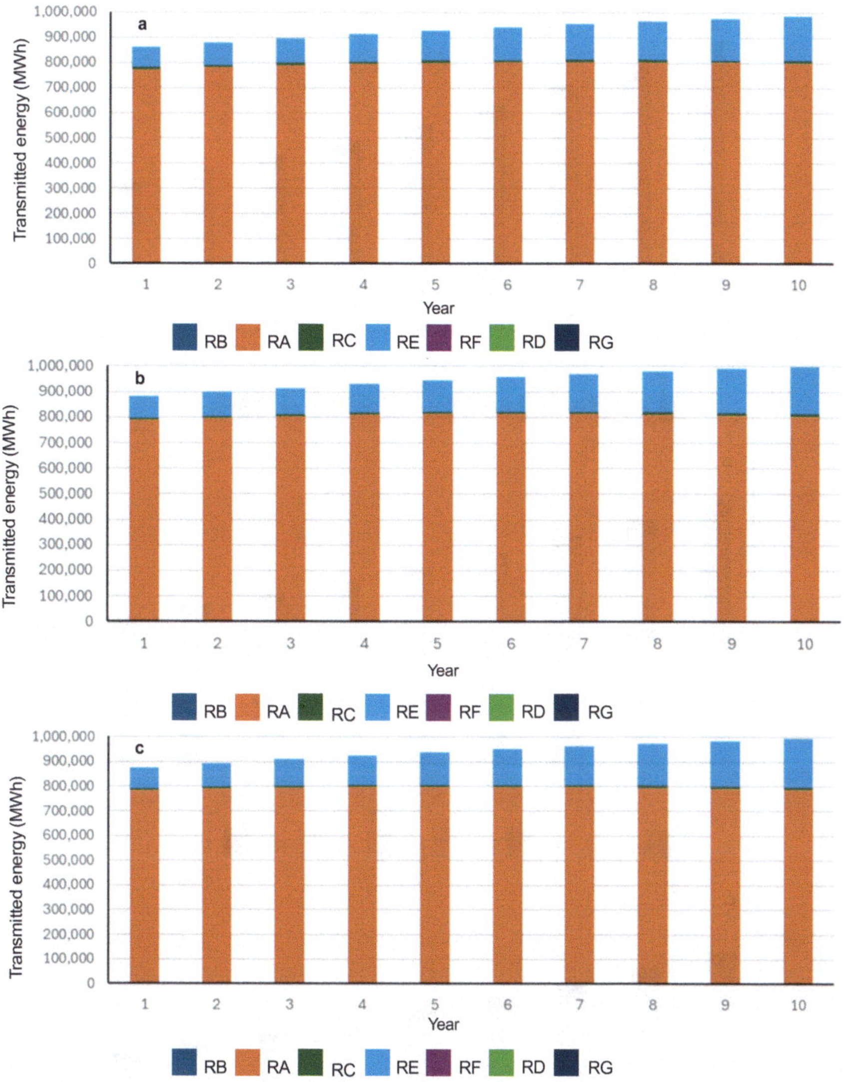

Fig. 6.15 Electricity transferred from one location to another

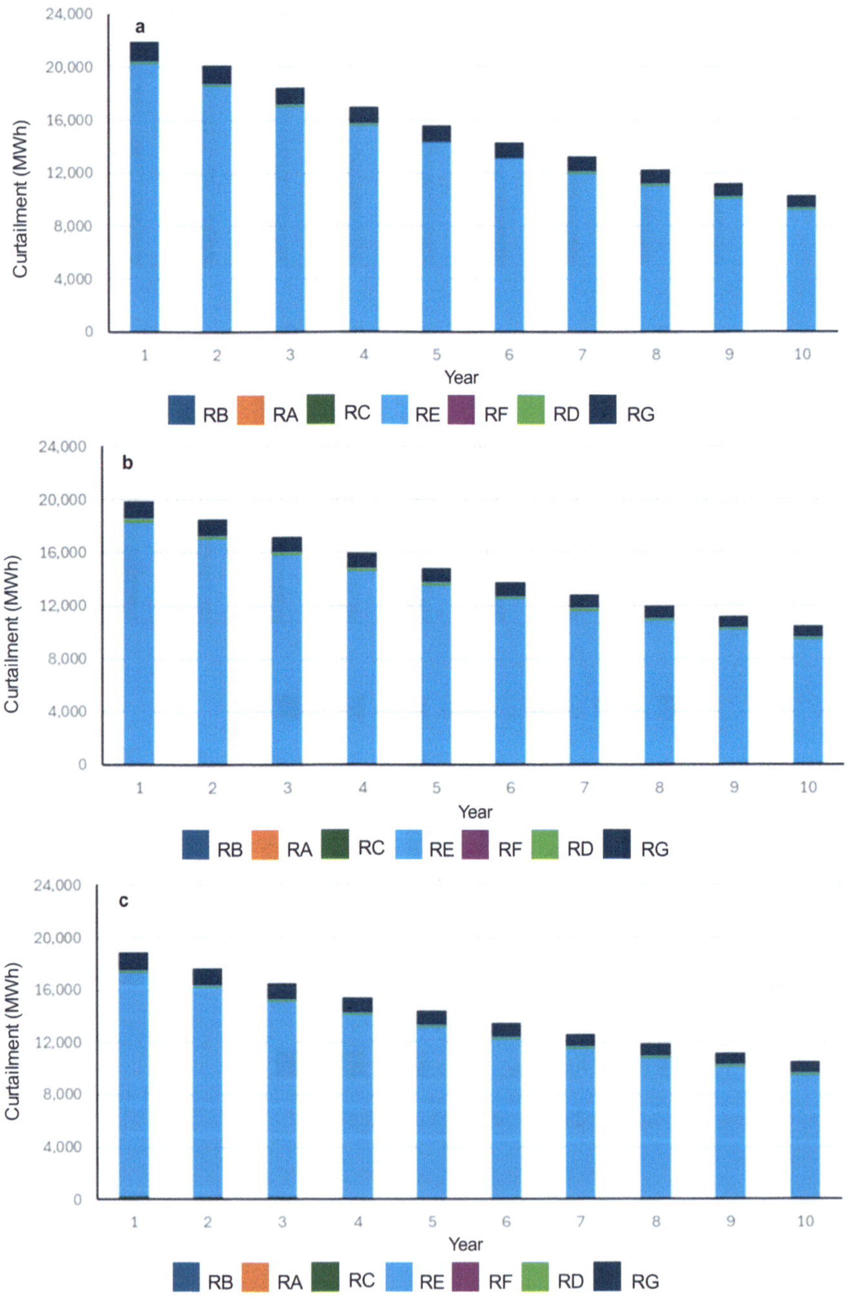

Fig. 6.16 Curtailment of each region

Table 6.8 Electricity production in areas RB, RA, and RC during the representative weeks of set A

Year1	Generation technology								
	Turbogas	Combined cycle	Efficient cogeneration	Internal combustion	Conventional thermoelectric	Wind	Solar	Hydroelectric	Total energy (MWh)
Tabasco region (RB)									
	14,582.4	0.0	25,530.9	3,172.5	0.0	0.0	0.0	0.0	43,285.8
2	14,588.0	0.0	25,875.6	3,193.0	0.0	0.0	0.0	0.0	43,656.7
3	14,620.8	0.0	27,078.9	3,198.4	0.0	0.0	0.0	0.0	44,898.1
4	14,714.9	0.0	28,726.7	3,223.5	0.0	0.0	0.0	0.0	46,665.1
5	14,885.9	0.0	31,453.9	3,255.0	0.0	0.0	0.0	0.0	49,594.8
6	15,169.2	0.0	35,780.1	3,348.7	0.0	0.0	0.0	0.0	54,298.0
7	15,619.6	0.0	41,158.1	3,508.2	0.0	0.0	0.0	0.0	60,285.8
8	16,624.9	0.0	47,309.4	3,811.1	0.0	0.0	0.0	0.0	67,745.4
9	18,048.2	0.0	53,510.8	4,145.9	0.0	0.0	0.0	0.0	75,704.9
10	20,423.4	0.0	60,459.4	4,779.5	0.0	0.0	0.0	0.0	85,662.2
Grijalva region (RA)									
1	8,131.2	0.0	0.0	67.2	0.0	5,470.2	0.0	1,176,916.9	1,190,585.5
2	8,131.2	0.0	0.0	67.5	0.0	5,470.2	0.0	1,194,636.5	1,208,305.4
3	8,143.3	0.0	0.0	67.8	0.0	5,470.2	0.0	1,211,009.6	1,224,690.9
4	8,145.6	0.0	0.0	68.0	0.0	5,470.2	0.0	1,226,016.5	1,239,700.4
5	8,168.4	0.0	0.0	69.2	0.0	5,470.2	0.0	1,239,675.8	1,253,383.6
6	8,287.4	0.0	0.0	70.8	0.0	5,470.2	0.0	1,251,662.1	1,265,490.4
7	8,467.4	0.0	0.0	74.1	0.0	5,470.2	0.0	1,262,100.5	1,276,112.1
8	8,796.2	0.0	0.0	79.0	0.0	5,470.2	0.0	1,270,775.2	1,285,120.6

(continued)

Table 6.8 (continued)

Tabasco region (RB)									
9	9,526.3	0.0	0.0	86.2	0.0	5,470.2	0.0	1,278,465.1	1,293,547.7
10	10,490.0	0.0	0.0	99.7	0.0	5,470.2	0.0	1,284,403.7	1,300,463.5
Lerma region (RC)									
1	3,158.4	18,394.7	7,926.6	0.0	0.0	31,976.9	0.0	0.0	61,456.6
2	3,176.2	18,940.8	8,090.0	0.0	0.0	31,976.9	0.0	0.0	62,183.9
3	3,186.0	19,702.8	8,235.5	0.0	0.0	31,976.9	0.0	0.0	63,101.2
4	3,186.1	20,810.8	8,373.4	0.0	0.0	31,976.9	0.0	0.0	64,347.1
5	3,186.6	22,020.8	8,566.5	0.0	0.0	31,976.9	0.0	0.0	65,750.8
6	3,191.0	23,083.9	8,927.2	0.0	0.0	31,976.9	0.0	0.0	67,179.0
7	3,203.9	24,842.6	9,143.6	0.0	0.0	31,976.9	0.0	0.0	69,167.0
8	3,211.1	26,743.6	9,440.0	0.0	0.0	31,976.9	0.0	0.0	71,371.5
9	3,230.9	28,489.9	9,692.5	0.0	0.0	31,976.9	0.0	0.0	73,390.2
10	3,252.4	31,272.0	10,125.1	0.0	0.0	31,976.9	0.0	0.0	76,626.4

Table 6.9 Electricity production in regions RE, RF, and RD during the representative weeks of set A

Merida region (RE)									
Year	Generation technology								
	Turbogas	Combined cycle	Efficient cogeneration	Internal combustion	Conventional thermoelectric	Wind	Solar	Hydroelectric	Total energy (MWh)
1	4,054.0	131,730.2	17,821.1	894.2	71.2	115,116.7	6,245.4	0.0	275,932.8
2	4,074.0	141,284.1	18,636.0	894.2	71.2	115,116.7	6,245.4	0.0	286,321.6
3	4,074.0	151,874.7	19,306.5	895.5	71.2	115,116.7	6,245.4	0.0	297,584.0
4	4,069.0	163,979.0	19,561.7	896.8	71.3	115,116.7	6,245.4	0.0	309,939.9
5	4,079.6	176,389.2	20,127.8	898.1	71.3	115,116.7	6,245.4	0.0	322,928.1
6	4,086.5	189,118.9	20,877.4	900.7	71.4	115,116.7	6,245.4	0.0	336,417.0
7	4,108.3	202,095.6	21,636.1	904.7	71.6	115,116.7	6,245.4	0.0	350,178.3
8	4,132.4	215,702.2	22,370.0	911.2	71.6	115,116.7	6,245.4	0.0	364,549.5
9	4,128.2	228,908.8	24,384.8	921.1	71.7	115,116.7	6,245.4	0.0	379,776.7
10	4,167.7	241,474.9	25,709.8	954.6	71.8	115,116.7	6,245.4	0.0	393,740.9
Cancun region (RF)									
1	12,892.3	0.0	0.0	0.0	215.0	14,410.9	0.0	0.0	27,518.2
2	12,922.6	0.0	0.0	0.0	215.0	14,410.9	0.0	0.0	27,548.5
3	12,943.5	0.0	0.0	0.0	215.0	14,410.9	0.0	0.0	27,569.5
4	12,962.7	0.0	0.0	0.0	215.4	14,410.9	0.0	0.0	27,588.9
5	12,986.3	0.0	0.0	0.0	215.4	14,410.9	0.0	0.0	27,612.6
6	13,000.6	0.0	0.0	0.0	215.7	14,410.9	0.0	0.0	27,627.2
7	13,066.3	0.0	0.0	0.0	216.0	14,410.9	0.0	0.0	27,693.2
8	13,129.5	0.0	0.0	0.0	216.4	14,410.9	0.0	0.0	27,756.9

(continued)

Table 6.9 (continued)

Merida region (RE)									
9	13,222.7	0.0	0.0	0.0	216.3	14,410.9	0.0	0.0	27,849.9
10	13,422.3	0.0	0.0	0.0	217.6	14,410.9	0.0	0.0	28,050.8
Chetumal region (RD)									
1	4,816.4	0.0	0.0	0.0	0.0	6,395.4	0.0	0.0	11,211.8
2	4,834.5	0.0	0.0	0.0	0.0	6,395.4	0.0	0.0	11,229.9
3	4,837.2	0.0	0.0	0.0	0.0	6,395.4	0.0	0.0	11,232.6
4	4,844.2	0.0	0.0	0.0	0.0	6,395.4	0.0	0.0	11,239.5
5	4,857.2	0.0	0.0	0.0	0.0	6,395.4	0.0	0.0	11,252.6
6	4,862.2	0.0	0.0	0.0	0.0	6,395.4	0.0	0.0	11,257.6
7	4,884.3	0.0	0.0	0.0	0.0	6,395.4	0.0	0.0	11,279.7
8	4,912.8	0.0	0.0	0.0	0.0	6,395.4	0.0	0.0	11,308.2
9	4,927.0	0.0	0.0	0.0	0.0	6,395.4	0.0	0.0	11,322.3
10	4,975.4	0.0	0.0	0.0	0.0	6,395.4	0.0	0.0	11,370.8

Table 6.10 Electricity production in region RG during the representative weeks of set A

Cozumel region (RG)									
Year	Generation technology								
	Turbogas	Combined cycle	Efficient cogeneration	Internal combustion	Conventional thermoelectric	Wind	Solar	Hydroelectric	Total energy (MWh)
1	3,597.9	0.0	0.0	0.0	0.0	3,997.1	0.0	0.0	7,595.0
2	3,598.9	0.0	0.0	0.0	0.0	3,997.1	0.0	0.0	7,596.0
3	3,605.1	0.0	0.0	0.0	0.0	3,997.1	0.0	0.0	7,602.2
4	3,611.2	0.0	0.0	0.0	0.0	3,997.1	0.0	0.0	7,608.3
5	3,623.3	0.0	0.0	0.0	0.0	3,997.1	0.0	0.0	7,620.4
6	3,631.3	0.0	0.0	0.0	0.0	3,997.1	0.0	0.0	7,628.4
7	3,658.9	0.0	0.0	0.0	0.0	3,997.1	0.0	0.0	7,656.0
8	3,668.0	0.0	0.0	0.0	0.0	3,997.1	0.0	0.0	7,665.1
9	3,710.6	0.0	0.0	0.0	0.0	3,997.1	0.0	0.0	7,707.7
10	3,801.3	0.0	0.0	0.0	0.0	3,997.1	0.0	0.0	7,798.4

Table 6.11 Electricity production in regions RB, RA, and RC for the designated representative weeks B

Year	Generation technology								
	Turbogas	Combined cycle	Efficient cogeneration	Internal combustion	Conventional thermoelectric	Wind	Solar	Hydroelectric	Total energy (MWh)
Tabasco region (RB)									
1	14,582.4	0.0	25,287.3	3,176.6	0.0	0.0	0.0	0.0	43,046.3
2	14,599.2	0.0	25,698.8	3,192.9	0.0	0.0	0.0	0.0	43,490.9
3	14,632.9	0.0	26,686.8	3,198.4	0.0	0.0	0.0	0.0	44,518.2
4	14,702.3	0.0	28,170.7	3,223.3	0.0	0.0	0.0	0.0	46,096.3
5	14,803.4	0.0	30,996.8	3,263.0	0.0	0.0	0.0	0.0	49,063.2
6	15,006.4	0.0	35,436.6	3,338.4	0.0	0.0	0.0	0.0	53,781.5
7	15,458.2	0.0	41,152.1	3,497.3	0.0	0.0	0.0	0.0	60,107.6
8	16,477.8	0.0	46,708.9	3,718.4	0.0	0.0	0.0	0.0	66,905.0
9	17,655.4	0.0	53,423.1	4,068.3	0.0	0.0	0.0	0.0	75,146.7
10	19,996.3	0.0	60,766.6	4,661.8	0.0	0.0	0.0	0.0	85,424.6
Grijalva region (RA)									
1	8,131.2	0.0	0.0	67.3	0.0	5,470.2	0.0	1,193,401.7	1,207,070.3
2	8,131.2	0.0	0.0	67.7	0.0	5,470.2	0.0	1,210,050.8	1,223,719.8
3	8,131.2	0.0	0.0	67.8	0.0	5,470.2	0.0	1,225,897.6	1,239,566.7
4	8,140.2	0.0	0.0	68.4	0.0	5,470.2	0.0	1,240,200.9	1,253,879.7
5	8,182.5	0.0	0.0	69.2	0.0	5,470.2	0.0	1,252,348.3	1,266,070.1
6	8,229.6	0.0	0.0	70.6	0.0	5,470.2	0.0	1,262,744.0	1,276,514.4
7	8,395.3	0.0	0.0	73.5	0.0	5,470.2	0.0	1,271,580.2	1,285,519.1
8	8,780.1	0.0	0.0	78.1	0.0	5,470.2	0.0	1,278,641.9	1,292,970.3

(continued)

Table 6.11 (continued)

Tabasco region (RB)									
Year	Generation technology								
	Turbogas	Combined cycle	Efficient cogeneration	Internal combustion	Conventional thermoelectric	Wind	Solar	Hydroelectric	Total energy (MWh)
9	9,416.3	0.0	0.0	85.3	0.0	5,470.2	0.0	1,284,576.8	1,299,548.6
10	10,345.2	0.0	0.0	97.5	0.0	5,470.2	0.0	1,288,932.9	1,304,845.8
Lerma region (RC)									
1	3,163.1	18,204.1	7,806.2	0.0	0.0	31,976.9	0.0	0.0	61,150.3
2	3,177.2	18,571.3	7,961.0	0.0	0.0	31,976.9	0.0	0.0	61,686.3
3	3,177.2	19,252.7	8,092.8	0.0	0.0	31,976.9	0.0	0.0	62,499.5
4	3,186.6	20,309.8	8,212.2	0.0	0.0	31,976.9	0.0	0.0	63,685.5
5	3,196.0	21,600.4	8,354.0	0.0	0.0	31,976.9	0.0	0.0	65,127.3
6	3,200.7	23,467.2	8,377.7	0.0	0.0	31,976.9	0.0	0.0	67,022.5
7	3,200.7	25,431.1	8,664.4	0.0	0.0	31,976.9	0.0	0.0	69,273.1
8	3,200.7	28,003.2	9,049.5	0.0	0.0	31,976.9	0.0	0.0	72,230.2
9	3,214.8	30,565.6	9,529.0	0.0	0.0	31,976.9	0.0	0.0	75,286.3
10	3,219.5	33,091.7	10,149.4	0.0	0.0	31,976.9	0.0	0.0	78,437.6

Table 6.12 Electricity production in regions RE, RF, and RD during the representative weeks of set B

Merida region (RE)									
Year	Generation technology								
	Turbogas	Combined cycle	Efficient cogeneration	Internal combustion	Conventional thermoelectric	Wind	Solar	Hydroelectric	Total energy (MWh)
1	4,062.0	136,774.6	17,543.7	890.2	71.2	115,116.7	6,245.4	0.0	280,703.9
2	4,068.0	148,584.5	17,953.5	891.5	71.2	115,116.7	6,245.4	0.0	292,930.9
3	4,068.0	160,990.1	18,344.7	892.8	71.2	115,116.7	6,245.4	0.0	305,729.0
4	4,079.7	174,075.8	19,058.2	901.0	71.2	115,116.7	6,245.4	0.0	319,548.0
5	4,083.1	188,064.6	19,937.3	906.0	71.2	115,116.7	6,245.4	0.0	334,424.3
6	4,111.8	201,875.8	21,062.6	909.9	71.2	115,116.7	6,245.4	0.0	349,393.5
7	4,113.1	216,087.8	22,147.9	908.6	71.2	115,116.7	6,245.4	0.0	364,690.7
8	4,108.1	231,311.8	23,262.7	909.9	71.2	115,116.7	6,245.4	0.0	381,025.8
9	4,110.0	246,299.5	24,714.9	910.0	71.2	115,116.7	6,245.4	0.0	397,467.7
10	4,113.9	261,436.7	25,711.0	912.5	71.2	115,116.7	6,245.4	0.0	413,607.5
Cancun region (RF)									
1	12,878.2	0.0	0.0	0.0	215.0	14,410.9	0.0	0.0	27,504.2
2	12,901.0	0.0	0.0	0.0	215.0	14,410.9	0.0	0.0	27,526.9
3	12,904.7	0.0	0.0	0.0	215.0	14,410.9	0.0	0.0	27,530.6
4	12,976.5	0.0	0.0	0.0	215.0	14,410.9	0.0	0.0	27,602.5
5	13,015.9	0.0	0.0	0.0	215.0	14,410.9	0.0	0.0	27,641.8
6	13,103.1	0.0	0.0	0.0	215.0	14,410.9	0.0	0.0	27,729.1
7	13,125.5	0.0	0.0	0.0	215.0	14,410.9	0.0	0.0	27,751.4
8	13,115.0	0.0	0.0	0.0	215.0	14,410.9	0.0	0.0	27,741.0

(continued)

Table 6.12 (continued)

Merida region (RE)									
9	13,123.9	0.0	0.0	0.0	215.0	14,410.9	0.0	0.0	27,749.9
10	13,137.7	0.0	0.0	0.0	215.0	14,410.9	0.0	0.0	27,763.6
Chetumal region (RD)									
1	4,820.2	0.0	0.0	0.0	0.0	6,395.4	0.0	0.0	11,215.6
2	4,828.6	0.0	0.0	0.0	0.0	6,395.4	0.0	0.0	11,224.0
3	4,833.9	0.0	0.0	0.0	0.0	6,395.4	0.0	0.0	11,229.3
4	4,848.7	0.0	0.0	0.0	0.0	6,395.4	0.0	0.0	11,244.1
5	4,853.6	0.0	0.0	0.0	0.0	6,395.4	0.0	0.0	11,249.0
6	4,896.3	0.0	0.0	0.0	0.0	6,395.4	0.0	0.0	11,291.7
7	4,907.7	0.0	0.0	0.0	0.0	6,395.4	0.0	0.0	11,303.1
8	4,901.9	0.0	0.0	0.0	0.0	6,395.4	0.0	0.0	11,297.2
9	4,902.3	0.0	0.0	0.0	0.0	6,395.4	0.0	0.0	11,297.7
10	4,901.5	0.0	0.0	0.0	0.0	6,395.4	0.0	0.0	11,296.9

Table 6.13 Electricity production in region RG over the representative weeks B

Cozumel region (RG)									
Year	Generation technology								
	Turbogas	Combined cycle	Efficient cogeneration	Internal combustion	Conventional thermoelectric	Wind	Solar	Hydroelectric	Total energy (MWh)
1	3,586.2	0.0	0.0	0.0	0.0	3,997.1	0.0	0.0	7,583.3
2	3,589.0	0.0	0.0	0.0	0.0	3,997.1	0.0	0.0	7,586.1
3	3,600.0	0.0	0.0	0.0	0.0	3,997.1	0.0	0.0	7,597.1
4	3,616.7	0.0	0.0	0.0	0.0	3,997.1	0.0	0.0	7,613.8
5	3,627.8	0.0	0.0	0.0	0.0	3,997.1	0.0	0.0	7,624.9
6	3,639.0	0.0	0.0	0.0	0.0	3,997.1	0.0	0.0	7,636.1
7	3,642.6	0.0	0.0	0.0	0.0	3,997.1	0.0	0.0	7,639.7
8	3,648.3	0.0	0.0	0.0	0.0	3,997.1	0.0	0.0	7,645.4
9	3,656.9	0.0	0.0	0.0	0.0	3,997.1	0.0	0.0	7,654.0
10	3,662.8	0.0	0.0	0.0	0.0	3,997.1	0.0	0.0	7,660.0

Table 6.14 Electricity production in regions RB, RA, and RC for the designated representative weeks C

Year	Generation technology								
	Turbogas	Combined cycle	Efficient cogeneration	Internal combustion	Conventional thermoelectric	Wind	Solar	Hydroelectric	Total energy (MWh)
Tabasco region (RB)									
1	14,600.9	0.0	25,669.6	3,181.9	0.0	0.0	0.0	0.0	43,452.4
2	14,604.1	0.0	26,080.5	3,191.0	0.0	0.0	0.0	0.0	43,875.6
3	14,626.7	0.0	26,969.6	3,202.7	0.0	0.0	0.0	0.0	44,798.9
4	14,685.2	0.0	28,895.8	3,219.5	0.0	0.0	0.0	0.0	46,800.5
5	14,770.8	0.0	32,518.5	3,269.8	0.0	0.0	0.0	0.0	50,559.1
6	14,963.4	0.0	37,815.5	3,329.8	0.0	0.0	0.0	0.0	56,108.7
7	15,397.7	0.0	43,818.0	3,436.5	0.0	0.0	0.0	0.0	62,652.1
8	16,209.7	0.0	50,574.2	3,622.5	0.0	0.0	0.0	0.0	70,406.4
9	17,461.1	0.0	57,536.8	4,021.0	0.0	0.0	0.0	0.0	79,018.9
10	19,885.8	0.0	63,432.8	4,757.0	0.0	0.0	0.0	0.0	88,075.6
Grijalva region (RA)									
1	8,131.2	0.0	0.0	67.5	0.0	5,470.2	0.0	1,189,573.5	1,203,242.4
2	8,131.2	0.0	0.0	67.5	0.0	5,470.2	0.0	1,204,028.6	1,217,697.5
3	8,136.5	0.0	0.0	68.0	0.0	5,470.2	0.0	1,217,129.8	1,230,804.4
4	8,145.9	0.0	0.0	68.1	0.0	5,470.2	0.0	1,228,788.9	1,242,473.1
5	8,171.4	0.0	0.0	68.9	0.0	5,470.2	0.0	1,239,036.1	1,252,746.5
6	8,222.5	0.0	0.0	70.6	0.0	5,470.2	0.0	1,247,812.4	1,261,575.6
7	8,376.2	0.0	0.0	72.7	0.0	5,470.2	0.0	1,255,685.6	1,269,604.6
8	8,694.3	0.0	0.0	76.2	0.0	5,470.2	0.0	1,262,621.4	1,276,862.1

(continued)

Table 6.14 (continued)

Tabasco region (RB)									
9	9,340.4	0.0	0.0	84.0	0.0	5,470.2	0.0	1,269,024.1	1,283,918.6
10	10,347.0	0.0	0.0	98.4	0.0	5,470.2	0.0	1,274,013.9	1,289,929.5
Lerma region (RC)									
1	3,172.5	19,172.3	8,073.2	0.0	0.0	31,976.9	0.0	0.0	62,394.9
2	3,172.5	19,529.4	8,165.9	0.0	0.0	31,976.9	0.0	0.0	62,844.7
3	3,191.3	20,601.4	8,274.0	0.0	0.0	31,976.9	0.0	0.0	64,043.6
4	3,189.6	21,979.1	8,774.9	0.0	0.0	31,976.9	0.0	0.0	65,920.5
5	3,196.0	23,144.0	8,962.2	0.0	0.0	31,976.9	0.0	0.0	67,279.0
6	3,205.4	24,370.7	8,936.0	0.0	0.0	31,976.9	0.0	0.0	68,489.0
7	3,210.8	26,414.9	9,227.6	0.0	0.0	31,976.9	0.0	0.0	70,830.2
8	3,219.5	29,567.4	9,622.8	0.0	0.0	31,976.9	0.0	0.0	74,386.6
9	3,238.3	31,880.0	10,049.7	0.0	0.0	31,976.9	0.0	0.0	77,144.9
10	3,269.0	34,878.5	10,451.5	0.0	0.0	31,976.9	0.0	0.0	80,575.9

Table 6.15 Electricity production in regions RE, RF, and RD for the designated representative weeks C

Merida region (RE)									
Year	Generation technology								
	Turbogas	Combined cycle	Efficient cogeneration	Internal combustion	Conventional thermoelectric	Wind	Solar	Hydroelectric	Total energy (MWh)
1	4,062.6	138,404.1	19,069.2	893.3	71.2	115,116.7	6,245.4	0.0	283,862.6
2	4,074.0	152,531.0	19,274.8	900.7	71.2	115,116.7	6,245.4	0.0	298,213.9
3	4,086.9	166,686.0	20,170.9	901.5	71.2	115,116.7	6,245.4	0.0	313,278.5
4	4,099.1	181,593.0	20,515.8	906.5	71.2	115,116.7	6,245.4	0.0	328,547.8
5	4,098.0	197,635.7	20,580.8	912.1	71.2	115,116.7	6,245.4	0.0	344,660.0
6	4,098.0	213,935.0	20,862.7	915.2	71.2	115,116.7	6,245.4	0.0	361,244.3
7	4,109.4	228,927.4	21,850.2	915.2	71.2	115,116.7	6,245.4	0.0	377,235.5
8	4,136.0	242,962.7	22,806.2	921.6	71.6	115,116.7	6,245.4	0.0	392,260.2
9	4,146.6	257,552.6	24,108.1	933.5	71.4	115,116.7	6,245.4	0.0	408,174.4
10	4,232.4	272,226.9	25,319.4	965.4	72.0	115,116.7	6,245.4	0.0	424,178.2
Cancun region (RF)									
1	12,920.0	0.0	0.0	0.0	215.0	14,410.9	0.0	0.0	27,545.9
2	12,988.1	0.0	0.0	0.0	215.0	14,410.9	0.0	0.0	27,614.0
3	13,004.5	0.0	0.0	0.0	215.0	14,410.9	0.0	0.0	27,630.4
4	13,060.9	0.0	0.0	0.0	215.0	14,410.9	0.0	0.0	27,686.8
5	13,103.4	0.0	0.0	0.0	215.0	14,410.9	0.0	0.0	27,729.3
6	13,126.4	0.0	0.0	0.0	215.0	14,410.9	0.0	0.0	27,752.4
7	13,152.7	0.0	0.0	0.0	215.0	14,410.9	0.0	0.0	27,778.7
8	13,228.2	0.0	0.0	0.0	216.0	14,410.9	0.0	0.0	27,855.1

(continued)

Table 6.15 (continued)

Merida region (RE)									
9	13,358.7	0.0	0.0	0.0	216.0	14,410.9	0.0	0.0	27,985.6
10	13,652.4	0.0	0.0	0.0	217.3	14,410.9	0.0	0.0	28,280.6
Chetumal region (RD)									
1	4,831.8	0.0	0.0	0.0	0.0	6,395.4	0.0	0.0	11,227.2
2	4,843.3	0.0	0.0	0.0	0.0	6,395.4	0.0	0.0	11,238.7
3	4,860.9	0.0	0.0	0.0	0.0	6,395.4	0.0	0.0	11,256.3
4	4,893.8	0.0	0.0	0.0	0.0	6,395.4	0.0	0.0	11,289.1
5	4,890.0	0.0	0.0	0.0	0.0	6,395.4	0.0	0.0	11,285.4
6	4,881.4	0.0	0.0	0.0	0.0	6,395.4	0.0	0.0	11,276.8
7	4,906.4	0.0	0.0	0.0	0.0	6,395.4	0.0	0.0	11,301.8
8	4,931.6	0.0	0.0	0.0	0.0	6,395.4	0.0	0.0	11,327.0
9	4,944.0	0.0	0.0	0.0	0.0	6,395.4	0.0	0.0	11,339.4
10	5,051.2	0.0	0.0	0.0	0.0	6,395.4	0.0	0.0	11,446.6

Table 6.16 Electricity production in area RG over the typical weeks C

Cozumel region (RG)									
	Generation technology								
Year	Turbogas	Combined cycle	Efficient cogeneration	Internal combustion	Conventional thermoelectric	Wind	Solar	Hydroelectric	Total energy (MWh)
1	3,595.1	0.0	0.0	0.0	0.0	3,997.1	0.0	0.0	7,592.2
2	3,601.4	0.0	0.0	0.0	0.0	3,997.1	0.0	0.0	7,598.5
3	3,605.2	0.0	0.0	0.0	0.0	3,997.1	0.0	0.0	7,602.3
4	3,615.1	0.0	0.0	0.0	0.0	3,997.1	0.0	0.0	7,612.2
5	3,630.0	0.0	0.0	0.0	0.0	3,997.1	0.0	0.0	7,627.1
6	3,632.2	0.0	0.0	0.0	0.0	3,997.1	0.0	0.0	7,629.3
7	3,640.1	0.0	0.0	0.0	0.0	3,997.1	0.0	0.0	7,637.2
8	3,664.7	0.0	0.0	0.0	0.0	3,997.1	0.0	0.0	7,661.9
9	3,712.3	0.0	0.0	0.0	0.0	3,997.1	0.0	0.0	7,709.4
10	3,799.4	0.0	0.0	0.0	0.0	3,997.1	0.0	0.0	7,796.5

References

Albadi, M. H., & El-Saadany, E. F. (2010). Overview of wind power intermittency impacts on power systems. *Electric Power System Research, 80*(6), 627–632. https://doi.org/10.1016/J.EPSR.2009.10.035

Almaimouni, A., Ademola-Idowu, A., Kutz, J. N., Negash, A., & Kirschen, D. (2018). Selecting and evaluating representative days for generation expansion planning. *20th Power Systems Computation Conference (PSCC)*. https://doi.org/10.23919/PSCC.2018.8442580

de la Cruz, J., & Castellanos, S. (2024). Battery energy storage systems' integration in Baja California Sur, Mexico: A long-term electrical grid assessment. *Journal of Energy Storage, 76*, Article 109703. https://doi.org/10.1016/J.EST.2023.109703

De Sisternes, F. J., & Webster, M. D. (2013). *Optimal Selection of Sample Weeks for Approximating the Net Load in Generation Planning Problems*. MIT lib. https://dspace.mit.edu/handle/1721.1/102959

Domínguez-Muñoz, F., Cejudo-López, J. M., Carrillo-Andrés, A., & Gallardo-Salazar, M. (2011). Selection of typical demand days for CHP optimization. *Energy Building, 43*(11), 3036–3043. https://doi.org/10.1016/J.ENBUILD.2011.07.024

Ehrgott, M., Ide, J., & Schöbel, A. (2014). Minmax robustness for multi-objective optimization problems. *European Journal of Operational Research, 239*, 17–31. https://doi.org/10.1016/j.ejor.2014.03.013

Gao, S., Hu, B., Xie, K., Niu, T., Li, C., & Yan, J. (2021). Spectral clustering based demand-oriented representative days selection method for power system expansion planning. *International Journal of Electrical Power, 125*, Article 106560. https://doi.org/10.1016/J.IJEPES.2020.106560

Helistö, N., Kiviluoma, J., & Reittu, H. (2020). Selection of representative slices for generation expansion planning using regular decomposition. *Energy, 211*, Article 118585. https://doi.org/10.1016/J.ENERGY.2020.118585

Heuberger, C. F., Staffell, I., Shah, N., & Dowell, N. M. (2017). A systems approach to quantifying the value of power generation and energy storage technologies in future electricity networks. *Computers & Chemical Engineering, 107*, 247–256. https://doi.org/10.1016/J.COMPCHEMENG.2017.05.012

INECC. (2020). *Technology Roadmap and Mitigation Potential for Utility-Scale Electricity Storage in Mexico*. National Institute of Ecology and Climate Change. https://www.gob.mx/inecc/documentos/hoja-de-ruta-tecnologica-y-potencial-de-mitigacion-del-almacenamiento-de-electricidad-a-escala-de-servicios-publicos-en-mexico. Accessed 20 January 2024.

Islam, S., & Roy, N. K. (2023). Renewables integration into power systems through intelligent techniques: Implementation procedures, key features, and performance evaluation. *Energy Reports, 9*, 6063–6087. https://doi.org/10.1016/J.EGYR.2023.05.063

Kate, R. J. (2016). Using dynamic time warping distances as features for improved time series classification. *Data Mining and Knowledge Discovery, 30*, 283–312. https://doi.org/10.1007/s10618-015-0418-x

Koltsaklis, N. E., & Dagoumas, A. S. (2018). State-of-the-art generation expansion planning: A review. *Applied Energy, 230*, 563–589. https://doi.org/10.1016/J.APENERGY.2018.08.087

Kotzur, L., Markewitz, P., Robinius, M., & Stolten, D. (2018). Impact of different time series aggregation methods on optimal energy system design. *Renewable Energy, 117*, 474–487. https://doi.org/10.1016/J.RENENE.2017.10.017

Kotzur, L., Nolting, L., Hoffmann, M., Groß, T., Smolenko, A., Priesmann, J., Büsing, H., Beer, R., Kullmann, F., Singh, B., Praktiknjo, A., Stolten, D., & Robinius, M. (2021). A modeler's guide to handle complexity in energy systems optimization. *Advance Applied Energy, 4*, Article 100063. https://doi.org/10.1016/j.adapen.2021.100063

Lara, C. L., Mallapragada, D. S., Papageorgiou, D. J., Venkatesh, A., & Grossmann, I. E. (2018). Deterministic electric power infrastructure planning: Mixed-integer programming model and nested decomposition algorithm. *European Journal of Operational Research, 271*(3), 1037–1054. https://doi.org/10.1016/J.EJOR.2018.05.039

Li, C., Conejo, A. J., Siirola, J. D., & Grossmann, I. E. (2022). On representative day selection for capacity expansion planning of power systems under extreme operating conditions. *International Journal of Electrical Power, 137*, Article 107697. https://doi.org/10.1016/J.IJEPES.2021.107697

Li, Z., Xia, Y., Bo, Y., & Wei, W. (2024). Optimal planning for electricity-hydrogen integrated energy system considering multiple timescale operations and representative time-period selection. *Applied Energy, 362*, Article 122965. https://doi.org/10.1016/J.APENERGY.2024.122965

Lozano, M. A., Ramos, J. C., Carvalho, M., & Serra, L. M. (2009). Structure optimization of energy supply systems in tertiary sector buildings. *Energy Building, 41*(10), 1063–1075. https://doi.org/10.1016/J.ENBUILD.2009.05.008

Mavrotas, G., Diakoulaki, D., Florios, K., & Georgiou, P. (2008). A mathematical programming framework for energy planning in services' sector buildings under uncertainty in load demand: The case of a hospital in Athens. *Energy Policy, 36*(7), 2415–2429. https://doi.org/10.1016/J.ENPOL.2008.01.011

Moradi-Sepahvand, M., & Amraee, T. (2021). Integrated expansion planning of electric energy generation, transmission, and storage for handling high shares of wind and solar power generation. *Applied Energy, 298*, Article 117137. https://doi.org/10.1016/J.APENERGY.2021.117137

Nahmmacher, P., Schmid, E., Hirth, L., & Knopf, B. (2016). Carpe diem: A novel approach to select representative days for long-term power system modeling. *Energy, 112*, 430–442. https://doi.org/10.1016/J.ENERGY.2016.06.081

Niu, G., Ji, Y., Zhang, Z., Wang, W., Chen, J., & Yu, P. (2021). Clustering analysis of typical scenarios of island power supply system by using cohesive hierarchical clustering based K-Means clustering method. *Energy Reports, 7*, 250–256. https://doi.org/10.1016/J.EGYR.2021.08.049

Pfenninger, S. (2017). Dealing with multiple decades of hourly wind and PV time series in energy models: A comparison of methods to reduce time resolution and the planning implications of inter-annual variability. *Applied Energy, 197*, 1–13. https://doi.org/10.1016/J.APENERGY.2017.03.051

Poncelet, K., Hoschle, H., Delarue, E., Virag, A., & Drhaeseleer, W. (2017). Selecting representative days for capturing the implications of integrating intermittent renewables in generation expansion planning problems. *IEEE Transactions on Power Systems, 32*(3), 1936–1948. https://doi.org/10.1109/TPWRS.2016.2596803

Raya-Tapia, A. Y., López-Flores, F. J., Tovar-Facio, J., & Ponce-Ortega, J. M. (2025). Comparative framework of representative weeks selection methods for the optimization of power systems. *Computers & Chemical Engineering, 194*, Article 108985. https://doi.org/10.1016/j.compchemeng.2024.108985

Ratanamahatana, C. A., & Keogh, E. (2004). *Making time-series classification more accurate using learned constraints*, Proceedings 11–22. https://doi.org/10.1137/1.9781611972740.2

Renner, S. (2018). Poverty and distributional effects of a carbon tax in Mexico. *Energy Policy, 112*, 98–110. https://doi.org/10.1016/j.enpol.2017.10.011

Scikit learn, S. sklearn.preprocessing.StandardScaler. (2023). Obtenido de https://scikit-learn.org/stable/modules/generated/sklearn.preprocessing.StandardScaler.html. Accessed 21 November 2023.

SENER. (2018). *National Electric System Development Program 2018–2032 "PRODESEN"*. National Secretariat of Energy. https://www.gob.mx/cms/uploads/attachment/file/331770/PRODESEN-2018-2032-definitiva.pdf. Accessed 16 February 2024.

Serrano-Arévalo, T. I., Tovar-Facio, J., & Ponce-Ortega, J. M. (2023). Optimal incorporation of intermittent renewable energy storage units and green hydrogen production in the electrical sector. *Energies, 16*, 2609. https://doi.org/10.3390/en16062609

Serrano-Arévalo, T. I., Díaz-Alvarado, F. A., Tovar-Facio, J., & Ponce-Ortega, J. M. (2024). The impact of circular economy indicators in the optimal planning of energy systems. *Sustainability Production Consumer, 44*, 234–249. https://doi.org/10.1016/j.spc.2023.12.018

Short, W., Sullivan, P., Mai, T., Mowers, M., Uriarte, C., Blair, N., Heimiller, D., & Martinez, A. (2011). *Regional Energy Deployment System (ReEDS)*. National Renewable Energy Lab. (NREL), Golden, CO (United States). https://doi.org/10.2172/1031955

Tahir, H., Park, D. H., Park, S. S., & Kim, R. Y. (2022). Optimal ESS size calculation for ramp rate control of grid-connected microgrid based on the selection of accurate representative days. *International Journal of Electrical Power, 139*, Article 108000. https://doi.org/10.1016/J.IJEPES.2022.108000

Teichgraeber, H., & Brandt, A. R. (2022). Time-series aggregation for the optimization of energy systems: Goals, challenges, approaches, and opportunities. *Renewable and Sustainable Energy Reviews, 157*, Article 111984. https://doi.org/10.1016/J.RSER.2021.111984

Tejada-Arango, D. A., Domeshek, M., Wogrin, S., & Centeno, E. (2018). Enhanced representative days and system states modeling for energy storage investment analysis. *IEEE Transactions on Power Systems, 33*(6), 6534–6544. https://doi.org/10.1109/TPWRS.2018.2819578

Wang, X., Liu, X., Jian, S., Peng, X., & Yuan, H. (2021). A distribution network reconfiguration method based on comprehensive analysis of operation scenarios in the long-term time period. *Energy Reports, 7*, 369–379. https://doi.org/10.1016/J.EGYR.2021.01.057

Yeganefar, A., Amin-Naseri, M. R., & Sheikh-El-Eslami, M. K. (2020). Improvement of representative days selection in power system planning by incorporating the extreme days of the net load to take account of the variability and intermittency of renewable resources. *Applied Energy, 272*, Article 115224. https://doi.org/10.1016/J.APENERGY.2020.115224

Zhang, G., Lu, J., & Gao, Y. (2015). Optimization models. In *Multi-level decision making, intelligent systems reference library*, Springer. https://doi.org/10.1007/978-3-662-46059-7

Zhou, H., Lu, L., Shen, L., Zhang, H., Jiang, L., & Liao, K. (2022). Integrated location and capacity coordination planning scheme for multi-power complementary generation system. *Energy Reports, 8*, 10–18. https://doi.org/10.1016/J.EGYR.2022.05.102

Chapter 7
The Energy-Food Nexus: Geopolitical and Health Crisis Analysis

7.1 Introduction

Increased market fluctuations, global supply chain disruptions, travel bans, and investor unease contributed to the pandemic's impact on worldwide financial systems (Rubbaniy et al., 2022; Tuna & Tuna, 2022). In the post-pandemic period, the global economy entered a phase of gradual revival, but now they are dealing with a fresh systemic shock brought on by the geopolitical confrontation between Russia and Ukraine. Compared to COVID-19, it is anticipated to be a more permanent economic shock since the global supply chain has been adversely impacted by the uncertainty, which has resulted in systematic political, economic, and humanitarian issues worldwide. A number of penalties on particular commercial activities and export and import flows have made the economic situation worse (Chishti et al., 2023).

However, countries that are geographically close to Russia and Ukraine, face elevated levels of uncertainty in economic policymaking as theyrely heavily on Russian energy, and have robust guarantees for political liberties and open political discourse are more vulnerable to geopolitical risks (Hossain et al., 2024).

According to Inacio et al. (2023), this is due to the fact that Russia is one of the world's leading energy suppliers and that any interruption in the sale of its energy resources plays a crucial role in shaping global energy dynamics, mostly hurting supplies to European member states. The EU cannot become energy self-sufficient due to its limited fossil fuel supplies and must import a significant quantity of energy. Given their more advantageous geographic location compared to other energy providers worldwide, the European Union's energy dependence on Russia is a result of Russia's enormous energy supplies, lower prices, and inexpensive transit costs. Additionally, by constructing gas and oil pipelines, energy shipping costs have been cut by the leading economic powers (Chen et al., 2023a).

A. Y. Raya-Tapia et al., *Machine Learning and Clustering for a Sustainable Future*,
Studies in Computational Intelligence 1233,
https://doi.org/10.1007/978-3-032-03876-0_7

7.1.1 Energy Sector Repercussions

The EU is heavily reliant on Russian energy goods, and Russia's international energy commerce is equally reliant on the Asian and European markets, according to an analysis of the EU's energy use and the structure of its energy trade. This creates mutual vulnerabilities encompassing Russia's energy position, EU members' societal/political health, and Russia's domestic economic/social balance, all of which are directly impacted by the EU's energy boycott on Russia (Chen et al., 2023b).

Because there was less demand for crude oil during the COVID-19 pandemic, prices and returns decreased (Liu et al., 2020). On the other hand, prices rose quickly as a result of the sanctions and the start of the geopolitical confrontation between Russia and Ukraine. These occurrences have led to a pattern of significant volatility and abrupt fluctuations in crude oil prices (Ali et al., 2023; Zhang et al., 2023).

Energy market turbulence complicates Europe's green transition, especially because natural gas uniquely facilitates rapid structural changes in sustainable energy supply chains (Chen et al., 2023a; Liang et al., 2021). In the near term, it continues to be an essential and irreplaceable component, but the EU hopes to lessen its reliance on this fossil fuel in order to reach "net zero" in coming decades (Chen et al., 2023b).

The energy supply in Europe is now more unpredictable as a result of this geopolitical confrontation. The EU, Asia, and other areas have realized how urgent it is to discover alternate energy sources due to the volatility of the gas and oil markets. Renewable energy investments are more appealing since society today relies heavily on the use of fossil fuels and because natural gas and oil costs are high (Zhang et al., 2023). But during crises, fossil fuels are making a comeback, and governments are more focused on bringing down the price of gas and oil than they are on drastically cutting emissions (Ben Hassen & El Bilali, 2022). Furthermore, because renewable energy sources depend on metals like aluminum and battery-grade nickel, their costs are increasing due to growing metal prices (World Bank, 2022).

Global energy supply is significantly impacted by geopolitical upheaval in energy-exporting regions through supply chains including transportation and energy production (Yang et al., 2014). Global geopolitical conflicts have been significantly impacted by the change in the energy system.

7.1.2 Impacts on the Food System

The crisis has affected commodities other than energy. Disruptions to crop production and distribution channels caused by the conflict resulted in both food insecurity concerns and rising prices across multiple commodity markets, including grain and fertilizer, which are major commodities produced and exported by Ukraine and Russia (Wang et al., 2023; Yang et al., 2014). Consequently, market uncertainty was increased by the potential of penalties and trade disruptions.

Ukraine ranks fifth in the world for wheat exports and ninth in terms of production. International wheat markets were rocked by the conflict between Russia and Ukraine, and concerns were expressed about how it may affect Ukraine's capacity to harvest and export its available supply of this essential crop. Due to the disruption of this grain's transportation caused by the loss of bridges, roads, and trains, many European logistics systems have collapsed, which has raised the cost of staple foods (Ghorbali et al., 2023; Saâdaoui et al., 2022). Consumers suffer greatly as a result of the conflict's damaging and pervasive impacts, which include increased inflation of wheat prices and a decrease in worldwide availability. Given that 20% of the world's calories come from wheat (Devadoss & Ridley, 2024; Shiferaw et al., 2013), these shocks might have catastrophic effects on the world's food security (Azani et al., 2022). The prices of wheat have stayed generally above their pre-war levels, despite a significant decrease from their first peak at the start of the conflict. The pressure on product pricing has only been made worse by this.

During the COVID-19 pandemic, panic purchasing and stockpiling were major aspects of consumer behavior connected to crises and disasters that garnered a lot of media attention (Hall et al., 2021). At the national and personal levels, the geopolitical confrontation between Ukraine and Russia led to a panic purchasing trend. The problem was made worse by the fact that although several nations imposed export limits in order to protect their inhabitants and ensure local food supply and reduce inflation, some other nations similarly restricted exports in order to do the same (Ben Hassen & El Bilali, 2022). These policies may have unanticipated, catastrophic consequences for vulnerable individuals in nations that import food, raising costs and exacerbating issues of food insecurity that the epidemic has already made worse. However, more food waste can result from more food purchases and a rise in family food supplies (Jagtap et al., 2022).

The advancement of the Sustainable Development Goals (SDGs), particularly SDG 12 (Responsible Production and Consumption), SDG 2 (Zero Hunger), and SDG 1 (End Poverty), may be impacted by geopolitical and health crises. In many nations, the sustainable transformation of food systems will be delayed by geopolitical turmoil. Since the effects on food security are being made worse by a number of underlying rigidities, vulnerabilities, and inefficiencies in global food systems, as global food supplies tighten, several governments are lobbying Europe to delay eco-farming policies and maximize production. Therefore, immediate and long-term policies and reforms must be adopted to strengthen the shift to healthy, egalitarian, and environmentally sustainable food systems (Ben Hassen & El Bilali, 2022; Jamalimoghaddam et al., 2023).

Russia accounts for approximately 17% of global potash fertilizer exports, over 10% of nitrogen fertilizer shipments, and nearly 17% of mixed fertilizer supplies worldwide (Benton et al., 2022). Along with shortages of important grains, the conflict between Russia and Ukraine has also caused shortages of fertilizer, particularly in underdeveloped nations where price consequences might drastically restrict usage and lead to reduced crop outputs amid worldwide shortages and unprecedented price surges. Consequently, numerous agricultural producers worldwide are switching to crops like soybeans that require less fertilizer than others like wheat and corn, which demand expensive fertilizers. This might raise the cost of bread, cereals, and other goods and worsen the present supply constraints.

Future harvests are uncertain, Ukrainian exports have stopped, and agricultural product prices have skyrocketed globally, putting millions at risk of starvation and hardship. Geopolitical violence impacts national and individual purchasing power, which impacts financial access to food by impeding the post-COVID-19 economic recovery. Additionally, it diminishes the region's food supply at a time when governments lack the financial flexibility to mitigate the effects of growing food prices because of COVID-19-related economic restraints. Governments' capacity to sustain social safety net programs and consistently subsidize basic food items will be put to the test in several import-dependent low- and middle-income nations (FAO, 2023). (Bentley, 2022). Furthermore, as they spend the majority of their money on food, those with low incomes will be most impacted by rising food costs (Ramírez-Márquez & Ponce-Ortega, 2023). Humanitarian requirements are therefore at an all-time high, with millions of people on the verge of starvation due to COVID-19, rising prices, wars, and climatic shocks (WFP, 2023). Due to a potential change in consumption habits toward less expensive, but frequently worse quality, food goods, price rises brought on by decreased food supply are anticipated to be followed by poor product quality.

The whole agri-food sector will be constrained by logistical reliance on fuel, and the drop in gas supplies will make matters worse by reducing the amount of ammonia and urea produced for nitrogen fertilizers. The ability to move commodities and products to their intended locations is a problem that goes beyond simple production. In addition to reducing the availability of gas and fuel, the geopolitical struggle has damaged effective transportation networks that connect companies and farms. Theoretically, the markets for metals, energy, and agriculture are intertwined through several routes that encompass investment, consumption, and production (Adeleke et al., 2022; Jiang & Chen, 2024).

7.2 Description of the Addressed Problem

Since Russia is a major global supplier of fossil fuels, the geopolitical war between Ukraine and Russia has had a substantial influence on energy markets, causing another systemic shock following the COVID-19 epidemic. Because of the uncertainty, the penalties placed on Russia, and the effect on supply chains, the cost of food and energy has increased at a record rate. This impacts not just the environment and stock markets, but also threatens global food security. This study uses 26 features from 2019 to 2022, including two unusual events—the coronavirus pandemic and the geopolitical confrontation between Ukraine and Russia—to categorize groupings of nations based on their effects on the energy-food nexus. In order to identify commonalities across the groups and, as a result, create more suitable paths of action in the pursuit of resilient, sustainable, and equitable systems. Future systemic shocks and the transformation of the food and energy systems will be addressed by taking into account solutions that tackle social, economic, and political aspects.

7.3 Implemented Methodology

The approach used in this work is depicted in Fig. 7.1, whereby the raw data set is first gathered for each of the years under study, and then the pertinent computations are performed for the percentage variation data set. After normalizing each dataset, the k-means unsupervised machine learning algorithm creates groups for each year, which are then described for each year.

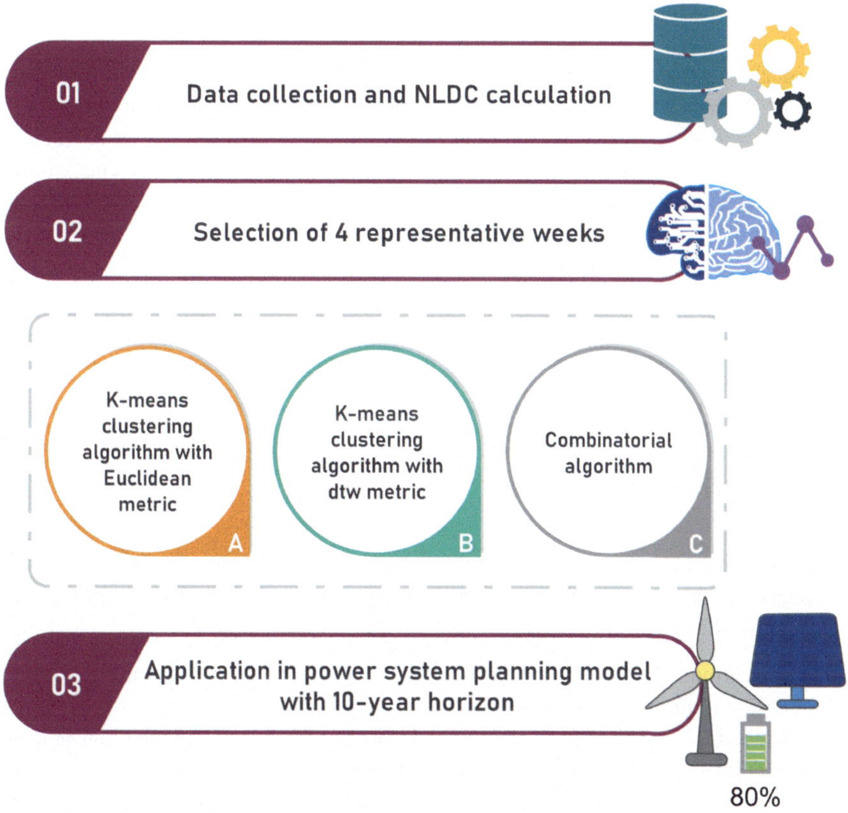

Fig. 7.1 Implemented general methodology

7.3.1 Data Collection

Data on energy was gathered, including fuel and power costs, imports, stockpiles, and consumption. Information on the food sector, including manpower, cultivated areas, meat output, pricing indexes for different food categories, agricultural input price indices, and per capita subsidies (EUROSTAT, 2023). Corresponds to 25 EU member states from 2019 to 2022, with each year representing a data set of 26 attributes (see Table 7.1). Using 2015 as the base year, the percentages of variance were computed using this data. As a result, four more data sets were created and subjected to the clustering technique K-means.

7.3.2 Data Preprocessing

The variables must be rescaled to ensure comparability, since each feature uses a distinct numerical range. The sklearn preprocessing package's Standard Scaler method (Scikit learn, 2023) was applied. Equation 7.1 illustrates how to normalize the features by removing the average and scaling to the unit variance. The sample x's standard score is determined as follows:

$$z = \frac{x - u}{s} \tag{7.1}$$

Here u corresponds to the average of the samples and s is the sample standard deviation.

7.3.3 K-Means Algorithm

As nations are categorized into several categories according to commonalities, the unsupervised machine learning method k-means forms clusters. The stages of the k-means algorithm are shown below:

1. Initialization: Randomly select k initial centroids m_k^t corresponding to the predefined number of clusters.
2. Distance Calculation: Each data point p_i (represented as d-dimensional feature vector) is assigned a Euclidean distance (Eq. 7.2) to every centroid.

$$d(p_{id}, m_{kd}) = \sqrt{\sum_d (p_{id} - m_{kd})^2}\ \forall i, \forall k \tag{7.2}$$

3. Cluster Assignment: Assign every point to the nearest centroid, forming k clusters.

Table 7.1 26 attributes applied in clustering in several periods

ID	Attribute
Energy metrics	
I1	Domestic consumption of natural gas (Mm^3)
I2	Volume of imported natural gas (Mm^3)
I3	Household gas tariffs for medium consumers (€ per gigajoule)
I4	Commercial/industrial gas rates for medium users (€ per gigajoule)
I5	Household electricity costs for medium consumption (€ per kilowatt-hour)
I6	Commercial/industrial electricity pricing for medium-scale users (€ per kilowatt-hour)
I7	Oil and refined product imports (thousands of tons)
I8	Initial oil and petroleum product inventories (thousands of tons)
I9	Hard coal import volumes (thousands of tons)
Economic indicators	
I10	HICP—All-items consumer price index
I11	Agricultural workforce (thousands of annual work units, AWU)
Land use and crop metrics	
I12	Total area under cereal production (thousand hectares)
I13	Land dedicated to wheat cultivation (thousand hectares)
I14	Area for fresh vegetable production (thousand hectares)
I15	Permanent crop acreage (thousand hectares)
Price indices (base year 2015 = 100)	
I16	Cereal
I17	Vegetable
I18	Fruit
I19	Livestock
I20	Milk
I21	Eggs
Production and agricultural inputs	
I22	Beef output volume (thousands of tons)
I23	Agricultural input price index: Electricity (Year 2015 = 100)
I24	Agricultural input price index: Fertilizers and soil amendments (Year 2015 = 100)
I25	Agricultural input price index: Livestock feed (Year 2015 = 100)
Subsidies	
I26	Per capita agricultural production subsidies (€ per inhabitant)

4. Centroid Update: Recompute each centroid as the mean of all points in its cluster (Eq. 7.3).

$$m_k^{t+1} = \frac{1}{\left|C_k^t\right|} \sum\nolimits_{x_i \in C_k^t} p_i \, \forall k \tag{7.3}$$

5. Iteration: Repeat steps 2–4 until convergence, achieved by minimizing the total within-cluster variance (sum of squared Euclidean distances, Eq. 7.4).

$$\sum\nolimits_i \sum\nolimits_k d(x_i - m_k)^2 \tag{7.4}$$

7.4 Results

7.4.1 Cluster Analysis: Raw Data

Figure 7.2 illustrates the annual clustering results of raw data across three distinct groups. The blue cluster maintained a consistent membership of four nations throughout 2019–2022. The pink cluster showed greater variability, containing six countries in 2019, decreasing to three in 2020–2021, then rising to five in 2022. Meanwhile, the yellow cluster expanded from fifteen members in 2019 to eighteen in 2020–2021, before slightly decreasing to sixteen in 2022.

The groups created using the mean value for each cluster across the 26 features between 2019 and 2022 are described in Tables 7.2 and 7.3. The defining energy attributes of the blue group include abundant oil and petroleum product inventories, substantial natural gas import volumes, and intensive natural gas utilization. Medium-sized families pay high prices for both electricity and natural gas; non-domestic electricity prices are high, with the exception of 2020; and medium-sized commercial

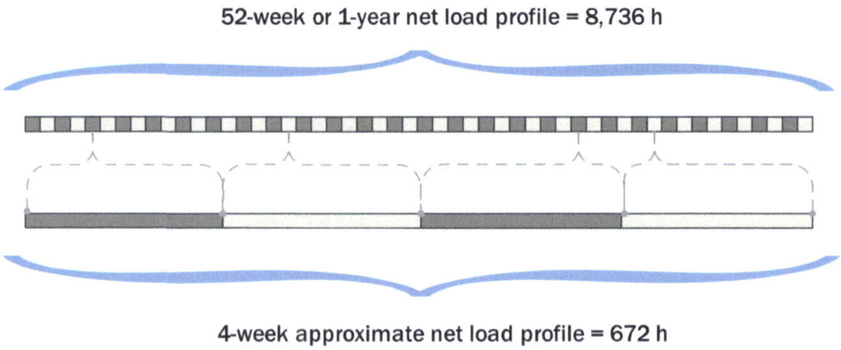

Fig. 7.2 Data clusters generated from raw datasets covering the period 2019–2022

Table 7.2 The normal ranges of variables present in the raw datasets across all groups in both 2019 and 2020

	2020			2019		
	Blue	Yellow	Pink	Blue	Yellow	Pink
I1	57896	6695	14809	61000	7791	7802
I2	75087	15732	18965	80417	19107	12656
I3	19	16	10	20	17	12
I4	7.0	6.8	7.7	8.1	8.3	8.3
I5	0.000	0.180	0.130	0.240	0.190	0.130
I6	0.000	0.080	0.090	0.090	0.080	0.080
I7	83168	19692	17716	96987	25282	12056
I8	272295	39406	61500	269115	42233	33258
I9	1496	8801	888	2991	11728	6942
I10	105	106	111	105	105	109
I11	773	112	948	818	125	610
I12	6021	722	5049	6204	697	3062
I13	2745	350	1828	3009	328	1209
I14	299	24	124	295	27	73
I15	2142	136	252	2106	158	135
I16	103.70	105.60	113.80	100.40	105.50	106.90
I17	110.50	113.80	142.20	107.10	115.80	130.70
I18	124.10	134.60	207.30	109.10	110.70	147.90
I19	101.80	104.60	111.60	105.30	104.40	111.10
I20	107.30	111.30	125.30	108.10	108.90	126.80
I21	107.20	100.80	112.80	107.80	100.10	106.90
I22	983.70	125.70	206.50	1002.40	146.20	116.60
I23	102.30	94.20	113.30	106.10	95.00	102.30
I24	92.50	81.10	94.70	96.40	86.90	95.80
I25	101.10	101.90	110.10	100.90	102.50	103.70
I26	8.90	22.30	16.70	9.10	15.10	33.80

and industrial gas users benefited from affordable natural gas rates during 2019 and again in 2022.

7.4.2 Results of Cluster Analysis: Percentage of Data Variation

The cluster analysis in Fig. 7.3 (2015 baseline year) categorizes the data into three groups using annual variation percentages from 2019 to 2022. There are two countries in the orange group in 2019, seventeen in 2020, sixteen in 2021, and nineteen in 2022.

Table 7.3 The normal ranges of variables present in the raw datasets across all groups in both 2021 and 2022

	2022			2021		
	Blue	Yellow	Pink	Blue	Yellow	Pink
I1	54467	6030	8334	61282	6725	15573
I2	75251	16140	7316	77390	15550	17123
I3	24.65	24.64	13.89	18.97	15.34	9.26
I4	16.52	20.82	21.50	6.65	7.39	6.61
I5	0.290	0.210	0.160	0.240	0.180	0.140
I6	0.170	0.159	0.159	0.090	0.084	0.079
I7	91333	22813	14121	86115	20128	18262
I8	247716	39554	42082	261925	38631	61987
I9	3130	13879	549	1747	10593	936
I10	116	120	133	107	109	116
I11	758	110	587	779	109	934
I12	5990	673	3349	6101	719	5057
I13	2970	323	1430	3018	363	1819
I14	287	24	70	307	25	122
I15	2143	145	157	2148	129	255
I16	190.37	198.75	218.85	135.16	134.92	149.14
I17	128.14	134.21	173.51	115.03	122.55	151.51
I18	129.77	135.62	200.72	132.05	132.94	195.34
I19	130.49	126.24	155.02	104.41	104.28	115.59
I20	146.65	155.89	206.30	112.58	119.43	135.60
I21	161.48	138.78	156.40	108.53	104.17	120.63
I22	956.88	134.22	131.74	990.53	122.92	206.75
I23	174.80	168.78	208.07	117.80	110.31	126.54
I24	201.43	217.80	264.30	113.35	110.17	127.06
I25	150.10	153.26	173.20	115.73	115.32	133.57
I26	8.96	15.15	47.64	8.84	22.63	17.29

In 2019, there were seven nations in the purple hue group; in 2020, there was just one; in 2021, there were five; and in 2022, there were three. The green cluster comprised 16 countries in 2019, decreasing to seven (2020), four (2021), and three (2022). Tables 7.4 and 7.5 present the group characteristics based on average values across 26 variables.

In the energy sector, the orange group's use of natural gas increased at the fastest rate in 2019 and 2020. Nonetheless, it has seen the biggest drops in household electricity rates, as well as in the prices of natural gas (until for 2020) and electricity prices (except for 2022). Additionally, 2019 and 2020 had the biggest drops in oil and oil product stocks (apart from 2019) and the import of hard coal and oil and derivatives.

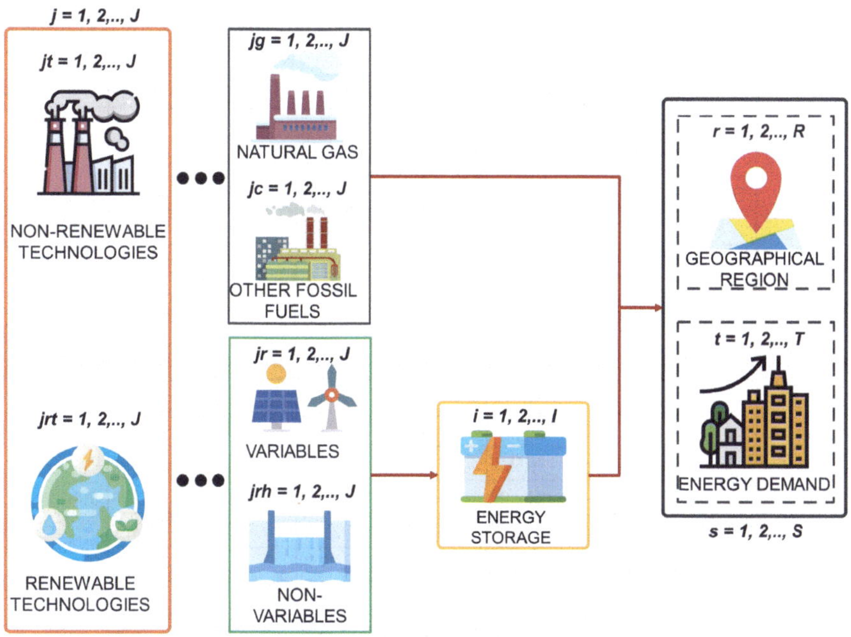

Fig. 7.3 Clustering patterns developed from percentage change data across the 2019–2022 period

In the food sector, the highest increases are recorded in the area dedicated to growing fresh vegetables (except from 2019), producing beef in 2020 and 2022, and providing per capita subsidies (apart from 2019). Reduced decline in the agricultural labor force (except from 2019). Greater declines in the price index of vegetables (unless in 2022), cereals, animals, fruits, eggs and milk (except in 2021), cereal and wheat crop area (except in 2022), and the HICP. Additionally, soil improvers and fertilizers (unless for 2019), feed for animals, and power prices in agriculture (except for 2020) are included in the index.

Imports of hard coal, natural gas, and petroleum and petroleum products have increased more than those of the purple cluster in the energy sector. Greater price rises for both residential and commercial consumers of natural gas (apart from 2022). As well as in the cost of power for non-household users (except from 2021) and households in 2020 and 2021. Expect further rise in oil and oil product inventories between 2019 and 2022.

Cereals, fruits, vegetables, animals, milk (until in 2021 and 2022), and eggs have all had higher price index rises in the food industry, with the exception of 2020. Additionally, there will be greater price rises for soil improvers and fertilizers (except from 2019), feed for animals, and power in agriculture in 2020 and 2021. Significant rise in the HICP index score, which represents the rate of inflation (except from 2022). In 2020 and 2022, there will be further declines in the sectors of Permanent Crops and Fresh Vegetable Crops (except from 2019). Higher declines in agricultural

Table 7.4 Cluster-wise mean values of percentage change metrics across all features (2019–2020)

	2020			2019		
	Orange	Green	Purple	Orange	Green	Purple
I1	11.9	9.0	5.4	47.6	7.5	8.4
I2	33.3	12.0	923.0	43.1	14.3	198.4
I3	-7.5	-11.6	4.0	-20.3	-3.2	0.7
I4	-24.5	-26.6	42.1	-18.5	-9.0	-2.7
I5	3.0	-0.4	12.0	-7.7	6.1	1.3
I6	-4.5	0.5	25.9	-17.0	0.3	5.2
I7	-9.3	-3.4	11.6	-3.3	2.7	6.3
I8	5.8	26.8	7.0	12.2	-1.3	19.6
I9	-29.0	-23.3	-10.7	-21.0	-16.8	21.5
I10	5.2	9.2	10.7	3.1	5.0	8.3
I11	-6.3	-19.3	-19.7	-14.1	-3.9	-8.6
I12	-6.6	1.8	-2.4	-20.6	-3.1	1.3
I13	-13.6	2.1	2.3	-28.2	-5.2	3.5
I14	18.5	2.4	-24.9	-8.3	17.7	0.8
I15	1.5	13.5	-4.1	4.2	-1.2	12.4
I16	3.8	11.1	14.7	-0.5	4.5	7.8
I17	12.3	20.7	62.7	12.7	13.9	28.8
I18	28.3	72.8	50.2	9.0	9.9	43.9
I19	3.1	7.5	19.6	1.1	5.0	10.3
I20	6.4	24.6	27.4	-2.7	9.0	26.9
I21	2.0	3.4	23.7	-1.0	1.9	6.6
I22	1.9	1.7	-27.6	-8.2	4.1	-0.4
I23	6.2	3.0	18.2	1.6	7.5	5.8
I24	-12.7	-10.3	2.4	5.1	-7.6	-3.5
I25	1.1	5.1	15.4	-3.2	2.1	5.1
I26	113.8	-9.7	0.0	14.8	65.9	-9.3

labor force participation in 2020 and 2022, as well as per capita subsidies (except from 2020).

In the energy industry, the green group has seen a greater increase in home power prices between 2019 and 2022. However, there were greater declines in imports of natural gas, natural gas usage (except from 2020), home natural gas prices in 2021 and 2022, and household electricity prices in 2020 and 2021. As well as in the imports of oil, oil products, and hard coal in 2021 and 2022. In the food sector, the price index for agricultural electricity increased more in 2019 and 2022, and beef output increased more in 2019 and 2021.

The pandemic's effects caused the EU's energy usage to plummet in 2020 before significantly increasing in 2021. However, in Spain, Germany, France, and Italy government fiscal interventions—including energy subsidies, price controls,

Table 7.5 Normal scores of attributes in percentage variations within each cluster during 2021 and 2022

	2022			2021		
	Orange	Green	Purple	Orange	Green	Purple
I1	-2.6	-31.9	0.8	16.0	-0.6	24.5
I2	44.5	-2.5	403.8	47.3	-3.5	332.8
I3	29.8	58.2	29.4	-8.5	-25.4	-5.8
I4	114.3	190.0	153.7	-22.8	-20.7	-11.2
I5	16.6	28.2	22.3	3.0	0.9	10.8
I6	93.3	91.7	109.3	0.7	10.3	2.8
I7	-4.7	-7.3	16.4	-9.2	-13.5	3.1
I8	-3.6	17.8	37.2	0.1	26.8	19.6
I9	-3.8	-28.6	3.7	-6.8	-28.6	-0.2
I10	18.2	35.4	31.9	7.5	13.0	14.2
I11	-9.9	-19.5	-27.0	-7.4	-25.3	-17.8
I12	-5.3	6.5	-8.7	-6.4	7.0	-5.6
I13	-4.3	13.3	1.1	-9.2	12.0	-2.2
I14	19.4	3.7	-19.8	24.7	0.1	-6.5
I15	-0.5	48.9	-7.3	-1.6	28.8	-1.2
I16	96.9	96.0	135.8	33.0	33.8	50.6
I17	32.4	47.7	89.4	16.8	30.1	46.3
I18	33.8	61.7	121.6	31.8	34.3	72.2
I19	26.6	43.7	60.1	3.5	5.4	12.8
I20	51.8	116.6	92.8	11.6	38.3	33.6
I21	42.4	47.9	66.4	5.4	3.8	14.1
I22	1.3	-11.2	-2.6	3.4	3.4	-0.9
I23	74.0	148.2	96.5	18.4	18.7	33.5
I24	123.2	127.9	162.7	13.6	15.0	28.0
I25	52.3	57.1	84.6	14.6	14.9	29.4
I26	63.3	21.6	7.0	82.3	-1.4	-6.2

and resulting taxation measures—represent policy responses to escalating energy expenses, the dispute between Russia and Ukraine has reached a climax.

Production, supply, manufacturing, processing, and logistical interruptions are to blame for recent spikes in food costs. Agricultural sectors in countries with critical dependencies—whether on Russian/Ukrainian food exports or Russian/Belarusian fertilizer supplies—show the highest systemic fragility (Borrell, 2023).

Electricity, fertilizer, and animal feed are examples of agricultural inputs that have increased significantly since the conflict started and Russia was subject to sanctions. Food costs have increased as a result, and when combined with energy prices, this has led to severe inflation in a number of nations.

The likelihood of food and energy networks being physically and financially disrupted grows as the battle rages on. It seems unlikely that the dispute will be settled very soon; it will continue to have an impact on the world's resource markets and raise the potential for extremely large "domino effects" or "risk cascades" in societies and economies all over the world. In areas and sectors distant from the war, these impacts may have immediate and serious repercussions (Benton et al., 2022).

7.5 Conclusions

Three groups were found when the raw data was grouped. The blue group is made up of nations that import more oil, oil products, and natural gas. These are developed nations that must import energy from other countries because their own output is insufficient. It is evident that they are committed to lowering emissions through the importation of natural gas in order to achieve their environmental objectives. Its exposure to fossil fuel market instability and potential energy supply disruptions poses a significant risk due to its heavy reliance on foreign energy. When it comes to the food industry, they have the most land for growing crops, producing meat, and employing agricultural workers. They also have the lowest costs for animal feed. This keeps the cost of foods like cereals, milk, fruits, and vegetables low. From 2019 to 2022, they have the lowest inflation and the fewest agricultural subsidies per capita.

Despite having the lowest household bills for electricity and natural gas, the pink group has the greatest rates of inflation and the most subsidies to lower the high cost of food. In the energy sector, the yellow group has lower non-domestic electricity costs, lower NG usage, and a significant coal import. The largest per capita subsidies are found in the food industry. Reduced agricultural labor and area dedicated to growing vegetables and cereals, as well as a lower index of animal and egg prices. Low index ratings for fertilizer prices and agricultural electricity.

Three groups were created using the percentage variation data. The orange group is distinguished by a rise in vegetable crop land and a fall in grains, such as wheat, which may have contributed to the decline in agricultural labor. The price index for a number of food categories and agricultural inputs has decreased. This could have resulted from agriculture receiving the biggest increases in per capita subsidies. It can cause inflation (HICP) to decline. There are a lot of differences in the energy business.

Oil, coal, petroleum products, and natural gas imports have increased the most, as indicated by the purple group. The price index for grains, animals, vegetables, animal feed, eggs, and fertilizers has increased the most in the food sector. With the exception of 2022, this is evidenced by the largest surge in inflation rates. With the biggest drops in natural gas usage and imports of oil, coal, and petroleum products, the green group had the most effects on the energy-food nexus in 2022. Together with the biggest drop in non-domestic electricity prices. However, the cost of electricity for homes and natural gas for both residential and non-domestic customers has increased more. The price index for meat production and cereals decreased more in 2022, while milk and agricultural electricity increased more. This category may have the most increase in inflation in 2022 due to all of the aforementioned factors.

The findings highlight the grave consequences of the geopolitical dispute between Russia and Ukraine, two nations crucial to the world's food and energy systems. Consequently, to safeguard global food security and protect both agricultural producers and consumers, international conflicts must not be allowed to disrupt farming operations and commodity exports. Global food security was immediately and significantly impacted by the geopolitical turmoil. As a result, some immediate measures are needed, such abolishing trade restrictions that limit food availability, sustaining or growing social programs by allocating funds for feeding the most needy, and raising more money for humanitarian assistance. To create a more resilient and sustainable food system that allows people to consume nutritious food, stakeholders from the government, society, and industry must cooperate. To mitigate and avert future worldwide food shortages, we must tackle the root systemic factors behind hunger and malnutrition. Climate change is compounding the dual challenge facing global agriculture: meeting food demands while reducing ecological impacts within increasingly fragile systems.

References

Adeleke, M. A., Awodumi, O. B., & Adewuyi, A. O. (2022). Return and volatility connectedness among commodity markets during major crises periods: Static and dynamic analyses with asymmetries. *Resources Policy, 79*, https://doi.org/10.1016/J.RESOURPOL.2022.102963.

Ali, M. K., Zahoor, M. K., Saeed, A., Nosheen, S., & Thanakijsombat, T. (2023). Impact of vertical integration strategies on environmental, social, and governance sustainability: Policy implication for oil and gas energy sector. *Process Integration and Optimization for Sustainability, 1–15*, https://doi.org/10.1007/S41660-023-00375-2.

Azani, M., Shaerpour, M., Yazdani, M. A., Aghsami, A., & Jolai, F. (2022). A novel scenario-based bi-objective optimization model for sustainable food supply chain during the COVID-19: A case study. *Process Integration and Optimization for Sustainability, 6*(1), 139–159. https://doi.org/10.1007/S41660-021-00203-5.

Ben Hassen, T., & El Bilali, H. (2022). Impacts of the Russia-Ukraine war on global food security: Towards more sustainable and resilient food systems? *Foods, 11*(15), 2301. https://doi.org/10.3390/FOODS11152301.

Bentley, A. (2022). Broken bread—avert global wheat crisis caused by invasion of Ukraine. *Nature, 603*(7902), 551. https://doi.org/10.1038/D41586-022-00789-X.

Benton, T. G., Froggatt, A., Wellesley, L., Grafham, O., King, R., Morisetti, N., & Schröder, P. (2022). The Ukraine war and threats to food and energy security. *Chatham House—International Affairs Think Tank* 4.

Borrell, J. (2023). *Food Insecurity: The Time to Act Is Now*. https://www.eeas.europa.eu/eeas/food-insecurity-time-act-now_en. Accessed 2 December 2023.

Chen, J., Xiao, Z., Bai, J., & Guo, H. (2023a). Predicting volatility in natural gas under a cloud of uncertainties. *Resources Policy, 82*, https://doi.org/10.1016/J.RESOURPOL.2023.103436.

Chen, S., Bouteska, A., Sharif, T., & Abedin, M. Z. (2023b). The Russia-Ukraine war and energy market volatility: A novel application of the volatility ratio in the context of natural gas. *Resources Policy, 85*, Article 103792. https://doi.org/10.1016/J.RESOURPOL.2023.103792.

Chishti, M. Z., Khalid, A. A., & Sana, M. (2023). Conflict vs sustainability of global energy, agricultural and metal markets: A lesson from Ukraine-Russia war. *Resources Policy, 84*, Article 103775. https://doi.org/10.1016/J.RESOURPOL.2023.103775.

Devadoss, S., & Ridley, W. (2024). Impacts of the Russian invasion of Ukraine on the global wheat market. *World Development, 173*, Article 106396. https://doi.org/10.1016/J.WORLDDEV.2023.106396.

EUROSTAT. (2023). *European Union*. Retrieved from https://ec.europa.eu/eurostat.

FAO. (2023). *Impact of the Ukraine-Russia Conflict on Global Food Security and Related Matters under the Mandate of the Food and Agriculture Organization of the United Nations*. https://www.fao.org/3/ni734en/ni734en.pdf. Accessed 11 December 2023.

Ghorbali, B., Kaabia, O., Naoui, K., Urom, C., & Slimane, I. B. (2023). Wheat as a hedge and safe haven for equity investors during the Russia-Ukraine war. *Finance Research Letter, 58*, Article 104534. https://doi.org/10.1016/J.FRL.2023.104534.

Hall, C. M., Fieger, P., Prayag, G., & Dyason, D. (2021). Panic buying and consumption displacement during COVID-19: Evidence from New Zealand. *Economies, 9*(2), 46. https://doi.org/10.3390/ECONOMIES9020046.

Hossain, A. T., Masum, A. A., & Saadi, S. (2024). The impact of geopolitical risks on foreign exchange markets: Evidence from the Russia-Ukraine war. *Finance Research Letter, 59*, Article 104750. https://doi.org/10.1016/J.FRL.2023.104750.

Inacio, C. M. C., Kristoufek, L., & David, S. A. (2023). Assessing the impact of the Russia-Ukraine war on energy prices: A dynamic cross-correlation analysis. *Physica a: Statistical Mechanics and Its Applications, 626*, Article 129084. https://doi.org/10.1016/J.PHYSA.2023.129084.

Jagtap, S., Trollman, H., Trollman, F., Garcia-Garcia, G., Parra-López, C., Duong, L., Martindale, W., Munekata, P. E. S., Lorenzo, J. M., Hdaifeh, A., Hassoun, A., Salonitis, K., & Afy-Shararah, M. (2022). The Russia-Ukraine conflict: Its implications for the global food supply chains. *Foods, 11*(14), 2098. https://doi.org/10.3390/FOODS11142098.

Jamalimoghaddam, E., Yazdani, S., Farajzadeh, Z., & Mahoozi, H. (2023). Investigating the environmental and economic sustainability of crop subsector. *Process Integration and Optimization for Sustainability, 7*(4), 765–784. https://doi.org/10.1007/S41660-023-00321-2.

Jiang, W., & Chen, Y. (2024). Impact of Russia-Ukraine conflict on the time-frequency and quantile connectedness between energy, metal and agricultural markets. *Resources Policy, 88*, Article 104376. https://doi.org/10.1016/J.RESOURPOL.2023.104376.

Liang, C., Ma, F., Wang, L., & Zeng, Q. (2021). The information content of uncertainty indices for natural gas futures volatility forecasting. *Journal of Forecasting, 40*(7), 1310–1324. https://doi.org/10.1002/FOR.2769.

Liu, L., Wang, E. Z., & Lee, C. C. (2020). Impact of the COVID-19 pandemic on the crude oil and stock markets in the US: A time-varying analysis. *Energy Research Letter, 1*(1), 13154. https://doi.org/10.46557/001c.13154.

Ramírez-Márquez, C., & Ponce-Ortega, J. M. (2023). Macro-index for the assessment of the water-energy-food nexus: Toward accessibility and sustainability of resources. *Process Integration and Optimization for Sustainability, 1–6*, https://doi.org/10.1007/S41660-023-00364-5.

Rubbaniy, G., Khalid, A. A., Syriopoulos, K., & Samitas, A. (2022). Safe-haven properties of soft commodities during times of Covid-19. *Journal of Commodity Markets, 27*, Article 100223. https://doi.org/10.1016/J.JCOMM.2021.100223.

Saâdaoui, F., Ben Jabeur, S., & Goodell, J. W. (2022). Causality of geopolitical risk on food prices: Considering the Russo-Ukrainian conflict. *Finance Research Letter, 49*, Article 103103. https://doi.org/10.1016/J.FRL.2022.103103.

Scikit learn S. (2023). sklearn.preprocessing.StandardScaler. Retrieved from https://scikit-learn.org/stable/modules/generated/sklearn.preprocessing.StandardScaler.html.

Shiferaw, B., Smale, M., Braun, H. J., Duveiller, E., Reynolds, M., & Muricho, G. (2013). Crops that feed the world 10. Past successes and future challenges to the role played by wheat in global food security. *Food Security, 5*(3), 291–317. https://doi.org/10.1007/S12571-013-0263-Y.

Tuna, G., & Tuna, V. E. (2022). Are effects of COVID-19 pandemic on financial markets permanent or temporary? Evidence from gold, oil and stock markets. *Resources Policy, 76*, Article 102637. https://doi.org/10.1016/J.RESOURPOL.2022.102637.

Wang, S., Xu, L., Yu, S., & Wang, S. (2023). Russia-Ukraine war perspective of natural resources extraction: A conflict with impact on sustainable development. *Resources Policy, 85*, https://doi.org/10.1016/J.RESOURPOL.2023.103689.

WFP. (2023). *Food Security Implications of the Ukraine Conflict*. https://docs.wfp.org/api/documents/WFP-0000137463/download/?_ga=2.48605670.1346592276.1652513347-41481742.1648302216. Accessed 14 December 2023.

World Bank. (2022). *Commodity Markets Outlook: The Impact of the War in Ukraine on Commodity Markets.*

Yang, Y., Li, J., Sun, X., & Chen, J. (2014). Measuring external oil supply risk: A modified diversification index with country risk and potential oil exports. *Energy, 68*, 930–938. https://doi.org/10.1016/J.ENERGY.2014.02.091.

Zhang, W., He, X., & Hamori, S. (2023). The impact of the COVID-19 pandemic and Russia-Ukraine war on multiscale spillovers in green finance markets: Evidence from lower and higher order moments. *International Review Financial Analysis, 89*, Article 102735. https://doi.org/10.1016/J.IRFA.2023.102735.

Chapter 8
Clustering Urban Zones: A Study of Gentrification

8.1 Introduction

Gentrification remains one of the most debated and multifaceted concepts in urban studies, with scholars continuing to dispute its precise definition and implications. The term, first introduced by British sociologist Ruth Glass in 1964 to describe changes in London's neighborhoods, has evolved into a complex framework for understanding urban transformation. Glass's original formulation described a rapid process where "once this process of 'gentrification' starts in a district, it goes on rapidly until all or most of the original working-class occupiers are displaced and the whole social character of the district is changed." This foundational definition continues to inform contemporary discussions, though current scholarship recognizes gentrification as a more nuanced phenomenon involving simultaneous demographic, economic, spatial, and cultural transformations (Kennedy & Leonard, 2001; Smith, 1996).

Modern conceptualizations view gentrification as a comprehensive neighborhood metamorphosis where multiple dimensions of urban life undergo significant change. At its core, the process involves the influx of more affluent residents into previously working-class or economically disadvantaged areas, bringing with them new consumption patterns, housing preferences, and cultural norms. This demographic shift typically triggers a cascade of other changes: improved municipal services and infrastructure, upgraded housing stock, new commercial establishments catering to different tastes, and altered streetscapes and architectural aesthetics (Badcock, 2001; Raman, 2014). The physical transformation of neighborhoods often includes not just building renovations but complete changes in land use patterns, as former industrial spaces become loft apartments and local shops give way to boutique retailers (Bailey & Robertson, 1997; Bohl, 2010).

These visible changes reflect deeper socioeconomic transformations. Gentrifying neighborhoods typically experience rising median incomes (Beauregard, 1990), increased property values (Freeman, 2005), and shifts in occupational structures

A. Y. Raya-Tapia et al., *Machine Learning and Clustering for a Sustainable Future*, Studies in Computational Intelligence 1233,
https://doi.org/10.1007/978-3-032-03876-0_8

as professional workers replace blue-collar residents. The commercial landscape evolves to serve new residents, with changes in both the types of businesses and their pricing structures, which can make everyday goods and services less accessible to long-term residents (Bonotti & Barnhill, 2022). While these changes often bring measurable improvements in neighborhood conditions—better schools, reduced crime rates, improved public spaces—they simultaneously make the area unaffordable for existing lower-income residents, forcing displacement through both direct economic pressures and more subtle forms of cultural marginalization (Krieger, 2014; Pérez, 2004).

The psychological and social impacts on displaced or at-risk communities can be profound. Studies have documented increased stress, anxiety, and even sleep disorders among residents facing displacement pressures (Krieger, 2014). Long-term social networks that provided crucial support systems are disrupted, and cultural institutions that anchored community identity often disappear. This creates a paradox where neighborhood "improvement" actually decreases quality of life for vulnerable residents, illustrating how gentrification's impacts vary dramatically based on one's socioeconomic position (Raman, 2014).

Gentrification's unequal consequences extend to environmental justice issues as well. Research has shown that the burdens of environmental hazards—from pollution to inadequate sanitation—disproportionately affect poorer neighborhoods, and these disparities often persist or even worsen during gentrification processes (Brulle & Pellow, 2006). Meanwhile, the benefits of environmental improvements like park upgrades or air quality initiatives tend to accrue to newer, wealthier residents. Similar inequities appear in healthcare access, as gentrification can disrupt existing care networks while making services less affordable (Gibbons & Barton, 2016). The spatial mismatch between displaced residents and their healthcare providers creates additional barriers to access (Morello-Frosch & Shenassa, 2006).

The concept of urban vulnerability provides a valuable lens for understanding these multidimensional impacts. Defined as "the potential for urban populations to be affected by adverse circumstances or conditions of risk" (Alguacil & Camacho, 2014), urban vulnerability encompasses physical, social, economic, and environmental dimensions. Gentrification often increases vulnerability for some groups even as it decreases it for others, creating complex patterns of advantage and disadvantage within the same neighborhoods (Hernández-Aja, 2012). While some researchers argue that gentrification can reduce certain inequities through social mixing and increased investment (Musterd et al., 2016), evidence suggests it more often reshapes rather than resolves urban inequalities, frequently along racial and ethnic lines.

Measuring and analyzing gentrification presents significant methodological challenges. Traditional approaches relying on census data have focused on indicators like income changes, educational attainment shifts, housing price increases, and occupational structure transformations (Atuesta & Hewings, 2019; Aune et al., 2020). More recent studies have incorporated additional metrics like Airbnb penetration rates and property tax increases (Ilic et al., 2019; Wu & Rowe, 2022). Various methodological frameworks have emerged, from geodemographic change analyses (Reibel &

Regelson, 2013) to composite gentrification indices (Chapple & Zuk, 2016), each with strengths and limitations.

A fundamental limitation of many approaches is their reliance on single indicators or arbitrary thresholds, which often produce oversimplified or temporally inconsistent results (Hammel & Wyly, 1996; Slater, 2010). The common practice of using fixed percentage changes or percentile cutoffs (like neighborhoods below the regional median income) introduces artificial boundaries that may not reflect actual urban processes (Galster & Peacock, 1986). Moreover, the selection of indicators varies widely across studies, making comparisons difficult and sometimes leading to contradictory findings (Easton et al., 2020).

These measurement challenges reflect deeper conceptual issues in gentrification research. The process operates differently across cities and countries, influenced by local housing policies, economic conditions, and cultural factors (Finio, 2021). What constitutes significant change in one context may be negligible in another. Additionally, gentrification's multidimensional nature means that focusing solely on, say, economic indicators misses crucial cultural or political dimensions (Best et al., 2023).

Emerging computational methods offer promising solutions to these challenges. Unsupervised machine learning techniques like K-means clustering can identify patterns across multiple variables without imposing predetermined thresholds or categories (Zuk et al., 2018). By analyzing numerous socioeconomic and physical indicators simultaneously, these approaches can capture gentrification's complexity in ways traditional methods cannot (Hwang & Sampson, 2014). They also allow for more dynamic, temporal analyses that better reflect the processual nature of neighborhood change (Li et al., 2023).

The application of machine learning in gentrification research represents an important methodological advance, but also raises new questions about interpretation and policy application. As these techniques become more sophisticated, they offer the potential to develop more nuanced, context-sensitive understandings of urban transformation—understandings that could inform more equitable policy responses to the challenges gentrification presents (Reades et al., 2018). However, the ethical implications of data-driven urban analysis require careful consideration, particularly regarding how findings might be used to either mitigate or accelerate displacement pressures.

Ultimately, gentrification remains what Glass first recognized—a complex, multifaceted process that simultaneously creates and destroys, improves and excludes (Glass, 1964). Understanding its full impacts requires approaches as multidimensional as the phenomenon itself, combining quantitative precision with qualitative sensitivity to the lived experiences of urban residents across the socioeconomic spectrum. As cities continue to evolve, developing more sophisticated analytical frameworks will be crucial for creating urban policies that harness neighborhood change's positive potential while minimizing its human costs.

8.2 Machine Learning Approaches to Analyzing Gentrification

Machine learning, a critical subset of artificial intelligence, encompasses diverse data processing methodologies that can be broadly categorized into supervised and unsupervised learning techniques. These computational approaches are increasingly being applied to study complex urban phenomena like gentrification, offering new insights through advanced pattern recognition and predictive modeling. Supervised machine learning relies on labeled datasets containing both input data and corresponding correct outputs to train classification and regression algorithms. Classification algorithms excel at categorizing data into specific groups, while regression algorithms predict continuous outcomes by analyzing relationships between variables (Razzaghi et al., 2023; Tiwari, 2022). These methods have proven valuable in gentrification research, particularly in analyzing neighborhood transformations. For instance, decision tree-based algorithms have been successfully implemented in Sydney to predict gentrification patterns using the Socioeconomic Index of Advantage and Disadvantage, a composite measure developed by the Australian Bureau of Statistics through principal component analysis of 21 variables encompassing education, occupation, income, housing, and employment metrics. Zeng et al. (2022) employed XGBoost, an advanced decision tree algorithm that outperforms traditional methods like random forest, to examine how points of interest (POI) data—including art galleries, boutique shops, and specialty cafes—can serve as indicators of gentrification by revealing changes in local business landscapes.

The advent of big data and deep learning has enabled sophisticated image processing applications in urban studies. Thackway et al. (2023) developed a Siamese convolutional neural network that analyzes Google Street View imagery to detect property renovations as physical evidence of gentrification, effectively bridging the gap between social and physical indicators of neighborhood change. Similarly, Gray et al. (2023) compared three ensemble models—gradient boosting machine, extreme gradient boost, and bootstrap aggregation—to predict the spatial and temporal patterns of gentrification across national scales using variables including housing prices, occupational demographics, and racial composition.

Airbnb's role in gentrification has also been examined through machine learning approaches. Shan et al. (2023) developed indices and multiple linear regression models to analyze how short-term rental platforms interact with gentrification pressures in Beijing, finding differential impacts based on neighborhoods' initial gentrification levels by comparing housing market transactions with Airbnb data.

Unsupervised learning techniques, particularly clustering algorithms, offer alternative approaches for analyzing unlabeled urban data. These methods group similar data points based on inherent patterns without predefined categories (Chander & Gopalakrishnan, 2023; Tyagi et al., 2022). K-means clustering has been widely applied in gentrification studies, with Knorr et al. (2019) using it to track neighborhood changes through six census variables (housing values, rents, income, race, education, and multi-unit buildings) between 2000–2016. While this approach captures the natural progression of neighborhood change regardless of initial conditions, the limited variable selection may overlook important aspects of urban transformation.

Other studies have expanded variable selection in clustering analyses. Liu et al. (2019) incorporated seven indicators including property ownership data, occupational patterns, income levels, and ethnic composition in their K-means analysis, though they noted challenges in explicitly interpreting results in gentrification terms. Best et al. (2023) employed K-method clustering to examine climate gentrification, incorporating livability indices, socioeconomic factors, minority status, housing values, population data, and disaster frequency metrics.

The current study applies K-means clustering to analyze gentrification patterns across 46 demographic, social, educational, urban, environmental, and healthcare variables. Using three distinct datasets—including point data from 2015 and 2020 for each municipality plus variance measures between these periods—this approach enables comprehensive typology development of municipal groupings. This multidimensional analysis not only identifies gentrification patterns but also reveals associated vulnerabilities, providing a nuanced understanding of urban transformation processes across different community contexts. The methodology offers advantages over traditional threshold-based approaches by allowing natural patterns to emerge from complex, high-dimensional data without imposing arbitrary classification boundaries.

8.3 Problem Statement

Gentrification is a complex urban phenomenon that triggers significant transformations across demographic, economic, social, architectural, and infrastructural dimensions as higher-income residents displace lower-income populations. Given its multifaceted nature, there is no standardized approach to measuring gentrification globally, making it challenging to develop targeted policies that protect vulnerable communities. Traditional methods often rely on assessing structural or compositional neighborhood changes through composite indices, which, while useful for identifying gentrification patterns, often sacrifice interpretability. However, accurately capturing gentrification requires evaluating multiple indicators to distinguish it from other forms of neighborhood change. Conventional approaches frequently employ arbitrary threshold values to classify gentrification, which can introduce bias due to their limited dimensionality and subjective selection criteria. In contrast, machine learning techniques such as K-means clustering offer a more robust alternative by simultaneously analyzing a wide range of socioeconomic, demographic, educational, health, and environmental variables. This method allows for a data-driven assessment of neighborhood change without imposing restrictive thresholds, enabling a more nuanced understanding of gentrification dynamics. Rather than predicting gentrification likelihood, this study focuses on mapping the spatial distribution of municipal vulnerabilities to the process. While many existing studies concentrate solely on social dimensions, this research adopts a more comprehensive framework by integrating urban design, climate factors, healthcare access, and housing conditions. The analysis incorporates three distinct datasets: two raw datasets (untreated by scores or transformations) and a third dataset tracking percentage variations in key features between 2015 and 2020. This approach helps determine whether significant compositional shifts occurred during this period, providing insights into gentrification's temporal and spatial progression. By clustering municipalities based on these variables, the study identifies stable groupings with varying degrees of gentrification susceptibility while highlighting correlations among different urban characteristics. This neighborhood typification aids in assessing gentrification patterns and informing equitable urban development strategies. Additionally, statistical and correlation analyses ensure accurate variable selection, while principal component analysis (PCA) reduces dimensionality, identifying the most influential factors driving gentrification trends. Ultimately, this methodology offers policymakers a more reliable, data-driven tool to recognize at-risk areas and implement measures that mitigate displacement and promote inclusive urban growth.

8.4 Case Study Addressed

The urban transformation of Mexico City's central districts began in the late nineteenth century when low-income populations migrated to the city core, leading to the proliferation of multi-family dwellings and the iconic "vecindades" (traditional tenement housing). This migration pattern created significant urban densification while simultaneously contributing to area depreciation. The housing landscape became further complicated in 1942 when presidential decrees implemented rent control policies, maintaining artificially low housing and commercial rental prices. While these measures initially stimulated lower-income housing construction by 1950, they ultimately discouraged property maintenance and real estate investment as landlords found their returns limited (Hanakata et al., 2022). This policy environment gradually degraded the central zone's housing stock and urban fabric.

A pivotal moment occurred with the devastating 1985 earthquake that killed hundreds and destroyed thousands of homes, triggering mass depopulation of the central city. The post-disaster period saw significant policy shifts, including the 1992 deregulation of rental markets, which attracted upper-class housing developments but faced strong resistance from working-class communities (Anson, 2022). Subsequent revitalization programs launched in 2000, initially framed as pro-poor housing initiatives, were ultimately co-opted by private developers who predominantly constructed middle- and high-income residences (Delgadillo, 2016; McDermott, 2019). This pattern demonstrates how real estate interests have increasingly influenced urban policy to facilitate gentrification processes.

As Mexico's political, economic, and cultural capital, Mexico City (CDMX) represents a prime case study for examining contemporary gentrification dynamics

(Fig. 8.1). The city faces multiple intersecting pressures accelerating neighborhood transformation, including what scholars term "architectural heritage dispossession"—where real estate capital revalues historic districts while displacing traditional residents and their socioeconomic practices (Delgadillo, 2016; Janoschka, 2016). The proliferation of short-term rental markets, particularly through platforms like Airbnb, has exacerbated these trends by incentivizing property owners to convert long-term housing into tourist accommodations. This phenomenon intensified during the COVID-19 pandemic when Mexico's relaxed mobility restrictions attracted thousands of higher-earning "digital nomads," primarily from the United States, further distorting local housing markets.

Recent municipal agreements with Airbnb, projected to generate $3 billion annually from digital nomad tourism, have sparked community concerns about rising living costs and cultural displacement. The income disparity between these temporary residents and local populations has driven inflation in housing prices, goods, and services. This economic pressure coincides with problematic real estate development practices, including illegal construction expansions and violations of heritage protection laws. Official records from 2019–2023 document 469 unauthorized additional floors across 299 buildings, with 66 cases involving protected historic structures developed without proper approvals. These practices not only violate urban planning

Fig. 8.1 Geographical place of Mexico City

regulations but also increase seismic vulnerability in a city already prone to devastating earthquakes, demonstrating how gentrification pressures can compromise both social equity and physical safety in urban environments.

Gentrification in Mexico City has led to severe socioeconomic challenges, including skyrocketing rental prices that have become unaffordable for many residents. This has resulted in forced displacements—sometimes through violent evictions—pushing vulnerable populations into more economically precarious areas, including high-risk zones and protected natural lands, often through informal settlements. Despite being displaced, many of these individuals continue to work and access essential services in their original neighborhoods, facing significant mobility challenges due to long commutes and rising transportation costs.

Despite these visible effects, Mexico City lacks official studies or established indicators to systematically measure gentrification. This gap presents an opportunity to leverage existing data to analyze and visualize the phenomenon, raising awareness among policymakers and the public to implement protective measures for affected communities. Our research approaches gentrification as a complex, multidimensional process encompassing economic, demographic, social, educational, environmental, ethno-racial, healthcare, and urban development factors. By examining these interconnected dimensions, we aim to categorize municipal clusters and assess their relationship with gentrification dynamics. The evidence of gentrification in Mexico City is undeniable, marked by urban renewal projects that rehabilitate neighborhoods while simultaneously displacing and marginalizing low-income residents. The city's 16 territorial demarcations (see Table 8.1) each exhibit distinct patterns of this phenomenon, reinforcing the need for a comprehensive, data-driven approach to understand its mechanisms and mitigate its consequences.

Table 8.1 Regions of Mexico City (CDMX)

N°	Municipality	N°	Municipality
1	Álvaro Obregón	9	Iztapalapa
2	Azcapotzalco	10	La Magdalena Contreras
3	Benito Juárez	11	Miguel Hidalgo
4	Coyoacán	12	Milpa Alta
5	Cuajimalpa de Morelos	13	Tláhuac
6	Cuauhtémoc	14	Tlalpan
7	Gustavo A. Madero	15	Venustiano Carranza
8	Iztacalco	16	Xochimilco

8.5 Implemented Methodology

The research methodology, as illustrated in Fig. 8.2, follows a structured analytical process. Initially, two raw datasets from 2015 and 2020 were acquired and processed to calculate percentage variance. Subsequent analysis examined correlations between the datasets' variables, followed by data normalization during the preprocessing phase. The optimal number of clusters (K) was determined through application of both elbow and silhouette methods. Using this K value, the K-means clustering algorithm was implemented, with subsequent evaluation of the resulting clusters across both datasets. The analytical process concluded with principal component analysis (PCA), a dimensionality reduction technique employed to facilitate the creation of composite indices and identify key patterns within the data. This comprehensive approach enabled systematic examination of the multidimensional aspects of gentrification across different temporal periods.

8.5.1 Data Collection

Given the predominant focus on social dimensions in existing gentrification research, this study adopts a more holistic methodological framework to examine the phenomenon's multiple facets. The analysis incorporates 46 distinct indicators spanning ten critical dimensions: demographic composition, healthcare access, ethnic-racial characteristics, economic conditions, educational attainment, urban environment, service availability, security levels, housing patterns, and environmental factors. The research utilizes official datasets from multiple authoritative sources. Economic, demographic, health, and educational metrics were extracted from INEGI's comprehensive surveys (INEGI, 2023), including the Population

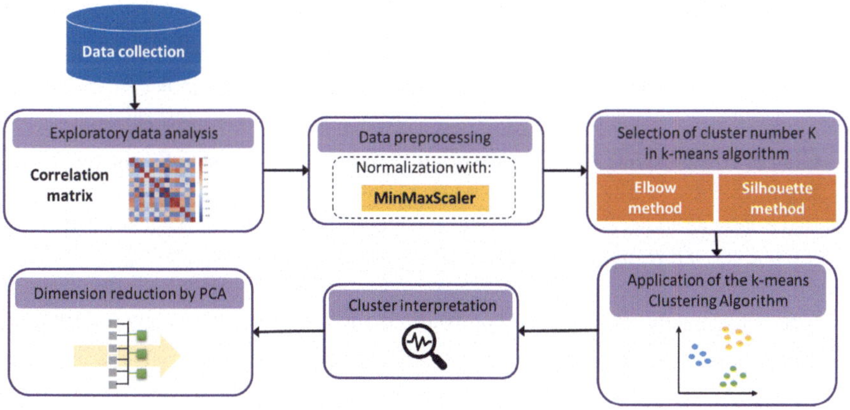

Fig. 8.2 Overall implemented methodology

and Housing Census, Occupation and Employment Survey (ENOE), National Housing Survey (ENVI), and Economic Census. Crime statistics were obtained from CDMX's Attorney General's Office (FGJ, 2023), while tourism data came from the city's Tourism Department (CDMX Tourism, 2023). Environmental indicators were derived from SEDEMA's annual air quality reports (SEDEMA, 2023). All data were collected at the municipal level across CDMX's 16 administrative divisions. Table 8.2 presents the complete set of 46 indicators collected for each municipality between 2015–2020, along with their calculated percentage variations across this five-year period. This comprehensive dataset enables a multidimensional analysis of gentrification patterns and their temporal evolution throughout Mexico City.

For all variables, the 2015–2020 percentage variation was derived through the following Eq. (8.1).

$$\Delta\% = \frac{V_1 - V_0}{V_0} \times 100 \tag{8.1}$$

where V_1 signifies the value of the attribute in 2020, and V_0 corresponds to the value of the attribute in 2015.

8.5.2 *K-Means Clustering Procedure*

The K-means algorithm operates through the following computational steps (Fig. 8.3).

1. Cluster Initialization: First determine optimal cluster count K using elbow and silhouette validation methods. Then randomly initialize K centroids μ_k^t, $\forall k \in K$ for the first iteration (t = 0) from dataset I.
2. Distance Computation: Calculate n-dimensional Euclidean distances between each data point x_i and all centroids μ_k using Eq. 8.2.

$$D(x_i, \mu_k)^t = \sqrt{\sum\nolimits_{n=1}^{p} (x_{i,n} - \mu_{k,n}^t)^2} \, \forall i \in I, k \in K \tag{8.2}$$

3. Cluster Assignment: Assign each point x_i to its nearest centroid's cluster μ_k^t via Eq. 8.3.

$$D(x_i, \mu_k^t) = min\{D(x_i, \mu_k^t), \forall i \in I, k \in K\} \rightarrow x_i \in C_k^t \tag{8.3}$$

4. Centroid Update: Recompute centroids as mean vectors of assigned points by Eq. 8.4.

$$\mu_k^{t+1} = \frac{1}{C_k^t} \sum\nolimits_{x_i \in C_k^t} x_i \, \forall k \in K \tag{8.4}$$

Table 8.2 The K-means clustering analysis employed 46 distinct variables across all 16 municipal districts

ID	Features
Demographic indicators	
F1	Fertility rate per thousand residents
F2	Mortality rate per thousand residents
F3	Resident population count
F4	Yearly population increase percentage
F5	Residents per square kilometer
F6	Residential mobility percentage (changed address in 5 years)
Healthcare access	
F7	Health insurance coverage rate
F8	Doctors per thousand residents
F9	Healthcare facilities per 100 k residents
Employment and diversity	
F10	Workforce participation rate
F11	Percentage of African-descended residents
F12	Indigenous language speakers percentage
F13	Poverty and extreme poverty prevalence
Education metrics	
F14	Preschool enrollment (ages 3–5)
F15	Elementary school attendance (ages 6–11)
F16	Middle school enrollment (ages 12–14)
F17	High school/college attendance (ages 15–24)
F18	Adult literacy rate (15+ years)
F19	Mean years of education (15+ population)
Urban services	
F20	Hotel rooms per 100 k residents
F21	Restaurants and food services per 100 k
F22	Number of commercial bank facilities per 100,000 eligible residents (18+)
F23	State development bank branches per 100 k adult inhabitants (18+)
F24	Law enforcement officers per 100 k
F25	Sports facilities per 100 k residents
F26	Retail stores per 100 k population
F27	Vehicle ownership rate per thousand adults
Public safety	
F28	Reported crimes per 100 k adult population
Housing conditions	
F29	Electrified households percentage

(continued)

Table 8.2 (continued)

ID	Features
F30	Plumbing access percentage
F31	Sewage system coverage
F32	Daily waste generation per person (kg)
F33	Persons per room ratio
F34	Homeownership rate
F35	Dwellings with solid flooring
F36	Homes with sturdy roofing
Real estate	
F37	Property price per square meter
Environmental factors	
F38	Total yearly rainfall (mm)
F39	Mean annual humidity levels
F40	Average yearly temperature (°C)
F41	Sulfur dioxide levels (ppb)
F42	Nitrogen dioxide concentration (ppb)
F43	Carbon monoxide measurements (ppm)
F44	Ozone exceedance hours
F45	Particulate matter (PM10) levels
F46	Fine particulate (PM2.5) concentrations

5. Convergence Evaluation: Calculate sum of squared errors (SSE) by Eq. 8.5.

$$SSE_{C_k^{t+1}} = \sum_{k=1}^{K} \sum_{x_i \in C_k^{t+1}} (x_i - \mu_k^{t+1})^2 \tag{8.5}$$

Terminate if $\left| SSE_{C_k^{t+1}}^{t+1} - SSE_{C_k^t}^{t} \right| \leq \xi$ (convergence threshold).

Otherwise repeat from step 2. This iterative process continues until centroid positions stabilize within the specified tolerance ξ, indicating optimal cluster formation. The algorithm minimizes intra-cluster variance while maximizing inter-cluster separation through successive approximation.

8.5.3 *Determining the Optimal Number of Clusters (K)*

This section details the methodology for determining the appropriate number of clusters (K) for our analysis. The optimal K value can be established through two approaches: when prior knowledge of data distribution exists, researchers may define custom clustering criteria; alternatively, empirical methods like elbow and silhouette

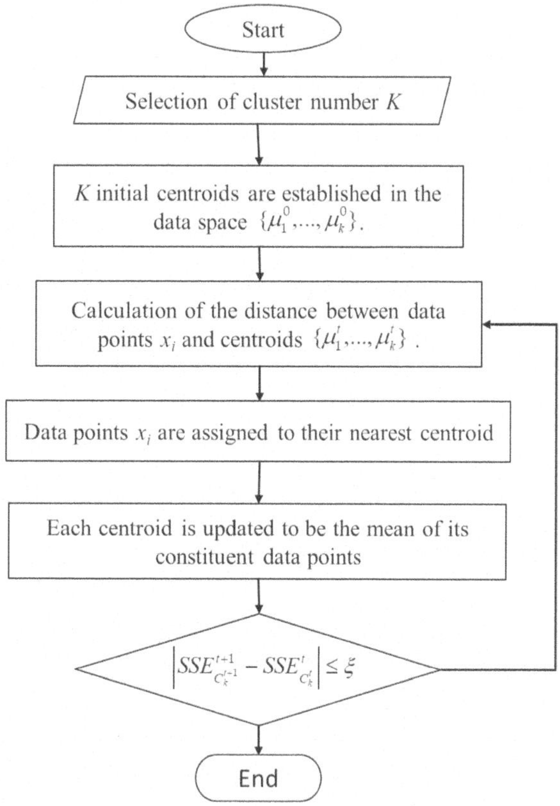

Fig. 8.3 K-means clustering algorithm

analysis provide data-driven estimates of cluster quantity. Our study employed the latter approach, utilizing these validation techniques to identify the most suitable cluster configuration. Complete methodological details and graphical results of both the elbow and silhouette analyses appear in the Appendix section.

8.5.3.1 Dimensionality Reduction Through Principal Component Analysis

The analysis implemented Principal Component Analysis (PCA) to streamline the dataset while maintaining critical dimensional relationships. This technique operates by:

1. Identifying orthogonal projection axes (principal components) that optimally preserve data structure
2. Utilizing Euclidean distance metrics to assess observation dissimilarities
3. Sequentially capturing variance through components:

- The primary component (PC1) accounts for maximum data variance.
- Subsequent components (PC2, PC3, etc.) each capture residual variance while remaining orthogonal to previous components.

As demonstrated by Vlachos (2017), PCA offers dual advantages: it reduces storage requirements through data compression while simultaneously revealing feature significance through component loading analysis. This approach enables researchers to both minimize dataset size and understand how original variables contribute to each derived component, providing valuable insights into the underlying data structure. The component loadings effectively map how different features influence each principal dimension, facilitating interpretation of the reduced-dimensional space.

8.6 Results

8.6.1 Amount of Clusters

To establish robust clustering parameters, we implemented complementary validation techniques—the elbow method and silhouette analysis. This dual approach was necessary because:

1. The ideal cluster count varies by dataset characteristics
2. Temporal differences between the 2015 and 2020 datasets warranted independent evaluation
3. The derived percentage variation dataset required separate assessments

All three analyses (2015 data, 2020 data, and their percentage change) converged on $k = 3$ as the optimal cluster number. This consistency across different data configurations suggests strong underlying patterns in the municipal characteristics. The Appendix provides comprehensive visualizations and quantitative results supporting this determination.

8.6.2 Cluster Interpretation and Characterization

The output from the K-means algorithm requires systematic interpretation to properly characterize each identified group. This analytical phase involves examining the distinctive properties of each cluster and assigning them meaningful labels based on their defining characteristics. Our interpretation methodology combined two approaches: examination of summary statistics for all municipalities within each cluster, and theoretical understanding of gentrification dynamics. The cluster centroids -representing the mean values of all variables within each group- served as the foundation for this analysis, as they mathematically define each cluster's core attributes.

Demographic Distinctions Between Clusters: The analysis revealed striking demographic variations across the gentrification spectrum.

1. High-gentrification areas showed exceptional population mobility, with the proportion of recent migrants (those changing residence within five years) exceeding levels in medium- and low-gentrification zones by more than 100%
2. Medium-gentrification municipalities demonstrated distinctive population scales, maintaining approximately twice the resident count of both high- and low-gentrification areas
3. Low-gentrification communities exhibited distinctive vital statistics, with per capita birth rates substantially surpassing those in other classification tiers

Temporal Evolution of Population Patterns: Comparative analysis of the 2015–2020 period uncovered significant spatial and temporal dynamics.

- Population density leadership shifted from medium-gentrification areas in 2015 to high-gentrification zones by 2020
- Medium-gentrification municipalities consistently showed the most sluggish population growth rates throughout the study period
- The growth rate leadership transitioned from low-gentrification areas in 2015 to high-gentrification regions by 2020.

The high-gentrification cluster emerged as particularly distinctive, with 29 variables registering elevated values in the baseline year (2015), as comprehensively documented in Table 8.3. These quantitative findings enable robust characterization of gentrification patterns across different municipal contexts.

In both years studied, highly gentrified municipalities show improved health outcomes due to a larger proportion of residents enrolled in health services, over twice as many medical staff compared to less gentrified areas, and a higher ratio of medical facilities per person. Although moderately gentrified communities have fewer medical units per capita than those with low gentrification, this could be due to their larger population size. Gentrification reshapes the type and geographic availability of health care, often benefiting wealthier residents while marginalizing vulnerable groups, such as minorities, low-income individuals, and uninsured patients. However, the proximity of health services is just one factor in accessibility and equitable distribution (Levesque et al., 2013). As neighborhoods gentrify, long-term residents may lose consistent access to care if they or their providers are displaced, forcing them to travel farther for treatment. Health care providers also face difficulties in addressing patients' needs amid complex social challenges—such as homelessness, unsafe housing, or exposure to violence—as well as threats of displacement and the logistical hurdles of providing follow-up care to patients who have been forced out of their communities. Table 8.4 highlights that the medium gentrification group has the highest values, whereas the low gentrification group has the lowest.

Table 8.3 Cluster analysis revealed characteristic attribute scores for each group in 2015, particularly showing higher values in high-gentrification zones

ID	Feature	Gentrification		
		Low	Medium	High
F2	Mortality rate per thousand residents	5.3	7.4	8.4
F6	Residential mobility percentage (changed address in 5 years)	6.6	7.5	16.6
F7	Health insurance coverage rate	79.3	79.3	79.8
F8	Doctors per thousand residents	1.4	1.7	3.7
F10	Workforce participation rate	43.6	44.6	51.9
F14	Preschool enrollment (ages 3-5)	69.5	73.8	78.8
F16	Middle school enrollment (ages 12-14)	95.9	96.8	97.0
F17	High school/college attendance (ages 15-24)	53.6	57.9	64.6
F18	Adult literacy rate (15+ years)	97.2	97.9	98.9
F19	Mean years of education (15+ population)	10.5	11.1	12.9
F20	Hotel rooms per 100k residents	201.5	205.9	2569.3
F22	Number of commercial bank facilities per 100,000 eligible residents (18+)	18.0	20.6	65.9
F24	Law enforcement officers per 100k	5.3	8.7	40.4
F26	Retail stores per 100k population	11.8	19.1	21.1
F27	Vehicle ownership rate per thousand adults	630.3	721.6	1317.7
F28	Reported crimes per 100k adult population	1644.9	2433.1	4646.9
F29	Electrified households percentage	99.6	99.9	100.0
F30	Plumbing access percentage	96.1	99.4	99.8
F32	Daily waste generation per person (kg)	1.3	2.1	2.1
F35	Dwellings with solid flooring	98.3	98.8	99.3
F36	Homes with sturdy roofing	87.8	94.3	97.3
F41	Sulfur dioxide levels (ppb)	3.2	4.6	5.1
F46	Fine particulate (PM2.5) concentrations	20.2	25.9	26.1
F9	Healthcare facilities per 100k residents	7.6	5.3	10.0
F11	Percentage of African-descended residents	2.0	1.5	2.4
F21	Restaurants and food services per 100k	4.9	4.3	53.3
F23	State development bank branches per 100k adult inhabitants (18+)	0.6	0.5	1.2
F31	Sewage system coverage	98.78	98.75	99.20
F37	Property price per square meter	32126.0	22873.0	51644.0

Table 8.4 Average attribute values per group in 2015 (Medium-level gentrification with elevated scores)

		Gentrification		
ID	Feature	Low	Medium	High
F3	Resident population count	398378.4	803099.3	438136.0
F5	Residents per square kilometer	3560.9	13689.9	13287.4
F40	Average yearly temperature (°C)	16.0	17.6	17.2
F42	Nitrogen dioxide concentration (ppb)	21.1	32.1	31.7
F43	Carbon monoxide measurements (ppm)	0.5	0.9	0.8
F45	Particulate matter (PM10) levels	36.3	45.0	43.7

Municipalities experiencing high gentrification tend to have distinct socioeconomic and ethnic profiles: a greater proportion of employed individuals and Afro-descendant residents, alongside fewer people speaking indigenous languages or living below the poverty line. In contrast, areas with minimal gentrification exhibit the opposite trends. The demographic shifts in indigenous and impoverished populations may suggest displacement in some cases—or at least a slowing decline of low-income groups in heavily gentrified towns. Table 8.5 highlights the key characteristics most prevalent in low-gentrification areas.

Across both years studied, highly gentrified communities exhibit higher school attendance rates among children aged 3–5 and youth aged 15–24 compared to other areas. These regions also demonstrate elevated literacy rates among residents over 15. On average, populations in high-gentrification zones complete over one additional year of schooling than those in moderately gentrified areas and two more years than residents in low-gentrification towns. By 2020, the high-gentrification group encompassed 24 high-value indicators, detailed in Table 8.6.

Urban areas with high gentrification exhibit markedly different characteristics compared to less gentrified communities. Data from both years show that per capita, these neighborhoods contain ten times the number of hotel rooms, triple the banking branches, double the development banks, and approximately twice as many vehicles as medium- or low-gentrification areas. They also host more restaurants and public safety personnel, yet experience crime rates nearly double those of other areas. Research (Lee, 2010) indicates that while gentrification may lower crime over time, it initially elevates it—likely due to socioeconomic disparities between new and long-term residents, as well as the increased appeal of targeting affluent newcomers. Additionally, heavily gentrified areas show a pronounced concentration of hospitality businesses, reflecting tourism's role in economic revitalization (Santos-Izquierdo et al., 2023), driven by expanded lodging and visitor-focused services. Conversely, low-gentrification communities boast more sports facilities per capita, and four consistent high-value traits were identified in moderately gentrified towns (Table 8.7).

Table 8.5 Standardized attribute scores per group in 2015 (Low gentrification but elevated values)

		Gentrification		
ID	Feature	Low	Medium	High
F1	Fertility rate per thousand residents	16.3	15.1	14.1
F12	Indigenous language speakers percentage	1.9	1.3	1.2
F13	Poverty and extreme poverty prevalence	54.6	39.2	14.9
F25	Sports facilities per 100k residents	10.5	8.5	4.6
F33	Persons per room ratio	0.9	0.8	0.7
F34	Homeownership rate	60.2	51.0	46.4
F38	Total yearly rainfall (mm)	684.3	619.8	600.3
F39	Mean annual humidity levels	58.6	55.5	52.7
F44	Ozone exceedance hours	294.3	204.3	138.7
F4	Yearly population increase percentage	0.6	-0.2	0.4
F15	Elementary school attendance (ages 6-11)	98.2	97.6	97.9

Housing characteristics vary significantly by gentrification level. High-gentrification areas produce more daily waste per resident and contain more homes with sturdy roofs, while property values average over double those in medium-gentrification zones—the most affordable market segment. These conditions contribute to reduced homeownership rates and lower household density. Utility access patterns shifted between 2015 and 2020: initially leading in electricity, water, and sewage services, high-gentrification areas were surpassed by medium-gentrification municipalities by 2020, revealing that lower-income residents often experience "housing poverty"—home ownership without adequate infrastructure.

Environmental conditions show similar disparities. Medium-gentrification towns consistently exhibited poorer air quality, with elevated NO_2, CO, and particulate matter (PM_{10} and $PM_{2.5}$) levels in both study years. Their higher vehicle ownership rates correlated with increased SO_2 and $PM_{2.5}$ in 2015, and SO_2 and NO_2 by 2020. This demonstrates unequal pollution distribution across gentrification levels, with high-gentrification areas experiencing cleaner air than their medium-gentrification counterparts. Low-gentrification localities maintained better air quality but recorded higher rainfall and humidity levels (Table 8.8). These findings support

Table 8.6 Standardized attribute values across groups in 2020 (High gentrification areas showing elevated scores)

		Gentrification		
ID	Feature	Low	Medium	High
F5	Residents per square kilometer	3024.3	13199.6	13990.4
F6	Residential mobility percentage (changed address in 5 years)	6.0	7.0	14.7
F7	Health insurance coverage rate	71.7	73.4	76.9
F8	Doctors per thousand residents	0.9	1.5	3.4
F10	Workforce participation rate	52.9	53.6	60.3
F11	Percentage of African-descended residents	1.8	2.0	2.8
F14	Preschool enrollment (ages 3-5)	70.8	73.7	79.4
F17	High school/college attendance (ages 15-24)	52.5	56.5	62.1
F18	Adult literacy rate (15+ years)	97.9	98.3	99.1
F19	Mean years of education (15+ population)	10.8	11.5	13.3
F20	Hotel rooms per 100k residents	202.5	227.3	2461.1
F21	Restaurants and food services per 100k	417.1	449.2	986.1
F22	Number of commercial bank facilities per 100,000 eligible residents (18+)	14.1	19.2	55.5
F27	Vehicle ownership rate per thousand adults	701.8	829.1	1461.5
F28	Reported crimes per 100k adult population	2001.6	2622.0	4673.7
F32	Daily waste generation per person (kg)	1.6	1.7	2.0
F36	Homes with sturdy roofing	89.4	95.4	97.8
F41	Sulfur dioxide levels (ppb)	2.2	3.3	3.8
F42	Nitrogen dioxide concentration (ppb)	15.9	25.4	26.2
F4	Yearly population increase percentage	1.3	0.5	1.3
F9	Healthcare facilities per 100k residents	6.5	4.5	7.9
F23	State development bank branches per 100k adult inhabitants (18+)	0.8	0.7	1.4
F24	Law enforcement officers per 100k	19.1	14.9	70.2
F37	Property price per square meter	27501.7	23402.4	45024.7

Table 8.7 Standardized attribute measurements across all groups in 2020 showed elevated values in moderately gentrified areas

		Gentrification		
ID	Feature	Low	Medium	High
F2	Mortality rate per thousand residents	9.5	12.9	11.1
F3	Resident population count	358735.3	809003.6	464835.7
F26	Retail stores per 100k population	14.1	19.3	19.2
F30	Plumbing access percentage	95.2	99.8	99.4
F31	Sewage system coverage	99.4	99.9	99.5
F35	Dwellings with solid flooring	98.3	99.6	99.4
F40	Average yearly temperature (°C)	16.2	18.4	18.3
F43	Carbon monoxide measurements (ppm)	0.28	0.44	0.39
F44	Ozone exceedance hours	173.7	230.1	202.7
F46	Fine particulate (PM2.5) concentrations	20.9	23.6	23.1
F29	Electrified households percentage	99.8	99.9	99.5
F45	Particulate matter (PM10) levels	40.9	41.5	38.1

existing research (Wong et al., 2013) showing disproportionate pollution burdens in non-gentrified communities compared to gentrified or gentrifying neighborhoods.

The variations in high-value features across each group are detailed in Appendix Table 8.13. Environmental analysis reveals that towns with medium gentrification faced worse conditions, exhibiting higher concentrations of NO_2, CO, and PM_{10} in 2015, as well as CO, O_3, PM_{10}, and $PM_{2.5}$ in 2020. These municipalities also had the highest motorization rates, contributing to elevated SO_2 and $PM_{2.5}$ levels in 2015 and increased SO_2 and NO_2 by 2020. This indicates an unequal distribution of air pollution across gentrification levels: high-gentrification areas experienced lower pollution shares compared to medium-gentrification zones. Meanwhile, low-gentrification localities recorded lower pollution levels but higher cumulative precipitation and relative humidity (Table 8.8). These findings align with existing research (e.g., Wong et al., 2013), which shows that air pollution disproportionately affects non-gentrifying and partially gentrified areas.

Figure 8.4 presents the $K = 3$ clustering results for 2015 and 2020, classifying municipalities into high, medium, and low gentrification based on their characteristics. In both years, three districts—Benito Juárez, Cuauhtémoc, and Miguel Hidalgo—were consistently categorized as high-gentrification areas.

Table 8.8 Standard scores for each group's attributes in 2020 (low gentrification with high average values)

		Gentrification		
ID	Feature	Low	Medium	High
F1	Fertility rate per thousand residents	7.2	7.1	6.7
F12	Indigenous language speakers percentage	1.8	1.2	1.2
F13	Poverty and extreme poverty prevalence	70.8	56.8	24.5
F15	Elementary school attendance (ages 6-11)	95.8	95.6	94.4
F16	Middle school enrollment (ages 12-14)	93.6	93.6	93.2
F33	Persons per room ratio	0.9	0.8	0.7
F34	Homeownership rate	66.4	54.4	49.1
F39	Mean annual humidity levels	53.3	51.7	48.3
F25	Sports facilities per 100k residents	7.5	2.3	4.4
F38	Total yearly rainfall (mm)	369.3	338.7	340.0

8.6.3 *Cluster Analysis of Variation Patterns*

The analysis of percentage variation data reveals distinct patterns across clusters. Cluster 4 demonstrates concerning health system declines, showing reductions in health coverage (affiliated population), physician availability (doctors per 1,000), and medical infrastructure (units per 100,000 inhabitants), as documented in Table 8.9. This cluster displays minimal increases in poverty rates among African-descent populations and shows divergent housing trends—decreasing numbers of buildings with durable floors alongside negligible improvements in weather-resistant roofing. Demographic shifts include declining birth rates and increasing mortality, potentially leading to reduced household density. While workforce participation and educational attainment show positive trends, these are offset by rising housing costs and increased criminal activity. Environmental indicators present a mixed picture: decreasing annual precipitation, humidity levels, and concentrations of SO_2/NO_2 contrast with rising temperatures and increased exposure to ozone, $PM_{2.5}$, and PM_{10} pollutants. Geographic analysis reveals disproportionate ozone impacts in southern Mexico City municipalities, while northern and central areas—the primary sources of precursor chemicals—experience lower pollution levels due to predominant north–south morning wind patterns that transport these compounds southward.

Cluster 5 exhibits different characteristics, with falling housing prices coinciding with significant increases in homeownership rates, despite substantial reductions in basic utility access (water and electricity). This cluster shows accelerated population growth and densification, coupled with growing numbers of African-descendant residents and expanding poverty levels. Contrasting trends emerge in education and cultural indicators, with sharp declines in indigenous language speakers and school

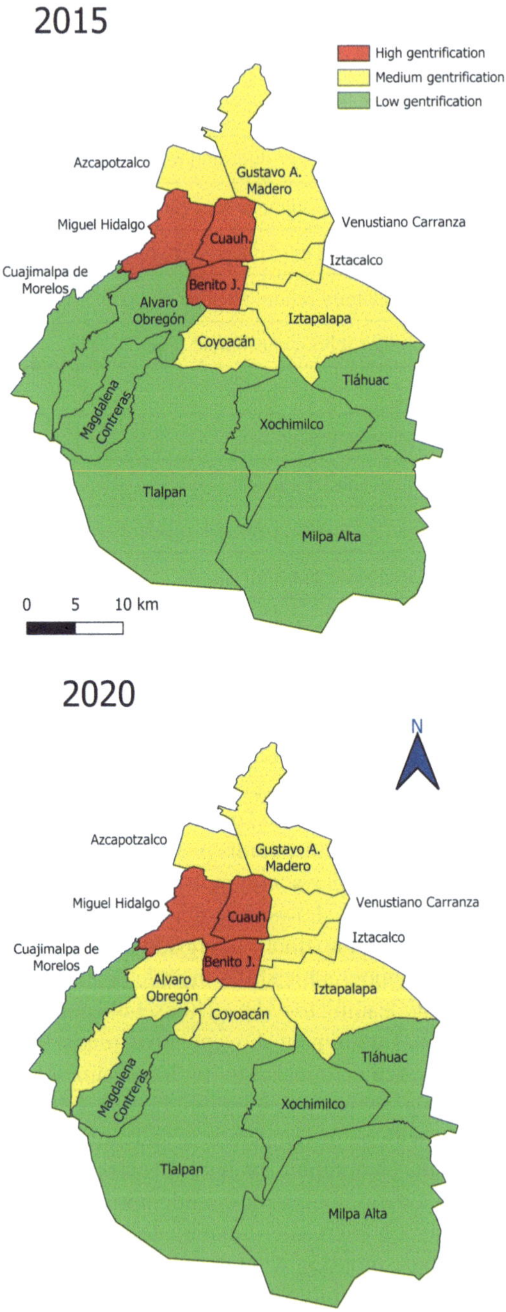

Fig. 8.4 Grouping of municipalities by the rank of gentrification in 2015 and 2020

Table 8.9 Average indicator values across all categories in the percentage variance clustering results

ID	Feature	Cluster		
		4	5	6
F1	Fertility rate per thousand residents	−57.60	−51.30	−54.00
F2	Mortality rate per thousand residents	95.00	63.70	71.70
F3	Resident population count	9.20	11.50	3.10
F4	Yearly population increase percentage	1.80	2.20	0.60
F5	Residents per square kilometer	9.60	11.70	3.10
F6	Residential mobility percentage (changed address in 5 years)	−9.40	−18.00	−5.30
F7	Health insurance coverage rate	−13.10	−4.50	−7.10
F8	Doctors per thousand residents	−40.20	−21.40	−2.90
F9	Healthcare facilities per 100 k residents	−21.70	−17.01	−17.03
F10	Workforce participation rate	29.90	15.20	19.40
F11	Percentage of African-descended residents	1.50	38.70	27.20
F12	Indigenous language speakers percentage	−13.30	−11.60	−5.70
F13	Poverty and extreme poverty prevalence	12.40	70.80	44.30
F14	Preschool enrollment (ages 3–5)	6.50	−1.60	1.00
F15	Elementary school attendance (ages 6–11)	−2.20	−4.00	−2.30
F16	Middle school enrollment (ages 12–14)	−2.40	−3.20	−3.00
F17	High school/college attendance (ages 15–24)	−2.70	−3.20	−1.70
F18	Adult literacy rate (15+ years)	0.57	0.16	0.59
F19	Mean years of education (15+ population)	5.20	0.40	3.50
F20	Hotel rooms per 100 k residents	2.80	4.70	1.30
F22	Number of commercial bank facilities per 100,000 eligible residents (18+)	−5.30	−12.30	−12.70
F23	State development bank branches per 100 k adult inhabitants (18+)	84.80	−47.00	29.20
F25	Sports facilities per 100 k residents	−0.70	−13.90	−37.10
F26	Retail stores per 100 k population	24.90	−2.40	25.50
F27	Vehicle ownership rate per thousand adults	12.60	8.60	15.50
F28	Reported crimes per 100 k adult population	39.00	6.60	13.60
F29	Electrified households percentage	−0.04	−0.60	0.10
F30	Plumbing access percentage	0.70	−0.60	−0.01
F31	Sewage system coverage	0.30	0.20	1.00
F32	Daily waste generation per person (kg)	−7.60	30.50	−2.20
F33	Persons per room ratio	−4.90	7.60	−1.90
F34	Homeownership rate	7.50	11.10	6.20

(continued)

Table 8.9 (continued)

ID	Feature	Cluster		
		4	5	6
F35	Dwellings with solid flooring	−0.55	−0.49	0.70
F36	Homes with sturdy roofing	1.10	2.20	1.70
F37	Property price per square meter	15.30	−9.50	−13.20
F38	Total yearly rainfall (mm)	−59.40	−43.70	−43.60
F39	Mean annual humidity levels	−13.80	−9.30	−7.20
F40	Average yearly temperature (°C)	8.10	7.60	3.40
F41	Sulfur dioxide levels (ppb)	−34.60	−17.40	−27.80
F42	Nitrogen dioxide concentration (ppb)	−38.80	−8.30	−19.30
F43	Carbon monoxide measurements (ppm)	−36.30	−51.30	−47.40
F44	Ozone exceedance hours	37.20	−45.00	13.10
F45	Particulate matter (PM10) levels	29.20	6.10	−5.10
F46	Fine particulate (PM2.5) concentrations	15.20	−0.02	−6.50

attendance among 3–24 year olds, resulting in slower improvements in literacy rates, educational achievement, and labor market participation. Additionally, this cluster experienced more modest increases in vehicle ownership and crime rates compared to other areas.

Cluster 6 exhibits distinct demographic patterns characterized by sluggish population growth, low density, and limited housing availability. Notably, this cluster shows minimal growth in homeownership rates despite declining property values, along with a smaller reduction in residents maintaining secondary residences. The most pronounced improvements appear in vehicle ownership rates, literacy levels, and housing infrastructure (including electricity access, drainage systems, and durable flooring). Environmental indicators reveal decreased $PM_{2.5}$ and PM_{10} levels alongside increased precipitation and humidity. These climatic factors significantly influence pollution dispersion, particularly through atmospheric washing—where rainfall effectively removes airborne particulates (>2.5 μm) and water-soluble gases from the atmosphere.

The analysis reveals an important housing market dynamic: Cluster 5's reduced prices enabled vulnerable populations to purchase more centrally-located properties, as opposed to Cluster 4's cheaper but more remote options. This finding underscores the value of combining clustering methodologies—using both point-in-time data and percentage variation analysis—to identify long-term gentrification patterns that might otherwise remain obscured when examining neighborhood changes within limited timeframes. Figure 8.5 presents the 2015–2020 change rate clustering, showing uneven municipal distribution (Clusters 4/5: 2 municipalities each; Cluster 6: 12 municipalities). The K-means approach employed here creates geographically-defined housing submarkets, enabling temporal comparison of cluster stability across years. Crucially, the variation-based clustering (Fig. 8.5) diverges from the raw data clustering (Fig. 8.4), suggesting characteristic changes among municipal groups

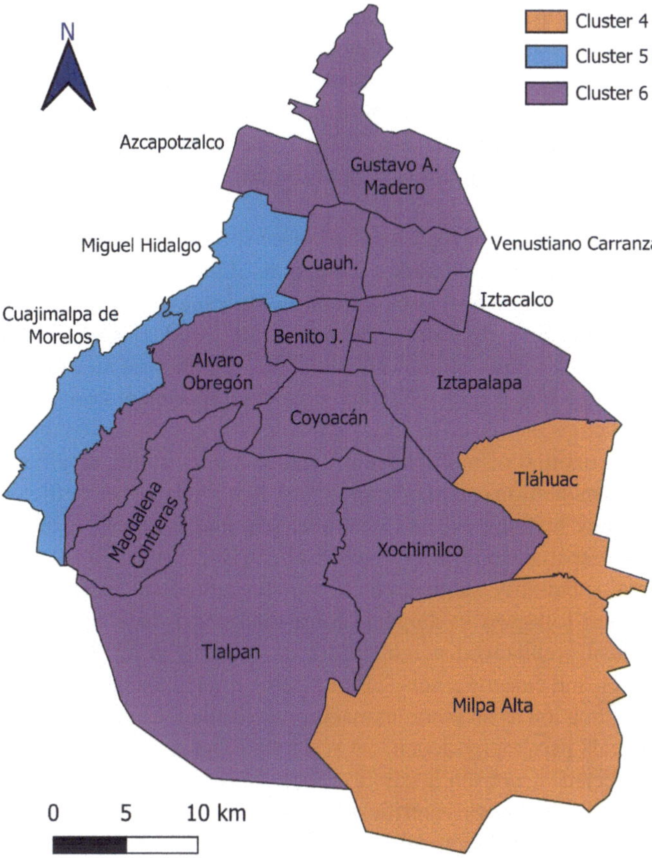

Fig. 8.5 Grouping of municipalities by utilizing percentage transformation among 2015 and 2020

predated the 2015–2020 period. Had these transformations occurred within the study period, both clustering methods would have produced more similar results.

Gentrification manifests differently across temporal and spatial contexts (Aalbers, 2019), with varying drivers and consequences depending on location (Cole et al., 2021; Anguelovski et al., 2021). Research demonstrates that housing price appreciation primarily occurs during early gentrification phases when incomes and property values remain relatively low, stabilizing in later stages (Atuesta & Hewings, 2019). This pattern suggests displacement pressures peak during initial gentrification before market conditions stabilize. Crucially, gentrification represents an evolutionary process rather than a discrete event, progressing through distinct phases at variable rates (Williams, 2015), with its impacts showing significant geographical and temporal variation (Wyly & Hammel, 1998).

In Mexico City, urban renewal policies have accelerated gentrification and spatial inequality through neighborhood redevelopment initiatives. These interventions enhance area desirability through improved infrastructure and transit, creating opportunities for real estate speculation. Developers capitalize on rising land values by catering to affluent buyers, exacerbating urban disparities (Anson, 2022; Delgadillo, 2016; McDermott, 2019). This investment concentration in established neighborhoods creates uneven development patterns, where privileged groups reclaim well-serviced central districts while marginalized populations are pushed to underserved peripheries (Janoschka et al., 2014). These dynamics underscore the importance of examining regional variations and multidimensional inequality within specific housing submarkets.

The study reveals pronounced housing price differentials across gentrification levels, suggesting distinct submarket formations. Land value fluctuations significantly influence urban transformation patterns, with developers responding to Mexico City's price surges through vertical expansion. By 2020, Benito Juárez (75.7%), Cuauhtémoc (73.3%), and Miguel Hidalgo (60.3%) showed the highest vertical growth rates (EVALÚA, 2021), reflecting intensified apartment development. These submarket variations enable differential gentrification speeds across the metropolis, warranting examination of multiple gentrification stages within single urban areas. Such market segmentation emerges from spatial–temporal market inefficiencies, critically determining housing access, locational advantages, wealth distribution, and ultimately socio-spatial segregation (Bolt et al., 2010).

This spatial inequality stems from a two-stage process: initial built environment devaluation followed by targeted infrastructure investments in select areas. These strategically enhanced neighborhoods -featuring prime locations, accessibility, amenities, and cultural assets (Smith, 1996) attract concentrated capital flows, initiating gentrification as an inherent market mechanism. Cluster analysis of 2015–2020 data reveals geography-dependent real estate dynamics, with socioeconomic disparities reflected in uneven access to urban services. The findings demonstrate growing spatial divisions as gentrification advances unevenly across municipalities. This unbalanced development stems from land price differentials that persist despite recent housing price fluctuations (Aguilar-Velázquez et al., 2024), with spatial inequality remaining stubbornly high or continuing to grow.

The interplay between real estate markets and market-influenced governance amplifies Mexico City's gentrification by widening land value gaps and spatial disparities. Effective policy responses require reliable cluster identification, enabling targeted interventions. However, methodological challenges persist, including temporal data limitations and inadequate analytical tools for spatiotemporal analysis. Where longitudinal data exists, cluster stability can be assessed, allowing for methodology adaptation to local contexts while maintaining comparative analytical value.

8.6.4 Principal Component Analysis (PCA)

Figure 8.6 presents the variance distribution and cumulative variance curves from our dimensionality reduction analysis. The results demonstrate that 16 principal components (PCs) capture the complete variance (cumulative variance = 1), effectively preserving all information from the original 46 variables without data loss.

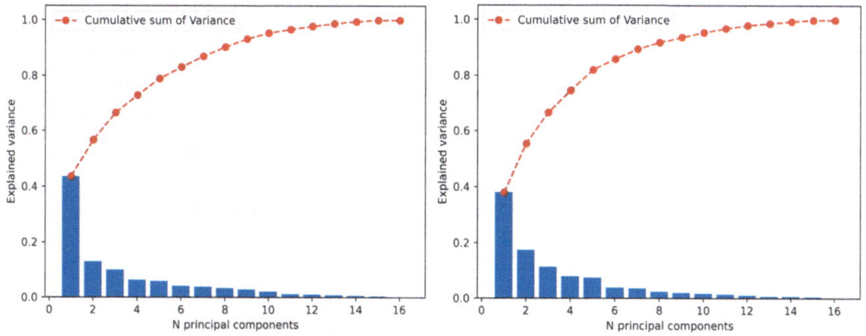

Fig. 8.6 Variance and cumulative variance plotted against principal component count for 2015 (left panel) and 2020 (right panel)

These PCs represent weighted linear combinations of the initial features, where each variable contributes differentially to each component.

The selection of optimal components involves balancing information retention with dimensionality reduction. For our analysis, we retained 12 PCs as they collectively explained 98% of the total variance in both study years. As illustrated in Fig. 8.6, the first principal component dominates the explained variance (accounting for 43% in 2015 and 39% in 2020), with the second component contributing substantially less (13% and 16% for 2015 and 2020 respectively).

The first principal component for 2015, represented in Eq. 8.2, emerges as a linear composite of the original 46 variables. A closer examination of the component loading matrices (provided in Appendix Tables 8.14 and 8.15) reveals the specific contribution patterns of each variable, offering deeper insight into the informational content captured by individual principal components.

$$PC1 = 0.139F1 - 0.187F2 - 0.011F3 + 0.05F4 \cdots - 0.110F45 - 0.175F46 \tag{8.6}$$

Each variable is weighted by its respective loading factor, though precisely quantifying each attribute's contribution proves challenging. Figure 8.7 display the ten most influential variables (with extreme positive/negative loadings) in the dominant first two components. These strongly correlated features include F13 (showing the strongest PC1 association) and F33. The supplementary materials provide complete loading and correlation matrices. Positive loadings indicate variables that positively influence the component, while negative loadings show inverse relationships, with magnitude reflecting relative importance.

For 2015 (Fig. 8.7), PC1 combines key socioeconomic and environmental indicators suitable for index development. It positively correlates with poverty rates, household density, homeownership, humidity levels, and indigenous language use, while inversely relating to durable roofs, employment, education metrics, literacy, and crime rates—suggesting high PC1 scores mark low-gentrification zones. PC2 integrates demographic, urban, and environmental measures, showing positive associations with temperature, population density, PM10/NO2 levels, low property values, medical service availability, school attendance (ages 6–11), and vehicle ownership. These patterns indicate medium-gentrification areas typically score high on PC2.

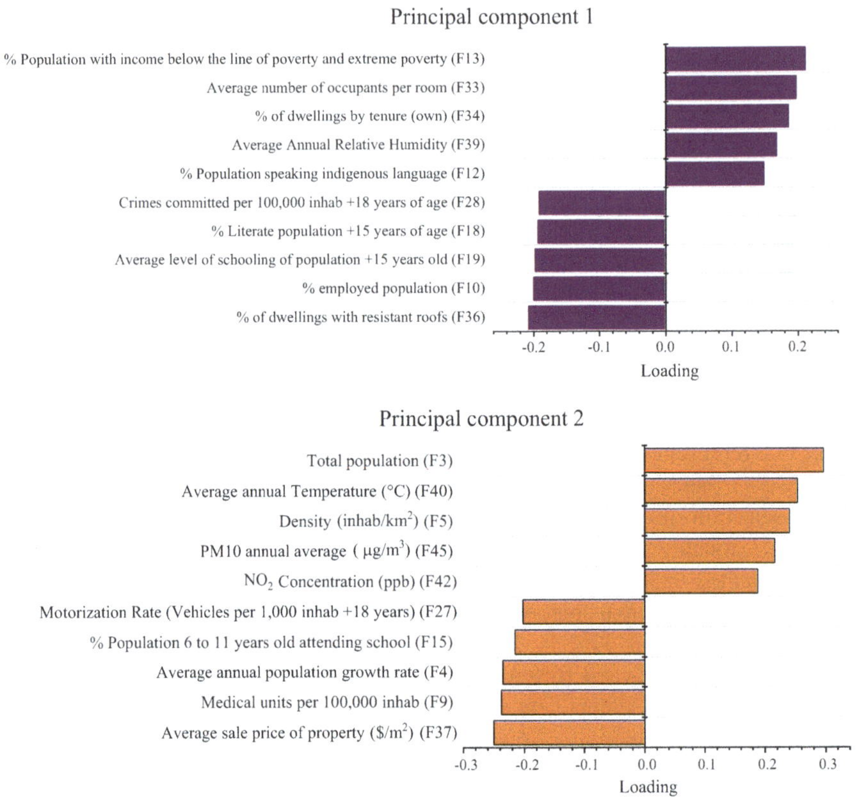

Fig. 8.7 The first two principal components' loading values (positive and negative) for the 2015 dataset

The loading patterns for 2020, illustrated in Fig. 8.8, reveal both continuities and evolutionary shifts in the principal component structure when compared to 2015. For the first principal component (PC1), we observe the persistence of three negatively-loaded features from the previous period (F19, F36, and F18), suggesting these variables maintain their inverse relationship with the component's underlying construct. Notably, two new indicators emerge as significant contributors—the motorization rate (F27) and the proportion of population maintaining multiple residences (F6)—reflecting evolving urban dynamics. The diagnostic pattern for low-gentrification areas remains robust, characterized by depressed values in these core features coupled with elevated scores in school attendance metrics (F15) and the recurring positive features (F13, F33, F34, and F12). This consistent signature across temporal periods strengthens the validity of these composite indicators for identifying neighborhood trajectories. More dramatically, the second component (PC2) demonstrates several polarity reversals in feature loadings: F3 transitioned from positive to negative valence, while F4, F37, and F9 exhibited the opposite shift from negative to positive loadings. These transformed associations now position PC2 as capturing

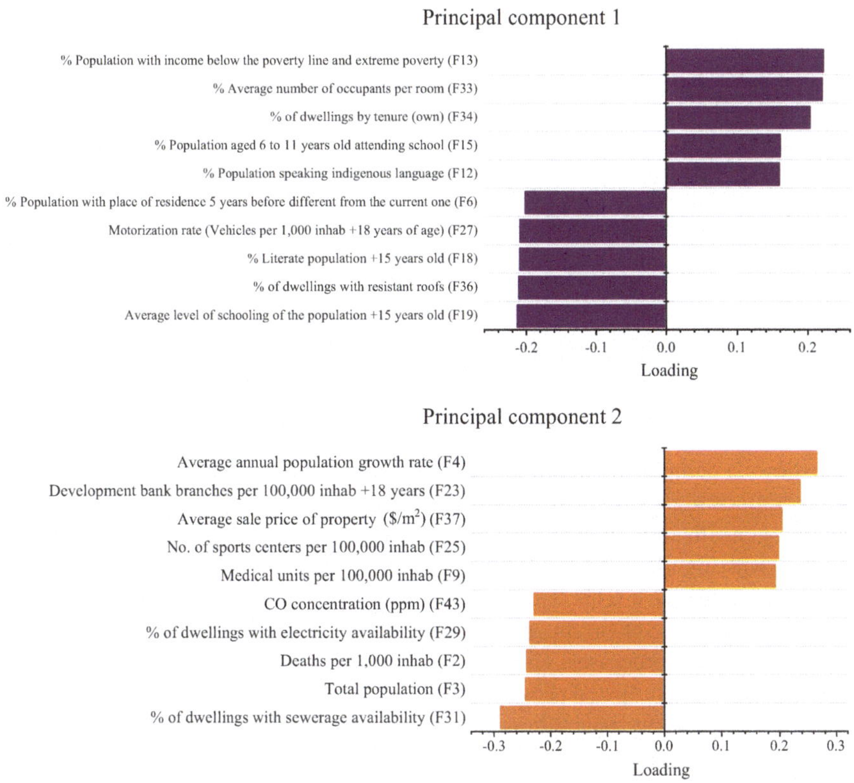

Fig. 8.8 The first two principal components' loading values (positive and negative) for the 2020 dataset

an emergent urban phenotype marked by accelerated demographic growth, financial sector expansion, housing market inflation, enhanced recreational infrastructure, and improved healthcare access. The inverse profile—featuring constrained utility access, elevated population density, and depressed CO_2 levels—may reflect either infrastructural deficits in rapidly growing areas or alternatively, more sustainable urban forms. These loading reversals suggest fundamental changes in the covariance structure of urban indicators between periods, possibly signaling shifts in the underlying socioeconomic processes driving urban transformation.

Figure 8.9 presents a comprehensive visualization of municipal gentrification patterns through principal component analysis. The scatter plot maps municipalities across PC1 and PC2 axes, with color-coding indicating gentrification levels (green = low, yellow = medium, red = high). PC1 reveals a consistent spatial hierarchy: strongly negative values identify three high-gentrification municipalities, intermediate values correspond to medium-gentrification areas, and positive values mark low-gentrification zones—a stable pattern across both years demonstrating PC1's reliability in detecting low-gentrification characteristics. More significantly,

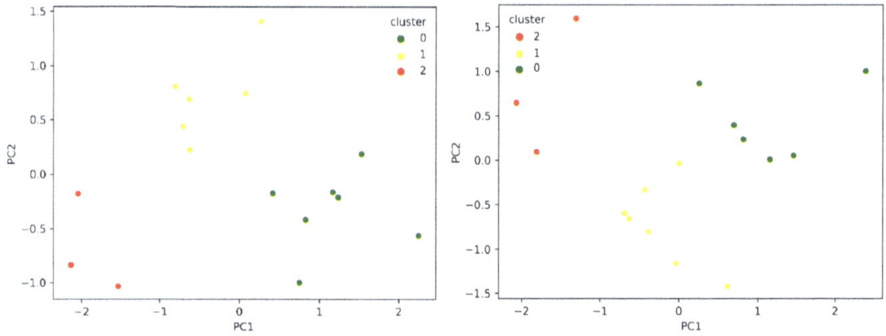

Fig. 8.9 Cluster visualization relative to principal components 1 and 2 for 2015 (left panel) and 2020 (right panel)

PC2 shows a complete temporal reversal in cluster positioning. The 2015 configuration grouped high and low gentrification areas together in negative territory, while medium-gentrification zones occupied positive space. By 2020, this relationship inverted, with high/low clusters shifting to positive values and medium-gentrification areas moving negative. This dramatic flip suggests evolving urban dynamics, where the demographic and environmental factors captured by PC2 now interact differently with gentrification processes. The visualization thus serves dual purposes: as a diagnostic tool for current spatial patterns and as a sensitive indicator of temporal changes in urban development trajectories, potentially reflecting shifts in housing markets, infrastructure investment, or population movements across the metropolitan area.

8.7 Conclusions

This study applies K-means clustering to analyze gentrification patterns across Mexico City's 16 municipalities using 46 socioeconomic, educational, urban infrastructure, health service, demographic, and environmental indicators. The investigation utilizes both raw data and percentage variance metrics to assess gentrification disparities between 2015 and 2020, establishing three distinct classification groups (low, medium, and high gentrification) based on characteristic attribute patterns. High-gentrification areas demonstrate population growth, improved service provision (medical, educational, hospitality), and economic development (increased businesses, banks, and vehicles), though accompanied by higher crime rates. While basic housing infrastructure (electricity, water, sewage) peaked in 2015, construction quality (durable flooring/roofing) improved by 2020 alongside significant property value increases. Medium-gentrification zones transitioned from having the highest population density in 2015 to being surpassed by high-gentrification areas by 2020, while maintaining the lowest housing prices despite infrastructure improvements. These areas also showed consistently higher pollution levels (CO, PM10) and temperatures. Low-gentrification regions were characterized by greater poverty rates, indigenous populations, birth rates, and homeownership percentages, along with distinctive environmental conditions of increased precipitation and humidity. Principal Component Analysis identified key differentiating factors: PC1 separated high/low gentrification areas primarily through socioeconomic indicators (poverty levels, housing occupancy, indigenous populations), while PC2 distinguished medium-gentrification zones through demographic and service accessibility metrics. The research provides crucial insights for urban policymakers to address spatial inequalities and displacement risks, though acknowledges the current municipal-scale analysis could be enhanced through future neighborhood-level investigations as more granular data becomes available, enabling more precise monitoring of gentrification dynamics across Mexico City's diverse urban landscape.

8.8 Appendix

8.8.1 Exploratory Data Analysis

The Pearson correlation coefficient quantifies the association between two variables (Choi et al., 2010; Edelmann et al., 2021). In this instance, it assesses the correlation among 46 variables, with the results shown via a heat matrix. This study aims to discover variables exhibiting a significant positive correlation (>0.9) and a strong negative correlation (<−0.9) with the purpose of eliminating them to prevent redundant information. Information is deemed redundant when certain variables may be deduced from others. This can affect the instability of the final centroids whenever the original centroids are altered.

The K-means clustering algorithm involves attributes measured on varying scales; consequently, employing the Euclidean distance formula directly could result in certain attributes being overshadowed by others with larger scales. Therefore, normalizing all attribute values is a common practice (Scikit-learn, 2023; Trebuňa et al., 2014). Equation A8.1 illustrates the correlation computation, utilizing variables y and z as an example.

$$r_{y,z} = \frac{\sum_{i=1}^{n}(y_i - \overline{y})(z_i - \overline{z})}{\sqrt{\sum_{i=1}^{n}(y_i - \overline{y})^2 \sum_{i=1}^{n}(z_i - \overline{z})^2}} \tag{8.6.8}$$

Normalization denotes the process of scaling characteristics to a range of [0, 1], representing a specific instance of minimum–maximum scaling. Minimum–maximum scaling is utilized to standardize the data across each feature column (Equation A8.2).

$$x_i^{norm} = \frac{x_i - x_{min}}{xmin_{max}} \tag{8.8}$$

x_i^{norm} corresponds to the normalized value, x_i is the actual value of the attribute, $x_{\max}$ and $x_{\min}$ are the maximum and minimum values of the attribute. This can be accomplished by using the MinMaxScaler scaler within the skleanr.preprocessing package (Scikit-learn, 2023).

8.8.1.1 Elbow Method

The elbow approach is a technique for interpreting and validating consistency in cluster analysis, aimed at determining the optimal number of clusters in a dataset. The SSE metric (Equation A8.6) facilitates the comparison of various K-means cluster configurations, with K ranging from 1 to 8 clusters in this instance (UPM, 2023) (Choi & Seong, 2020). The quantity of K clusters is represented on the x-axis, while the internal SSE corresponding to each K is depicted on the y-axis. The graph exhibits a substantial alteration in the slope, indicating the ideal number of K clusters for the dataset.

8.8.1.2 Silhouette Method

According to Choi and Seong (2020), the silhouette approach is employed to assess the quality of clustering and validate the outcomes of cluster analysis. It is also among the most often used techniques for choosing the ideal number of K clusters. According to Zhang et al. (2018), the silhouette coefficient is an assessment criterion that is determined exclusively by the intraclass and interclass distances of every pair of objects.

The silhouette coefficient is determined by utilizing the mean distance within the group (a) and the mean distance from the nearest group (b) for each sample x_i in the set X. For each x_i is calculated a_i as the mean distance to the rest of the points in its group. And b_i is calculated as the mean distance to the next nearest group. The silhouette value x_i is obtained from Equation A8.3.

$$S_i = \frac{b_i - a_i}{max\{a_i, b_i\}} \tag{8.9}$$

If $a_i = b_i$ is true then $S_i = 0$, which signifies that x_i can be an isolated point and/or is very close to or on the decision boundary between two clusters which may cause overlapping clusters.

If $a_i < b_i$, the value of S_i is equal or close to 1, representing that the data x_i are properly grouped and far from other clusters.

If $a_i > b_i$, the value of S_i is equal or close to -1, denoting that the data x_i is unsuccessfully clustered and near neighboring clusters, it should be more applicable to cluster it with a neighboring group.

The average of all values is the silhouette coefficient for a cluster in K groups (UPM, 2023). For a range of K values, the silhouette technique computes the observations' silhouette coefficient. According to Kaufman and Rousseeuw (1990), the ideal number of K clusters is the one that maximizes the silhouette coefficient across a range of potential values for K. Since this approach does not allow for the existence of a single cluster, K is changed between 2 and 6. The number of clusters, K, is shown on the x-axis of the silhouette technique plot, while the silhouette coefficient is shown on the y-axis (Table 8.10).

8.8.2 Results

8.8.2.1 Correlation

Intense red indicates a positive correlation (+1) and intense blue indicates a negative correlation (−1) between the 46 characteristics for the years 2015 and 2020, respectively, as shown in Figs. 8.10 and 8.11. The feature pairs were arranged using a correlation coefficient from highest to lowest in order to improve viewing and analysis. The feature pairs with positive correlation coefficients between 0.8 and 1 and negative correlation coefficients between −0.8 and −1 were analyzed.

Along with other factors, two features showed the most unfavorable connections in 2015. F13 and F33 are these characteristics. The population characteristics showing the strongest association with poverty include: elevated rates of residents below poverty and extreme poverty thresholds (F13), reduced educational attainment among adults, lower unemployment figures (F10), decreased vehicle ownership (F27), fewer individuals maintaining multiple residences (F6), reduced preschool enrollment (F14), limited banking infrastructure (F22), fewer homes with durable roofing, lower secondary/college attendance (F17), and higher illiteracy rates (F18). Household crowding demonstrates a positive relationship with poverty indicators F13 and F33, though F33 paradoxically shows negative correlations when considering its inverse associations with factors like employment (F19, F10), education access (F14, F17), transportation access (F27), residential mobility (F6), financial services (F22), and another key poverty metric (F36)—all of which typically decrease as household density increases.

F22 and F19 were the two traits with the most positive relationships. This indicates that, in addition to a higher percentage of employed people (F10), a higher rate of motorization (F27), a higher percentage of people with two or more places of residence (F6), a higher average level of education for people aged 15 and older (F19), more restaurants and food services per 100kpeople (F21), a higher average sale price ($\$/m^2$) (F37), and more crimes committed per 100,000 people aged 18 and older (F28), there are also more banking branches per 100,000 people over the age of 18 (F22).

Table 8.10 Statistical data for each attribute

Attribute	Mean	std	Min	P 0.25	P 0.50	P 0.75	Max
F1	15.4	1.7	11.5	14.9	15.3	16.2	18.3
F2	6.6	1.5	4.6	5.5	6.5	7.7	10.0
F3	557,603.3	418,202.2	137,927.0	364,554.5	416,674.5	625,635.3	1,827,868.0
F4	0.3	0.7	−0.7	−0.3	0.2	0.7	1.6
F5	9,183.0	5,748.1	462.5	3,795.4	9,570.0	13,843.3	16,901.5
F6	8.8	4.3	4.9	5.6	7.7	9.6	19.0
F7	79.4	2.7	75.6	77.4	79.5	81.1	84.1
F8	2.0	1.3	0.5	0.8	1.5	3.0	4.2
F9	7.2	2.9	3.3	5.1	6.9	7.8	13.6
F10	45.5	3.7	40.1	43.0	45.3	46.3	53.6
F11	1.9	0.8	0.8	1.3	1.8	2.4	3.6
F12	1.6	0.8	0.9	1.2	1.4	1.5	4.1
F13	41.4	17.7	8.2	30.9	42.7	52.2	69.2
F14	72.9	6.3	62.2	69.3	72.8	77.0	85.0
F15	97.9	0.9	95.4	97.6	98.1	98.4	98.9
F16	96.4	0.9	94.8	95.8	96.3	96.9	98.4
F17	57.3	5.8	48.5	53.1	56.5	59.0	71.7
F18	97.8	0.8	96.4	97.4	97.6	98.4	99.1
F19	11.2	1.1	9.4	10.4	11.2	11.6	13.8
F20	647.1	1,227.1	40.0	110.0	194.9	443.7	4,927.4
F21	13.7	22.5	0.8	2.1	5.6	10.1	86.1
F22	28.0	21.5	6.6	12.5	20.4	35.4	71.3
F23	0.7	0.5	0.0	0.4	0.6	0.9	2.2
F24	13.2	25.5	3.7	5.4	6.5	6.9	107.6
F25	8.6	5.1	1.8	4.4	9.1	12.2	18.6
F26	16.3	7.0	3.3	12.2	16.6	21.1	27.1
F27	793.4	308.1	428.7	565.7	733.9	900.8	1,529.4
F28	2,503.3	1,285.8	1,205.8	1,891.1	2,029.0	2,746.2	6,389.9
F29	99.8	0.2	99.3	99.7	99.8	99.9	100.0
F30	98.0	3.3	87.2	98.8	99.2	99.6	99.9
F31	98.8	0.3	98.2	98.7	98.9	99.0	99.5
F32	1.8	0.7	0.6	1.4	1.6	2.0	3.0
F33	0.8	0.1	0.6	0.7	0.8	0.9	1.1
F34	54.1	7.7	44.7	49.3	51.7	57.1	73.6
F35	98.7	0.6	97.4	98.6	98.8	98.9	99.7
F36	92.0	4.9	81.0	90.3	93.2	95.7	98.6
F37	32,316.4	15,367.7	13,000.0	19,632.8	30,161.5	44,097.0	66,458.0

(continued)

Table 8.10 (continued)

Attribute	Mean	std	Min	P 0.25	P 0.50	P 0.75	Max
F38	644.4	163.5	447.0	538.3	582.5	672.5	996.0
F39	56.3	3.4	51.0	55.0	56.0	58.3	63.0
F40	16.8	1.3	13.6	16.9	17.1	17.5	18.3
F41	4.1	1.1	2.6	3.5	3.9	5.0	6.1
F42	27.2	6.1	17.0	23.5	29.0	32.5	35.1
F43	0.7	0.2	0.4	0.6	0.7	0.9	1.1
F44	231.4	99.3	87.0	152.8	230.0	294.5	418.0
F45	40.9	6.0	31.0	37.8	38.5	47.0	51.0
F46	23.4	3.4	19.7	20.2	22.6	27.1	27.9

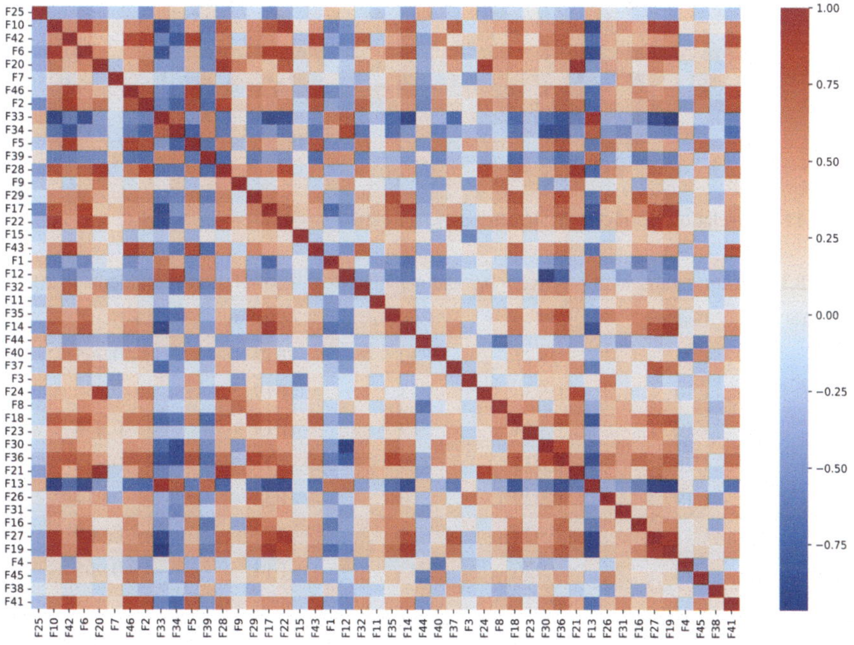

Fig. 8.10 Correlation matrix of the 46 attributes from 2015

Feature F19 indicates that a greater proportion of residents aged 15 and older have a higher level of education. Additionally, characteristics F6, F10, F17, F27, F14, and F22 all drop when F19 does.

The feature pairs with a correlation coefficient $r > 0.9$, or the greatest positive links, and the feature pairs with a correlation coefficient <-0.9, or the strongest negative relationships, for 2015 and 2020 are displayed in Tables 8.11 and 8.12.

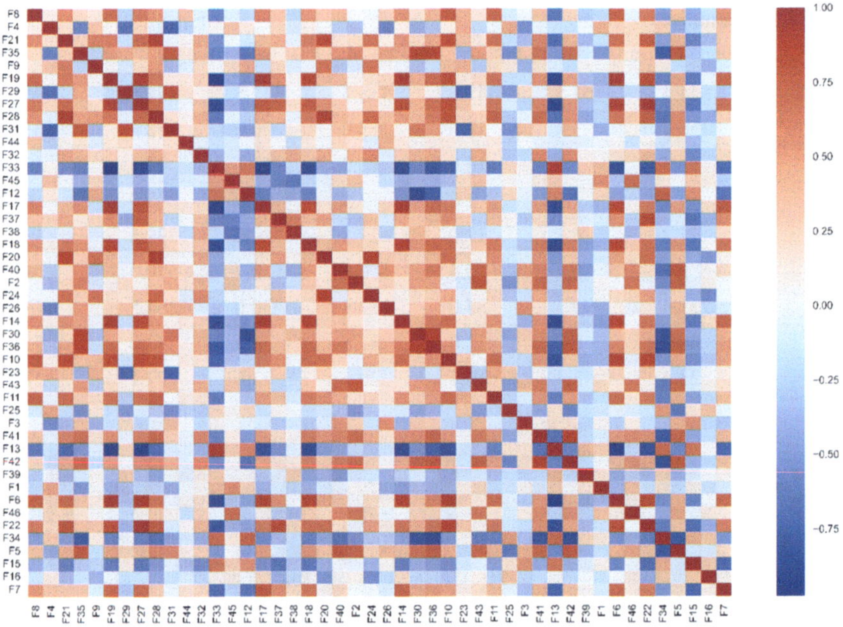

Fig. 8.11 Correlation matrix of the 46 attributes from 2020

Table 8.11 Feature pairs with positive > 0.9 and negative <−0.9 correlation coefficients in 2015

Feature	Feature	*r*	Feature	Feature	*r*
C20	C21	0.971	F30	F12	−0.965
C6	C19	0.936	F33	F19	−0.958
C19	C10	0.934	F19	F13	−0.936
C21	C28	0.931	F10	F13	−0.925
C17	C19	0.929	F10	F33	−0.914
C33	C13	0.929	F13	F27	−0.909
C10	C22	0.926	F33	F14	−0.905
C6	C27	0.925			
C19	C27	0.923			
C27	C22	0.923			
C10	C27	0.923			
C20	C24	0.917			
C20	C28	0.916			
C14	C17	0.915			
C6	C10	0.912			
C19	C14	0.904			

Table 8.12 Attribute pairs with positive correlation coefficients >0.9 and negative correlation coefficients <−0.9 in 2020

Feature	Feature	r	Feature	Feature	r
F27	F22	0.950	F6	F13	−0.951
F19	F6	0.943	F19	F13	−0.937
F17	F19	0.939	F33	F17	−0.930
F14	F17	0.924	F13	F27	−0.925
F19	F14	0.923	F27	F33	−0.903
F20	F28	0.917			
F13	F33	0.916			
F19	F27	0.915			
F28	F21	0.907			
F6	F27	0.906			

F20 and F21 showed the greatest positive association (0.971) in 2015. This indicates that there are more restaurants and food services per 100 k people in locations with a higher number of hotel rooms per 100,000 residents. F30 and F12 showed the biggest negative association. This suggests that fewer residences with piped water are located in locations where a larger proportion of the population speaks an indigenous language.

F13 and F33 were once again the two traits with the most negative associations for 2020. Accordingly, the attributes F6, F19, F27, F18, F17, F14, F22, and F10 all have lower values than in 2015 when a larger proportion of the population earns less than the poverty and extreme poverty level (F13). Similar to 2015, feature F33 likewise shows a negative association with F17, F27, F14, F6, F18, F36, and F22 when examining persons per room averages, F18 stands as the only outlier.

In 2020, the associated pair of characteristics F6 and F19 had the strongest positive correlations. As a result, there are more motorization rates (F27), a higher percentage of employed people (F10), a higher percentage of people aged 3 to 5 (F14) and 15 to 24 (F17) attending school, a higher literacy rate among people aged 15 and older (F18), and more banking branches per 100,000 people aged 18 and older (F22) when a larger percentage of the population owns more than one place of residence (F6). With the exception of F18, F19 has a similar relationship to 2015 in terms of characteristics F6, F17, F14, F27, F18, F10, and F22.

In 2020, a robust positive correlation of 0.95 was identified between F27 and F22, indicating that an increase in the number of banking branches per 100,000 residents aged 18 and older corresponds with a greater rate of motorization. The most pronounced negative correlation was −0.951 between F6 and F13, signifying that an increase in the percentage of the population owning multiple residences over the past five years corresponds with a decrease in the percentage of individuals living below the poverty and extreme poverty thresholds.

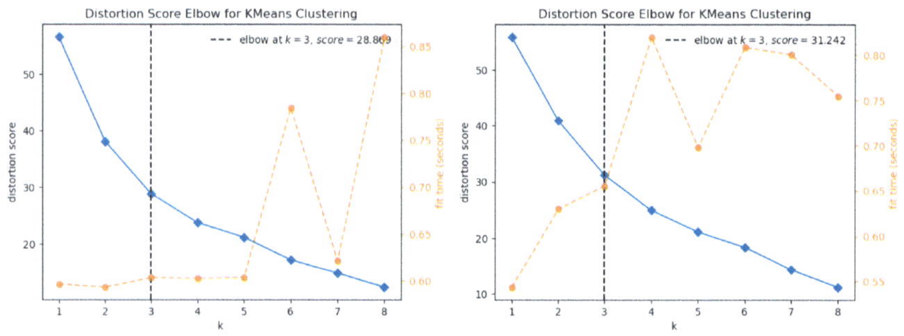

Fig. 8.12 Elbow graph for **a** 2015 (left) and **b** 2020 (right)

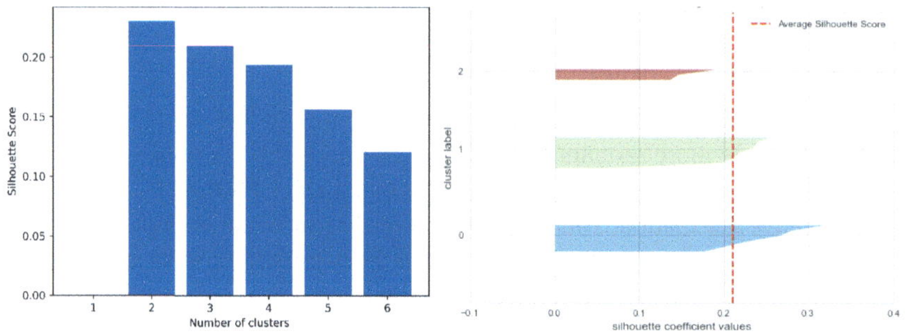

Fig. 8.13 Silhouette method graphs in 2015

8.8.2.2 Elbow Method

The elbow method results indicated that three groups are the optimal number of K clusters for the dataset in both 2015 and 2020. In Fig. 8.12, the x-axis represents the number of K clusters, while the y-axis displays the distortion score, which quantifies the sum of squared distances (SSE). The secondary y-axis illustrates the time expended to train the clustering model at each K.

8.8.2.3 Silhouette Method2015

The silhouette coefficient quantifies the density and spacing of clusters, hence assessing clustering quality. Figure 8.13 presents the outcomes of the silhouette approach for the year 2015. The bar chart displays the number of clusters K on the x-axis and the average silhouette coefficient on the y-axis. K = 2 is the best number of clusters, as evidenced by its maximum average silhouette coefficient of 0.23. Nonetheless, K = 3 yields a proximate value of 0.21; yet, for our classification objectives, K = 3 was selected as the two clusters are not suitable for our needs.

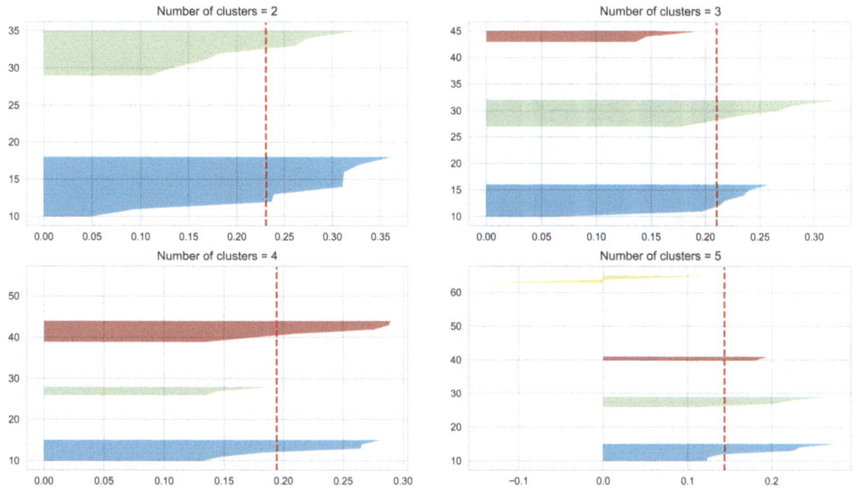

Fig. 8.14 K-means clustering silhouette plot for 16 samples or municipalities at K centers, $K = 2$, 3, 4 and 5 in 2015

The subsequent graph with three centers illustrates the silhouette coefficient for each group. The dashed line indicates the mean silhouette coefficient. One group is seen to be marginally below average, which may indicate a suboptimal quantity of K clusters. The graph illustrates the size of each group; for K = 3, groups 0 and 1 are bigger than group 2.

Figure 8.14 illustrates the silhouette diagrams for K-means clustering across 16 samples or municipalities, with K values of 2, 3, 4, and 5 represented by distinct colors in each diagram. The analysis for K = 4 and K = 5 reveals a reduced number of clusters, with those present exhibiting suboptimal conditions, as some clusters fall below the average silhouette coefficient. Additionally, the analysis indicates values below 0, suggesting potential inaccuracies in the data assigned to these clusters; therefore, K = 2 and K = 3 emerge as the most favorable clustering options.

8.8.2.4 2020

Figure 8.15 displays the results of the silhouette approach for 2020. The bar chart shows that the ideal number of clusters is K = 4 with 0.191 followed by K = 3 with 0.180.

The silhouette plot for K-means clustering with K = 3 indicates that all three clusters have values beyond zero, with just one cluster possessing a silhouette coefficient below the average silhouette coefficient. In contrast to the K = 4 and K = 5 clusters, which have negative values leading to poor distribution among the groupings (Fig. 8.16).

Table 8.13 displays each attribute alongside the corresponding color for the years 2015 and 2020, indicating the greatest average value for the respective cluster. The high gentrification cluster is shown in red, the medium gentrification cluster

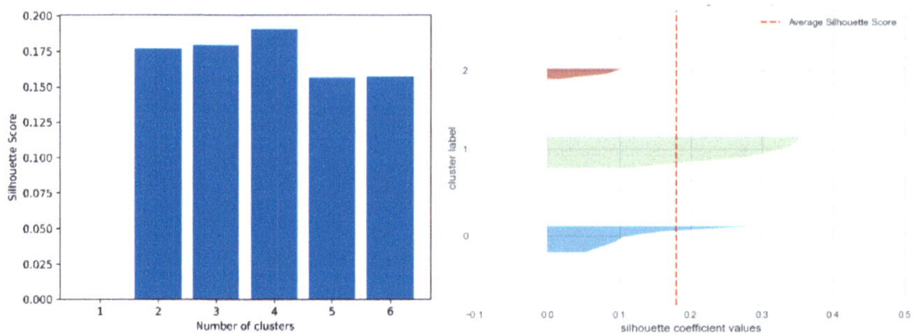

Fig. 8.15 Silhouette method graphs in 2020

in yellow, and the low gentrification cluster in green. For characteristic F1, the green hue observed in both years signifies that localities with little gentrification exhibit a higher birth rate per 1,000 inhabitants. For instance, in F4, the fastest population growth rate in 2015 was seen in the low gentrification cluster, however in 2020, it was noted in the high gentrification cluster. The cells with diagonals demonstrate that the values do not adhere to an ascending or descending sequence. For instance, in the feature of healthcare facilities per 100 k residents F9, the group with the highest average was the high gentrification group, followed by the low gentrification group, and subsequently the medium gentrification group. Although the cells lacking diagonals follow either an ascending or decreasing sequence, a notable exception arises in the Births per inhabitant feature F1. The low gentrification group had the greatest average for this attribute, succeeded by the medium gentrification group, and lastly, the high gentrification group (Tables 8.14, 8.15, 8.16, 8.17, 8.18, 8.19 and 8.20).

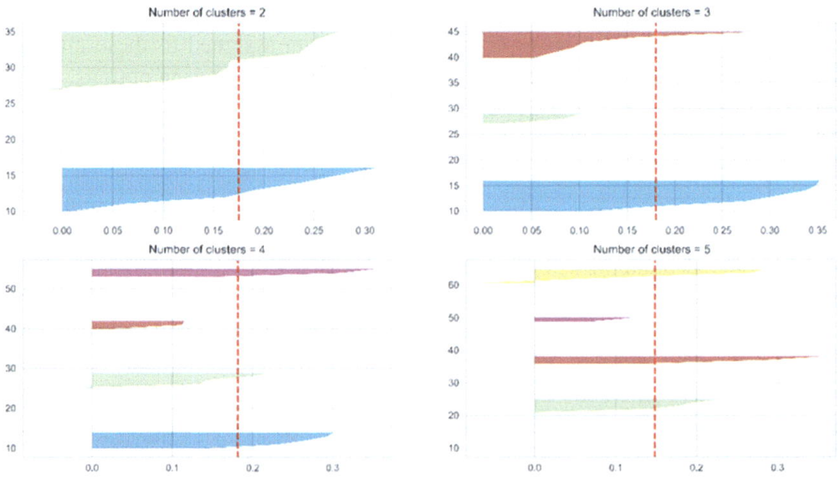

Fig. 8.16 K-means clustering silhouette plot for 16 samples or municipalities in K centers, $K = 2$, 3, 4 and 5 in 2020

Table 8.13 Modifications in the number of features in the groups

ID	Indicator	2015	2020
F1	Fertility rate per thousand residents		
F2	Mortality rate per thousand residents		
F3	Resident population count		
F4	Yearly population increase percentage		
F5	Residents per square kilometer		
F6	Residential mobility percentage (changed address in 5 years)		
F7	Health insurance coverage rate		
F8	Doctors per thousand residents		
F9	Healthcare facilities per 100k residents		
F10	Workforce participation rate		
F11	Percentage of African-descended residents		
F12	Indigenous language speakers percentage		
F13	Poverty and extreme poverty prevalence		
F14	Preschool enrollment (ages 3-5)		
F15	Elementary school attendance (ages 6-11)		
F16	Middle school enrollment (ages 12-14)		
F17	High school/college attendance (ages 15-24)		
F18	Adult literacy rate (15+ years)		
F19	Mean years of education (15+ population)		
F20	Hotel rooms per 100k residents		
F21	Restaurants and food services per 100k		
F22	Number of commercial bank facilities per 100,000 eligible residents (18+)		
F23	State development bank branches per 100k adult inhabitants (18+)		
F24	Law enforcement officers per 100k		
F25	Sports facilities per 100k residents		
F26	Retail stores per 100k population		
F27	Vehicle ownership rate per thousand adults		
F28	Reported crimes per 100k adult population		
F29	Electrified households percentage		

(continued)

Table 8.13 (continued)

F30	Plumbing access percentage		
F31	Sewage system coverage	X	
F32	Daily waste generation per person (kg)		
F33	Persons per room ratio		
F34	Homeownership rate		
F35	Dwellings with solid flooring		
F36	Homes with sturdy roofing		
F37	Property price per square meter	X	X
F38	Total yearly rainfall (mm)		X
F39	Mean annual humidity levels		
F40	Average yearly temperature (°C)		
F41	Sulfur dioxide levels (ppb)		
F42	Nitrogen dioxide concentration (ppb)		
F43	Carbon monoxide measurements (ppm)		
F44	Ozone exceedance hours		
F45	Particulate matter (PM10) levels		
F46	Fine particulate (PM2.5) concentrations		

Table 8.14 Loadings of each attribute on the principal components (2015)

	PC1	PC2	PC3	PC4	PC5	PC6	PC7	PC8	PC9	PC10	PC11	PC12
F1	0.14	−0.04	0.16	0.15	0.13	0.12	0.05	0.35	0.24	0.13	−0.17	0.15
F2	−0.19	0.10	0.16	0.04	0.09	−0.15	0.06	0.01	0.03	0.04	0.05	−0.25
F3	−0.01	0.30	−0.01	0.03	−0.23	0.33	−0.11	−0.19	−0.08	0.18	−0.06	−0.03
F4	0.05	−0.24	−0.05	−0.13	−0.10	0.15	−0.18	0.21	−0.47	−0.09	0.14	−0.08
F5	−0.17	0.24	0.04	−0.05	0.00	0.00	−0.07	0.05	−0.15	0.00	−0.13	−0.07
F6	−0.19	−0.20	−0.02	−0.09	−0.06	−0.01	0.01	−0.05	−0.02	−0.03	−0.13	0.09
F7	−0.02	−0.12	−0.11	−0.21	0.45	0.08	−0.06	−0.02	0.29	−0.10	−0.13	0.03
F8	−0.10	−0.12	0.26	−0.25	0.16	0.21	0.10	−0.14	−0.02	0.01	0.05	−0.17
F9	−0.01	−0.24	0.35	−0.03	0.09	−0.01	−0.06	0.04	−0.07	0.04	0.14	0.04
F10	−0.20	−0.15	−0.03	0.05	−0.06	0.07	0.02	−0.08	−0.12	−0.07	0.11	−0.03
F11	−0.06	−0.12	−0.08	−0.12	−0.26	−0.24	−0.40	−0.05	0.10	−0.32	−0.08	0.34
F12	0.15	−0.07	0.24	−0.20	−0.06	0.02	0.14	−0.11	−0.19	0.09	0.01	−0.01
F13	0.21	0.07	0.08	−0.02	−0.03	0.00	−0.11	0.01	0.10	0.02	0.07	0.08
F14	−0.17	−0.11	−0.21	−0.03	0.07	−0.17	0.09	−0.01	−0.11	0.13	0.08	−0.01
F15	0.00	−0.22	−0.13	−0.18	−0.08	−0.21	−0.01	0.38	0.20	0.17	0.45	−0.11
F16	−0.13	0.05	−0.16	−0.31	−0.18	−0.09	0.08	0.14	0.01	−0.08	−0.30	−0.08
F17	−0.18	−0.10	−0.15	−0.14	0.04	−0.13	0.00	−0.15	−0.05	0.16	0.04	0.07
F18	−0.19	−0.02	0.01	−0.10	0.02	0.07	0.15	0.19	0.07	−0.13	0.01	0.05
F19	−0.20	−0.14	−0.11	−0.03	−0.05	−0.05	0.08	−0.08	−0.10	0.06	0.07	0.11
F20	−0.14	−0.10	0.27	0.17	−0.14	−0.02	−0.11	0.07	0.09	−0.02	−0.08	−0.05
F21	−0.16	−0.15	0.24	0.09	−0.15	−0.01	−0.11	−0.02	0.04	−0.01	−0.07	0.03
F22	−0.18	−0.19	0.05	0.11	−0.04	0.13	0.03	−0.02	0.01	0.05	−0.06	0.15

(continued)

Table 8.14 (continued)

	PC1	PC2	PC3	PC4	PC5	PC6	PC7	PC8	PC9	PC10	PC11	PC12
F23	−0.07	−0.13	0.13	0.01	−0.02	0.20	0.38	−0.25	0.30	−0.25	0.11	0.04
F24	−0.10	−0.02	0.32	0.18	−0.10	−0.13	−0.23	0.09	0.03	0.05	−0.13	−0.09
F25	0.09	0.08	−0.05	−0.01	−0.15	0.44	0.13	0.25	−0.10	−0.34	0.13	0.26
F26	−0.12	0.15	0.10	−0.14	−0.28	0.22	−0.14	−0.07	0.01	0.27	0.10	−0.08
F27	−0.18	−0.20	−0.03	0.05	0.05	0.06	0.16	−0.08	−0.04	0.04	−0.07	0.07
F28	−0.19	−0.04	0.21	0.08	−0.09	−0.07	−0.04	0.02	−0.01	−0.01	−0.03	−0.04
F29	−0.17	0.04	−0.03	−0.21	−0.12	0.13	0.06	0.22	0.15	0.21	−0.24	−0.05
F30	−0.16	0.11	−0.21	0.17	0.02	0.06	−0.20	0.10	0.16	−0.03	0.06	0.12
F31	−0.11	−0.08	0.00	−0.05	0.18	0.26	−0.38	0.06	0.03	−0.35	0.18	−0.31
F32	−0.13	0.17	0.13	0.18	−0.02	−0.06	0.10	0.27	0.09	0.09	0.32	0.27
F33	0.20	0.11	0.14	−0.09	0.04	0.02	−0.03	0.03	0.04	−0.14	−0.08	−0.08
F34	0.19	−0.11	0.05	−0.18	−0.08	−0.16	0.10	−0.01	−0.16	0.00	−0.02	0.11
F35	−0.16	−0.02	−0.16	−0.16	−0.01	0.31	−0.10	−0.02	0.14	0.23	−0.01	−0.04
F36	−0.21	0.08	−0.10	−0.04	−0.04	0.08	−0.07	0.08	0.10	0.07	0.04	0.04
F37	−0.11	−0.25	−0.09	0.30	−0.07	0.13	0.08	−0.15	−0.01	0.00	−0.01	0.07
F38	0.02	−0.11	−0.19	0.36	0.26	0.10	−0.15	0.01	−0.16	0.12	−0.21	0.04
F39	0.17	−0.09	−0.02	−0.05	0.23	0.12	−0.19	−0.07	−0.08	0.31	0.11	0.17
F40	−0.11	0.25	−0.07	0.02	0.01	−0.17	−0.06	−0.30	0.16	−0.21	0.19	−0.15
F41	−0.18	0.08	0.05	0.05	0.24	−0.01	−0.09	0.10	−0.25	−0.10	0.06	−0.13
F42	−0.18	0.19	0.03	0.07	0.17	−0.04	0.12	0.05	−0.10	0.09	0.08	−0.04
F43	−0.16	0.18	0.01	0.02	0.12	−0.05	0.24	0.22	−0.20	−0.11	0.01	−0.02
F44	0.10	−0.03	−0.21	0.28	−0.27	0.04	0.17	−0.04	−0.06	−0.01	0.18	−0.21

(continued)

Table 8.14 (continued)

	PC1	PC2	PC3	PC4	PC5	PC6	PC7	PC8	PC9	PC10	PC11	PC12
F45	−0.11	0.22	0.16	−0.14	0.12	0.00	−0.12	−0.22	−0.06	0.06	0.30	0.44
F46	−0.18	0.15	0.04	−0.06	0.10	−0.06	0.07	0.06	−0.24	−0.13	−0.19	0.26

Table 8.15 Loadings of each attribute on the principal components (2020)

	PC1	PC2	PC3	PC4	PC5	PC6	PC7	PC8	PC9	PC10	PC11	PC12
F1	0.08	0.02	0.24	−0.03	0.25	0.16	0.34	0.02	−0.32	−0.24	0.09	0.07
F2	−0.09	−0.24	0.20	−0.04	0.10	−0.13	−0.02	−0.14	0.26	−0.22	−0.05	−0.19
F3	0.01	−0.24	0.04	−0.02	−0.18	0.43	0.05	0.06	−0.06	−0.02	0.29	0.10
F4	0.02	0.27	0.14	−0.14	0.23	0.00	−0.01	0.02	−0.24	0.14	0.10	−0.18
F5	−0.17	−0.21	0.11	−0.03	−0.06	0.00	−0.03	0.18	−0.02	−0.10	0.01	−0.24
F6	−0.20	0.14	−0.03	−0.11	−0.06	−0.06	−0.07	0.15	−0.11	0.04	0.01	0.03
F7	−0.16	0.10	−0.13	−0.17	0.17	−0.06	0.19	−0.10	−0.05	0.22	0.24	−0.10
F8	−0.15	0.11	0.04	−0.13	−0.22	0.01	0.28	0.15	0.08	0.16	0.07	0.09
F9	−0.02	0.19	0.22	0.28	−0.06	−0.16	−0.05	−0.05	0.15	0.13	0.31	−0.02
F10	−0.19	0.14	0.02	0.06	−0.17	−0.05	0.00	0.21	−0.05	0.00	0.02	−0.11
F11	−0.17	0.08	0.12	−0.03	−0.11	0.03	0.13	0.16	0.25	0.11	−0.31	0.49
F12	0.16	0.10	0.19	0.09	−0.22	−0.11	−0.01	0.09	0.00	−0.29	0.15	−0.04
F13	0.22	−0.07	0.01	0.08	−0.02	0.04	0.07	−0.04	0.12	0.04	−0.08	0.09
F14	−0.19	0.09	−0.12	−0.15	0.00	−0.20	−0.08	0.01	0.08	−0.16	0.04	−0.08
F15	0.16	−0.13	0.08	−0.09	−0.16	0.18	−0.23	0.26	0.08	−0.13	0.12	−0.05
F16	0.05	−0.01	−0.06	−0.36	0.05	−0.26	−0.29	−0.07	0.18	−0.22	0.09	0.04
F17	−0.19	0.03	−0.13	−0.20	−0.03	−0.11	−0.03	0.00	−0.10	−0.09	0.23	0.10
F18	−0.21	0.04	−0.02	−0.12	0.01	−0.16	−0.11	−0.06	−0.12	0.05	−0.25	0.01
F19	−0.21	0.09	−0.09	−0.09	−0.08	−0.07	0.00	0.09	−0.08	−0.08	0.10	0.11
F20	−0.14	0.09	0.19	0.28	−0.09	0.05	−0.13	−0.09	−0.02	−0.02	−0.01	0.05
F21	−0.17	0.13	0.18	0.15	−0.05	−0.14	−0.05	0.04	0.03	−0.02	−0.13	0.01
F22	−0.20	0.16	0.00	0.08	0.00	0.16	0.06	0.01	−0.10	−0.05	0.06	−0.07
F23	−0.02	0.24	0.19	−0.11	−0.10	0.13	0.13	−0.08	0.47	−0.05	−0.12	−0.18
F24	−0.10	0.03	0.16	0.38	0.04	−0.07	−0.12	−0.05	−0.18	0.00	0.07	−0.09
F25	0.09	0.20	−0.03	−0.13	−0.04	0.13	−0.33	−0.39	0.06	0.26	0.10	0.23

(continued)

Table 8.15 (continued)

	PC1	PC2	PC3	PC4	PC5	PC6	PC7	PC8	PC9	PC10	PC11	PC12
F26	−0.07	−0.14	0.01	0.05	−0.39	0.06	−0.01	−0.37	0.01	0.02	0.31	0.06
F27	−0.21	0.15	−0.02	−0.03	0.03	0.09	−0.03	0.07	0.00	−0.12	0.10	−0.03
F28	−0.20	0.04	0.17	0.17	−0.11	−0.02	−0.07	−0.01	0.05	0.05	−0.05	0.07
F29	0.02	−0.24	−0.13	0.21	−0.12	−0.34	0.04	0.13	−0.12	0.12	0.05	0.00
F30	−0.17	−0.18	−0.13	0.04	0.11	0.15	−0.01	−0.08	0.05	0.23	−0.09	−0.18
F31	−0.07	−0.29	−0.13	0.12	−0.06	−0.16	−0.04	0.01	−0.13	0.22	−0.13	0.12
F32	−0.12	−0.04	0.02	0.22	0.30	−0.11	−0.08	−0.21	0.01	−0.30	0.07	0.39
F33	0.22	−0.04	0.12	0.06	0.03	0.00	0.00	0.00	0.04	0.08	−0.10	−0.14
F34	0.20	0.13	−0.04	−0.01	−0.05	−0.18	−0.10	−0.04	−0.08	−0.15	0.02	0.01
F35	−0.17	−0.22	−0.12	0.07	−0.03	−0.03	−0.07	−0.02	0.15	0.04	0.09	−0.18
F36	−0.21	−0.11	−0.06	0.00	0.00	0.14	−0.05	−0.10	0.02	0.04	0.03	−0.04
F37	−0.13	0.20	−0.15	0.14	0.11	0.26	0.09	−0.11	0.00	−0.07	0.01	−0.04
F38	−0.03	0.03	−0.29	0.21	0.24	0.21	−0.08	−0.05	0.23	−0.13	−0.04	0.00
F39	0.09	−0.01	−0.13	0.10	0.23	−0.21	0.31	0.17	0.36	0.21	0.40	0.07
F40	−0.13	−0.19	0.16	−0.22	−0.11	0.10	0.03	−0.09	−0.04	0.02	0.03	0.04
F41	−0.19	−0.08	0.16	0.03	0.13	−0.02	−0.06	0.04	0.19	−0.03	0.17	−0.13
F42	−0.18	−0.15	0.09	−0.02	0.20	0.08	0.02	0.14	0.08	−0.08	−0.15	0.00
F43	−0.11	−0.23	0.17	−0.13	0.16	−0.01	0.08	0.00	−0.02	−0.14	0.13	0.33
F44	−0.04	−0.04	−0.03	−0.03	−0.20	−0.19	0.50	−0.48	−0.05	−0.18	−0.15	−0.15
F45	0.07	−0.04	0.39	−0.12	0.10	−0.12	0.04	−0.02	0.01	0.26	0.02	0.08
F46	−0.06	−0.11	0.32	−0.12	0.21	0.00	−0.15	−0.16	−0.08	0.23	0.00	−0.12

Table 8.16 Correlation of the attributes with each principal component (2015)

	PC1	PC2	PC3	PC4	PC5	PC6	PC7	PC8	PC9	PC10	PC11	PC12
F1	0.64	−0.09	0.36	0.26	0.22	0.17	0.07	0.46	0.29	0.14	−0.13	0.11
F2	−0.86	0.24	0.36	0.07	0.16	−0.22	0.09	0.01	0.04	0.05	0.04	−0.18
F3	−0.05	0.74	−0.02	0.04	−0.38	0.47	−0.15	−0.24	−0.09	0.19	−0.05	−0.02
F4	0.23	−0.58	−0.11	−0.23	−0.17	0.21	−0.26	0.27	−0.58	−0.10	0.12	−0.06
F5	−0.77	0.60	0.09	−0.09	0.01	−0.01	−0.09	0.07	−0.18	0.00	−0.11	−0.05
F6	−0.86	−0.50	−0.05	−0.16	−0.10	−0.01	0.02	−0.07	−0.03	−0.03	−0.10	0.07
F7	−0.10	−0.29	−0.25	−0.37	0.77	0.11	−0.08	−0.03	0.36	−0.11	−0.10	0.03
F8	−0.44	−0.31	0.59	−0.43	0.27	0.31	0.14	−0.18	−0.02	0.01	0.04	−0.12
F9	−0.05	−0.59	0.81	−0.06	0.15	−0.02	−0.09	0.06	−0.08	0.04	0.11	0.03
F10	−0.92	−0.38	−0.06	0.09	−0.11	0.11	0.03	−0.11	−0.15	−0.08	0.09	−0.02
F11	−0.25	−0.29	−0.19	−0.21	−0.44	−0.35	−0.57	−0.07	0.12	−0.34	−0.07	0.24
F12	0.69	−0.18	0.55	−0.35	−0.10	0.02	0.20	−0.14	−0.23	0.10	0.01	0.00
F13	0.97	0.18	0.18	−0.04	−0.05	0.00	−0.16	0.01	0.12	0.02	0.05	0.06
F14	−0.78	−0.27	−0.48	−0.06	0.11	−0.25	0.12	−0.02	−0.14	0.14	0.07	−0.01
F15	−0.01	−0.54	−0.29	−0.32	−0.14	−0.30	−0.02	0.49	0.25	0.19	0.37	−0.08
F16	−0.59	0.12	−0.36	−0.55	−0.31	−0.12	0.12	0.19	0.01	−0.09	−0.25	−0.05
F17	−0.83	−0.25	−0.34	−0.24	0.07	−0.19	0.00	−0.20	−0.06	0.17	0.04	0.05
F18	−0.89	−0.06	0.01	−0.17	0.03	0.09	0.22	0.24	0.09	−0.13	0.01	0.04
F19	−0.91	−0.34	−0.25	−0.05	−0.08	−0.07	0.11	−0.10	−0.12	0.06	0.06	0.08
F20	−0.64	−0.26	0.63	0.30	−0.24	−0.03	−0.15	0.09	0.11	−0.02	−0.07	−0.03
F21	−0.72	−0.36	0.55	0.16	−0.25	−0.02	−0.16	−0.03	0.05	−0.01	−0.06	0.02
F22	−0.84	−0.48	0.12	0.20	−0.07	0.18	0.04	−0.03	0.02	0.05	−0.05	0.11

(continued)

Table 8.16 (continued)

	PC1	PC2	PC3	PC4	PC5	PC6	PC7	PC8	PC9	PC10	PC11	PC12
F23	−0.32	−0.31	0.29	0.02	−0.04	0.30	0.54	−0.33	0.37	−0.27	0.09	0.03
F24	−0.46	−0.05	0.73	0.31	−0.16	−0.19	−0.32	0.11	0.04	0.05	−0.11	−0.07
F25	0.43	0.21	−0.11	−0.02	−0.25	0.64	0.19	0.33	−0.12	−0.37	0.10	0.19
F26	−0.57	0.38	0.23	−0.24	−0.47	0.32	−0.20	−0.09	0.01	0.28	0.08	−0.06
F27	−0.84	−0.50	−0.06	0.09	0.08	0.08	0.22	−0.10	−0.05	0.04	−0.06	0.05
F28	−0.88	−0.11	0.47	0.13	−0.15	−0.10	−0.05	0.03	−0.01	−0.02	−0.02	−0.03
F29	−0.78	0.10	−0.06	−0.38	−0.20	0.19	0.08	0.28	0.19	0.22	−0.19	−0.04
F30	−0.72	0.26	−0.48	0.31	0.03	0.08	−0.29	0.14	0.19	−0.03	0.05	0.09
F31	−0.51	−0.20	0.00	−0.08	0.31	0.37	−0.55	0.08	0.04	−0.37	0.15	−0.22
F32	−0.60	0.43	0.29	0.32	−0.03	−0.08	0.14	0.36	0.12	0.09	0.26	0.19
F33	0.91	0.27	0.33	−0.16	0.06	0.03	−0.04	0.04	0.05	−0.15	−0.07	−0.06
F34	0.85	−0.28	0.11	−0.31	−0.14	−0.23	0.15	−0.01	−0.20	0.00	−0.02	0.08
F35	−0.73	−0.04	−0.37	−0.28	−0.02	0.46	−0.14	−0.02	0.17	0.24	−0.01	−0.03
F36	−0.95	0.19	−0.22	−0.08	−0.06	0.12	−0.10	0.11	0.12	0.07	0.03	0.03
F37	−0.49	−0.62	−0.20	0.53	−0.12	0.18	0.11	−0.19	−0.01	0.00	−0.01	0.05
F38	0.09	−0.27	−0.44	0.63	0.44	0.14	−0.22	0.01	−0.19	0.13	−0.17	0.03
F39	0.77	−0.21	−0.04	−0.08	0.39	0.18	−0.27	−0.09	−0.09	0.33	0.09	0.12
F40	−0.50	0.63	−0.16	0.04	0.02	−0.25	−0.08	−0.39	0.19	−0.23	0.16	−0.10
F41	−0.80	0.21	0.11	0.09	0.41	−0.01	−0.13	0.13	−0.31	−0.11	0.05	−0.09
F42	−0.81	0.47	0.06	0.12	0.29	−0.06	0.16	0.06	−0.12	0.10	0.06	−0.03
F43	−0.74	0.43	0.03	0.04	0.20	−0.07	0.34	0.28	−0.25	−0.12	0.01	−0.01
F44	0.46	−0.06	−0.48	0.50	−0.46	0.05	0.24	−0.05	−0.07	−0.01	0.14	−0.15

(continued)

Table 8.16 (continued)

	PC1	PC2	PC3	PC4	PC5	PC6	PC7	PC8	PC9	PC10	PC11	PC12
F45	−0.50	0.54	0.36	−0.25	0.20	0.00	−0.17	−0.29	−0.08	0.06	0.24	0.32
F46	−0.80	0.38	0.10	−0.11	0.16	−0.09	0.11	0.08	−0.30	−0.14	−0.15	0.19

Table 8.17 Correlation of the attributes with each principal component (2020)

	PC1	PC2	PC3	PC4	PC5	PC6	PC7	PC8	PC9	PC10	PC11	PC12
F1	0.34	0.05	0.53	−0.05	0.48	0.22	0.45	0.02	−0.33	−0.21	0.08	0.05
F2	−0.40	−0.69	0.45	−0.09	0.19	−0.18	−0.02	−0.15	0.26	−0.20	−0.04	−0.14
F3	0.03	−0.70	0.10	−0.04	−0.34	0.61	0.07	0.06	−0.06	−0.02	0.25	0.07
F4	0.08	0.76	0.30	−0.29	0.44	−0.01	−0.02	0.02	−0.24	0.13	0.09	−0.13
F5	−0.72	−0.60	0.24	−0.07	−0.12	0.00	−0.04	0.19	−0.02	−0.09	0.01	−0.17
F6	−0.88	0.40	−0.06	−0.23	−0.12	−0.08	−0.10	0.15	−0.12	0.04	0.01	0.02
F7	−0.70	0.28	−0.29	−0.34	0.33	−0.09	0.25	−0.11	−0.05	0.19	0.20	−0.07
F8	−0.67	0.30	0.10	−0.27	−0.41	0.02	0.38	0.16	0.09	0.14	0.06	0.07
F9	−0.08	0.55	0.50	0.58	−0.11	−0.22	−0.07	−0.05	0.16	0.12	0.26	−0.02
F10	−0.83	0.41	0.05	0.13	−0.33	−0.08	0.01	0.22	−0.06	0.00	0.01	−0.08
F11	−0.75	0.22	0.26	−0.07	−0.21	0.04	0.17	0.17	0.25	0.10	−0.26	0.35
F12	0.70	0.29	0.41	0.18	−0.42	−0.16	−0.01	0.10	0.00	−0.26	0.13	−0.03
F13	0.97	−0.21	0.03	0.17	−0.03	0.06	0.09	−0.04	0.12	0.03	−0.07	0.07
F14	−0.83	0.25	−0.26	−0.31	0.00	−0.29	−0.11	0.01	0.08	−0.14	0.03	−0.06
F15	0.70	−0.36	0.18	−0.19	−0.31	0.26	−0.31	0.28	0.08	−0.12	0.10	−0.04
F16	0.22	−0.01	−0.13	−0.74	0.10	−0.36	−0.39	−0.07	0.18	−0.20	0.08	0.03
F17	−0.85	0.08	−0.28	−0.41	−0.05	−0.15	−0.04	0.01	−0.10	−0.08	0.20	0.07
F18	−0.91	0.11	−0.05	−0.24	0.02	−0.23	−0.15	−0.06	−0.12	0.04	−0.21	0.01
F19	−0.93	0.24	−0.20	−0.19	−0.14	−0.09	0.00	0.09	−0.09	−0.07	0.09	0.08
F20	−0.62	0.24	0.42	0.59	−0.17	0.07	−0.17	−0.09	−0.02	−0.02	−0.01	0.04
F21	−0.76	0.37	0.39	0.31	−0.10	−0.19	−0.06	0.04	0.03	−0.02	−0.11	0.01
F22	−0.87	0.44	−0.01	0.17	0.00	0.23	0.07	0.01	−0.10	−0.04	0.05	−0.05

(continued)

Table 8.17 (continued)

	PC1	PC2	PC3	PC4	PC5	PC6	PC7	PC8	PC9	PC10	PC11	PC12
F23	−0.09	0.67	0.41	−0.23	−0.19	0.18	0.18	−0.09	0.48	−0.04	−0.10	−0.13
F24	−0.43	0.09	0.36	0.79	0.08	−0.10	−0.16	−0.05	−0.18	0.00	0.06	−0.07
F25	0.38	0.56	−0.06	−0.27	−0.07	0.19	−0.45	−0.42	0.06	0.23	0.08	0.16
F26	−0.30	−0.39	0.02	0.10	−0.75	0.09	−0.02	−0.39	0.01	0.02	0.27	0.04
F27	−0.91	0.42	−0.05	−0.05	0.05	0.13	−0.04	0.08	0.00	−0.11	0.08	−0.02
F28	−0.85	0.12	0.38	0.35	−0.21	−0.02	−0.09	−0.02	0.05	0.04	−0.05	0.05
F29	0.10	−0.67	−0.28	0.44	−0.22	−0.48	0.05	0.14	−0.12	0.11	0.05	0.00
F30	−0.72	−0.51	−0.28	0.08	0.21	0.21	−0.02	−0.09	0.06	0.21	−0.08	−0.13
F31	−0.31	−0.82	−0.29	0.26	−0.11	−0.23	−0.05	0.01	−0.14	0.20	−0.11	0.09
F32	−0.50	−0.12	0.05	0.46	0.57	−0.16	−0.10	−0.22	0.01	−0.27	0.06	0.28
F33	0.96	−0.12	0.26	0.12	0.07	0.00	0.00	0.00	0.04	0.07	−0.08	−0.10
F34	0.89	0.37	−0.08	−0.02	−0.09	−0.26	−0.14	−0.04	−0.08	−0.13	0.02	0.01
F35	−0.73	−0.62	−0.26	0.15	−0.05	−0.04	−0.09	−0.02	0.15	0.04	0.08	−0.13
F36	−0.92	−0.32	−0.14	0.00	0.00	0.20	−0.07	−0.11	0.02	0.03	0.02	−0.03
F37	−0.56	0.58	−0.34	0.30	0.20	0.37	0.12	−0.12	0.00	−0.06	0.01	−0.03
F38	−0.14	0.08	−0.64	0.44	0.46	0.30	−0.11	−0.06	0.23	−0.12	−0.03	0.00
F39	0.38	−0.04	−0.28	0.21	0.44	−0.30	0.41	0.18	0.36	0.19	0.34	0.05
F40	−0.58	−0.54	0.36	−0.46	−0.22	0.14	0.04	−0.10	−0.04	0.02	0.03	0.03
F41	−0.82	−0.24	0.35	0.06	0.25	−0.03	−0.09	0.05	0.19	−0.03	0.15	−0.09
F42	−0.77	−0.44	0.20	−0.03	0.39	0.11	0.02	0.15	0.08	−0.07	−0.13	0.00
F43	−0.46	−0.65	0.38	−0.26	0.30	−0.01	0.11	0.00	−0.02	−0.12	0.11	0.23
F44	−0.16	−0.11	−0.07	−0.06	−0.38	−0.27	0.67	−0.52	−0.05	−0.17	−0.13	−0.11

(continued)

Table 8.17 (continued)

	PC1	PC2	PC3	PC4	PC5	PC6	PC7	PC8	PC9	PC10	PC11	PC12
F45	0.31	−0.12	0.88	−0.24	0.19	−0.17	0.06	−0.02	0.01	0.24	0.02	0.06
F46	−0.27	−0.32	0.72	−0.26	0.40	0.00	−0.20	−0.17	−0.08	0.21	0.00	−0.09

Table 8.18 Eigenvalues of the principal components of the year 2015 and 2020

Eigenvalues		
	2020	2015
PC1	18.990	21.016
PC2	8.098	6.166
PC3	4.988	5.220
PC4	4.307	3.079
PC5	3.619	2.846
PC6	2.043	2.113
PC7	1.833	2.042
PC8	1.139	1.711
PC9	1.027	1.505
PC10	0.805	1.139
PC11	0.726	0.654
PC12	0.504	0.509
PC13	0.379	0.452
PC14	0.366	0.369
PC15	0.244	0.245
PC16	0.000	0.000

Table 8.19 Scores of the municipalities in each principal component and the respective cluster to which it belongs 2015

	PC1	PC2	PC3	PC4	PC5	PC6	PC7	PC8	PC9	PC10	PC11	PC12	Cluster
1	1.16	−0.32	−1.70	0.86	−1.30	1.52	−0.77	−0.97	0.41	−0.80	0.34	−0.61	0
2	−2.08	1.58	−0.46	0.28	3.89	−0.77	−0.37	−0.84	−0.11	0.35	−0.82	−0.66	1
3	−7.77	−2.74	−1.40	−3.09	0.39	0.21	−1.29	−0.73	−2.07	−0.52	0.68	−0.06	2
4	−2.27	0.90	−1.94	−0.95	−1.60	−2.45	0.52	−1.45	−0.10	1.51	−0.30	1.29	1
5	2.33	−3.53	−2.16	2.33	0.55	0.89	−0.32	2.46	−0.84	1.44	−0.34	0.88	0
6	−7.31	−0.77	6.02	1.95	−1.65	−0.94	−1.66	0.71	0.21	0.08	−0.22	−0.11	2
7	0.25	3.00	−0.02	−0.36	0.47	0.46	−0.57	−0.19	1.43	2.54	0.84	−0.85	1
8	−1.91	2.29	−1.81	−0.49	−0.34	−1.69	1.44	2.49	−0.64	−0.91	−1.07	−0.97	1
9	1.52	5.24	0.58	0.42	−1.29	2.78	−0.45	−0.81	−1.49	−0.40	−0.98	0.65	1
10	4.27	−0.87	−0.77	3.09	2.02	−1.05	−1.40	−1.20	−0.59	−0.97	0.51	0.00	0
11	−5.94	−3.13	−0.46	0.97	0.27	1.69	2.79	−1.30	1.96	−0.37	−0.71	0.04	2
12	8.43	−2.80	4.27	−2.81	1.01	0.54	1.02	−0.25	−0.82	0.58	−0.57	−0.20	0
13	3.78	−0.46	−1.02	−2.37	−0.29	−0.32	−2.52	0.86	2.75	−1.15	−0.55	0.54	0
14	2.39	−1.31	−1.45	−0.02	−2.79	0.55	0.36	0.50	−0.40	0.30	0.81	−1.02	0
15	−2.27	2.54	1.33	−0.65	1.89	0.81	1.49	1.67	0.38	−0.81	1.75	0.94	1
16	5.42	0.37	0.99	0.85	−1.23	−2.23	1.73	−0.96	−0.07	−0.86	0.62	0.15	0

Table 8.20 Scores of the municipalities in each principal component and the respective cluster to which it belongs 2020

	PC1	PC2	PC3	PC4	PC5	PC6	PC7	PC8	PC9	PC10	PC11	PC12	Cluster
0	0.17	−0.45	−2.28	0.71	−1.48	1.16	3.08	−1.32	0.26	−0.02	−1.10	−0.84	0
1	−2.42	−1.92	1.43	−1.51	2.97	−1.48	0.38	0.22	−0.04	1.29	1.00	−0.53	0
2	−7.90	1.72	−2.59	−1.63	−2.71	−1.44	0.94	1.81	−0.57	0.46	0.28	0.45	1
3	−1.43	−1.33	−2.21	−1.06	−1.68	−0.69	−0.71	−0.82	−0.18	−0.78	0.48	−0.55	0
4	0.83	2.12	−3.12	3.42	3.75	0.60	0.40	0.32	−1.72	−0.16	0.38	−0.21	0
5	−6.70	0.69	3.76	5.39	−1.52	−0.26	−0.90	−0.42	−0.10	0.06	−0.02	0.04	1
6	0.27	−4.10	0.62	−0.48	0.74	0.03	1.36	−1.43	0.38	−0.67	1.61	0.98	0
7	−1.07	−2.79	−0.25	−1.37	1.46	−1.18	−2.05	0.30	−0.29	−1.49	−1.05	−0.93	0
8	2.90	−4.94	1.19	0.04	−1.60	3.31	−0.41	1.94	−0.38	0.43	0.33	−0.37	0
9	2.98	0.55	−2.75	1.57	1.53	−0.14	−0.55	1.10	2.58	0.52	−0.39	0.60	0
10	−5.32	5.61	1.88	−3.21	1.67	2.88	−0.28	−0.37	0.51	−0.44	−0.03	0.05	1
11	8.32	4.65	2.17	0.00	−1.60	−1.34	0.45	0.37	0.57	−0.16	1.01	−1.13	2
12	4.05	0.31	0.61	−1.47	−0.17	−0.32	−0.84	−1.21	−1.33	2.10	−1.00	0.35	0
13	2.36	1.06	−2.44	0.08	−1.89	0.51	−2.54	−1.16	0.10	−0.06	0.50	0.69	0
14	−2.17	−1.94	1.78	−0.02	0.49	−0.86	0.49	−0.24	1.24	0.25	−1.12	0.06	0
15	5.14	0.77	2.22	−0.46	0.04	−0.81	1.17	0.91	−1.04	−1.32	−0.86	1.34	2

References

Aalbers, M. B. (2019). Introduction to the forum: From third to fifth-wave gentrification. *Tijdschrift Voor Economische En Sociale Geografie, 110*(1), 1–11. https://doi.org/10.1111/tesg.12332

Aguilar-Velázquez, D., Rivera Islas, I., Romero Tecua, G., & Valenzuela-Aguilera, A. (2024). Gentrification and access to housing in Mexico City during 2000 to 2022. *Proceedings of the National Academy of Sciences, 121*(10), Article e2314455121.

Alguacil, J., Camacho, J., & Hernández Aja, A. (2014). Urban vulnerability in Spain. Identification and evolution of vulnerable neighborhoods. *Journal of Social Science Methodology, 27*, 73–94. https://doi.org/10.5944/empiria.27.2014.10863

Anguelovski, I., Cole, H. V., O'Neill, E., Baró, F., Kotsila, P., Sekulova, F., ... & Triguero-Mas, M. (2021). Gentrification pathways and their health impacts on historically marginalized residents in Europe and North America: Global qualitative evidence from 14 cities. *Health and Place*, 72, 102698. https://doi.org/10.1016/j.healthplace.2021.102698

Anson, K. (2022). Contesting the Neoliberal City: Place-based activism in the documentary films of barrio anti-gentrification movements in Bogotá and Mexico City. *Journal of Latin American Cultural Studies, 31*(3), 457–473. https://doi.org/10.1080/13569325.2022.2138700

Atuesta, L. H., & Hewings, G. J. D. (2019). Housing appreciation patterns in low-income neighborhoods: Exploring gentrification in Chicago. *Journal of Housing Economics, 44*, 35–47. https://doi.org/10.1016/j.jhe.2018.08.005

Aune, K. T., Gesch, D., & Smith, G. S. (2020). A spatial analysis of climate gentrification in Orleans Parish. *Louisiana Post-Hurricane Katrina. Environmental Research, 185*, Article 109384. https://doi.org/10.1016/j.envres.2020.109384

Badcock, B. (2001). Thirty Years On: Gentrification and class changeover in Adelaide's inner suburbs. *Urban Studies, 38*(9), 1559–1572. https://doi.org/10.1080/00420980120080441

Bailey, N., & Robertson, D. (1997). Housing renewal, urban policy and gentrification. *Urban Studies, 34*(4), 561–578. https://doi.org/10.1080/0042098975925

Beauregard, R. A. (1990). Trajectories of neighborhood change: The case of gentrification. *Environment and Planning A: Economy and Space, 22*(7), 855–874. https://doi.org/10.1068/A220855

Best, K. B., Jouzi, Z., Islam, M. S., Kirby, T., Nixon, R., Hossan, A., & Nyiawung, R. A. (2023). Typologies of multiple vulnerabilities and climate gentrification across the East Coast of the United States. *Urban Climate, 48*, Article 101430. https://doi.org/10.1016/j.uclim.2023.101430

Bohl, C. C. (2010). New urbanism and the city: Potential applications and implications for distressed inner-city neighborhoods. *Housing Policy Debate, 11*(4), 761–801. https://doi.org/10.1080/10511482.2000.9521387

Bolt, G., Phillips, D., & Van Kempen, R. (2010). Housing policy, (de) segregation and social mixing: An international perspective. *Housing Studies, 25*, 129–135. https://doi.org/10.1080/02673030903564838

Bonotti, M., & Barnhill, A. (2022). Food, gentrification and located life plans. *Food Ethics, 7*(1), 1–20. https://doi.org/10.1007/S41055-022-00101-7/METRICS

Brulle, R. J., & Pellow, D. N. (2006). Environmental justice: Human health and environmental inequalities. *Annual Review of Public Health, 27*, 103–124. https://doi.org/10.1146/annurev.publhealth.27.021405.102124

CDMX Tourism. (2023). Government of the city of Mexico. Retrieved from Tourism activity statistics: https://www.turismo.cdmx.gob.mx/actividad-turistica/estadisticas-de-la-actividad-turistica. Access May 20, 2023

Chander, B., & Gopalakrishnan, K. (2023). Data clustering using unsupervised machine learning. *Statistical Modeling in Machine Learning, 179–204.* https://doi.org/10.1016/b978-0-323-91776-6.00015-4

Chapple, K., & Zuk, M. (2016). Forewarned: The use of neighborhood early warning systems for gentrification and displacement. *Cityscape, 18*(3), 109–130. https://www.jstor.org/stable/26328275

Choi, J., Peters, M., & Mueller, R. O. (2010). Correlational analysis of ordinal data: From Pearson's r to Bayesian polychoric correlation. *Asia Pacific Education Review, 11*(4), 459–466. https://doi.org/10.1007/s12564-010-9096-y

Choi, M. K., & Seong, P. H. (2020). A methodology for evaluating human operator's fitness for duty in nuclear power plants. *Nuclear Engineering and Technology, 52*(5), 984–994. https://doi.org/10.1016/j.net.2019.10.024

Delgadillo, V. (2016). Selective modernization of Mexico City and its historic center. Gentrification without displacement? *Urban Geography, 37*(8), 1154–1174. https://doi.org/10.1080/02723638.2015.1096114

Easton, S., Lees, L., Hubbard, P., & Tate, N. (2020). Measuring and mapping displacement: The problem of quantification in the battle against gentrification. *Urban Studies, 57*(2), 286–306. https://doi.org/10.1177/0042098019851953

Edelmann, D., Móri, T. F., & Székely, G. J. (2021). On relationships between the Pearson and the distance correlation coefficients. *Statistics & Probability Letters, 169*, Article 108960. https://doi.org/10.1016/j.spl.2020.108960

EVALÚA. (2021). Evaluation of public housing policies in Mexico City 2021. Mexico City: Mexico City Evaluation Council. Retrieved from https://www.evalua.cdmx.gob.mx/storage/app/media/2022/evaluacion/Informe%20Final%20Evaluacion%20Vivienda%202021%20.pdf.

FGJ. (2023). Attorney General's Office. Retrieved from the Government of Mexico City: https://www.fgjcdmx.gob.mx/procuraduria/estadisticas-delictivas. Accessed May 20, 2023.

Finio, N. (2021). Measurement and definition of gentrification in urban studies and planning. *Journal of Planning Literature, 37*(2), 249–264. https://doi.org/10.1177/08854122211051603

Freeman, L. (2005). Displacement or succession?: Residential mobility in gentrifying neighborhoods. *Urban Affairs Review, 40*(4), 463–491. https://doi.org/10.1177/1078087404273341

Galster, G., & Peacock, S. (1986). Urban gentrification: Evaluating alternative indicators. *Social Indicators Research, 18*, 321–337. https://doi.org/10.1007/BF00286623

Gibbons, J., & Barton, M. S. (2016). The association of minority self-rated health with black versus white gentrification. *Journal of Urban Health, 93*(6), 909–922. https://doi.org/10.1007/s11524-016-0087-0

Glass, R. (1964). Introduction: Aspects of change. In Centre For Urban Studies (Ed) *London: Aspects of change* (pp.xiii-xlii). MacGibbon and Kee.

Gray, J., Buckner, L., & Comber, A. (2023). Predicting gentrification in England: A data primitive approach. *Urban Science, 7*(2), 64. https://doi.org/10.3390/urbansci7020064

Hammel, D. J., & Wyly, E. K. (1996). A model for identifying gentrified areas with census data. *Urban Geography, 17*(3), 248–268. https://doi.org/10.2747/0272-3638.17.3.248

Hernández-Aja, A. (2012). Atlas of urban vulnerability in Spain: Methodology, contents and credits. Retrieved from https://www.mitma.gob.es/recursos_mfom/pdf/37c8dd62-2236-49cf-8b4a-0bc3a8782478/111284/20120125_metodologia_atlas.pdf. Accessed August 15, 2023

Hwang, J., & Sampson, R. J. (2014). Divergent pathways of gentrification: Racial inequality and the social order of renewal in Chicago neighborhoods. *American Sociological Review, 79*(4), 726–751. https://doi.org/10.1177/0003122414535774

Ilic, L., Sawada, M., & Zarzelli, A. (2019). Deep mapping gentrification in a large Canadian city using deep learning and Google Street View. *PLoS ONE, 14*(3), Article e0212814. https://doi.org/10.1371/journal.pone.0212814

INEGI. (2023). National Institute of Statistics and Geography. Retrieved from Mexico in Numbers (Mexico City): https://www.inegi.org.mx/app/areasgeograficas/?ag=09#collapse-Tabulados. Accessed May 20, 2023

Janoschka, M. (2016). Gentrification, displacement, dispossession: Key urban processes in Latin American cities. *Revista Invi, 31*, 27–71.

Janoschka, M., Sequera, J., & Salinas, L. (2014). Gentrification in Spain and Latin America: A critical dialogue. *Journal of Geography Norte Grande, 58*, 7–40.

Kaufman, L., & Rousseeuw, P. J. (Eds.). (1990). *Finding Groups in Data.* https://doi.org/10.1002/9780470316801

Kennedy, M., & Leonard, P. (2001). Dealing with neighborhood change: a primer on gentrification and policy choices. California: The Brookings Institution Center on Urban and Metropolitan Policy. Retrieved from https://www.brookings.edu/wp-content/uploads/2016/06/gentrification.pdf. Accessed May 20, 2023

Knorr, D., Gilligan, J., & Camp, J. (2019). Using machine learning to identify and predict gentrification in Nashville, Tennessee. https://ir.vanderbilt.edu/handle/1803/13285

Krieger, N. (2014). Discrimination and health inequities. *International Journal of Health Services, 44*(4), 643–710. https://doi.org/10.2190/hs.44.4.b

Levesque, J. F., Harris, M. F., & Russell, G. (2013). Patient-centred access to health care: Conceptualising access at the interface of health systems and populations. *International Journal for Equity in Health, 12*, 1–9. https://doi.org/10.1186/1475-9276-12-18

Lee, Y. Y. (2010). Gentrification and crime: Identification using the 1994 Northridge earthquake in Los Angeles. *Journal of Urban Affairs, 32*(5), 549–577. https://doi.org/10.1111/j.1467-9906.2010.00506.x

Li, W., Zhang, H., Shibasaki, R., Chen, J., & Kobayashi, H. H. (2023). Mobility pattern clustering with big human mobility data. *Handbook of Mobility Data Mining, 27–48.* https://doi.org/10.1016/b978-0-443-18424-6.00006-4

Liu, C., Deng, Y., Song, W., Wu, Q., & Gong, J. (2019). A comparison of the approaches for gentrification identification. *Cities, 95*, Article 102482. https://doi.org/10.1016/j.cities.2019.102482

McDermott, J. (2019). Towards an icon model of gentrification: Global capitalism, policing, and the struggle for iconic spaces in Mexico City. *Urban Studies, 56*(16), 3522–3539. https://doi.org/10.1177/0042098018813065

Morello-Frosch, R., & Shenassa, E. D. (2006). The environmental "Riskscape" and social inequality: Implications for explaining maternal and child health disparities. *Environmental Health Perspectives, 114*(8), 1150–1153. https://doi.org/10.1289/ehp.8930

Musterd, S., Marcińczak, S., van Ham, M., & Tammaru, T. (2016). Socioeconomic segregation in European capital cities. Increasing separation between poor and rich. *Urban Geography, 38*(7), 1062–1083. https://doi.org/10.1080/02723638.2016.1228371

Hanakata, N. C., Streule, M., & Schmid, C. (2022). Incorporation of urban differences in Tokyo, Mexico City, and Los Angeles. *City, 26*(5–6), 791–819. https://doi.org/10.1080/13604813.2022.2126231

Pérez, G. (2004). *The near northwest side story: Migration, displacement, and Puerto Rican families.* University of California Press.

Raman, S. (2014). Gentrification. In A. C. Michalos (Eds.), *Encyclopedia of Quality of Life and Well-Being Research* (pp. 2509–2512). Springer, Dordrecht. https://doi.org/10.1007/978-94-007-0753-5_1157

Razzaghi, P., Abbasi, K., & Ghasemi, J. B. (2023). Multivariate pattern recognition by machine learning methods. *Machine Learning and Pattern Recognition Methods in Chemistry from Multivariate and Data Driven Modeling, 47–72.* https://doi.org/10.1016/b978-0-323-90408-7.00002-2

Reades, J., De Souza, J., & Hubbard, P. (2018). Understanding urban gentrification through machine learning. *Urban Studies, 56*(5), 922–942. https://doi.org/10.1177/0042098018789054

Reibel, M., & Regelson, M. (2013). Neighborhood racial and ethnic change: The time dimension in segregation. *Urban Geography, 32*(3), 360–382. https://doi.org/10.2747/0272-3638.32.3.360

Santos-Izquierdo, F., Blanco-Vílchez, M., Romero-Padilla, Y., & Navarro-Jurado, E. (2023). The touristification of historic centres through commercial gentrification in times of COVID-19. In *Urban Dynamics in the Post-pandemic Period: Tourist Spaces and Urban Centres* (pp. 47–62). Cham: Springer International Publishing. https://doi.org/10.1007/978-3-031-36017-6_4

SEDEMA. (2023). Air quality in Mexico City. Retrieved from Government of Mexico City: http://www.aire.cdmx.gob.mx/default.php?opc=Z6BhnmI=. Accessed May 20, 2023

Scikit-learn. (2023). Scaling features to a range. Retrieved from Preprocessing data: https://scikit-learn.org/stable/modules/preprocessing.html#preprocessing-scaler. Accessed April 20, 2023

Shan, L., He, S., & Wan, C. (2023). Unraveling the dynamic Airbnb-gentrification interrelation before and after the COVID-19 Pandemic: Evidence from Beijing. *China. Cities, 137*, Article 104270. https://doi.org/10.1016/j.cities.2023.104270

Slater, T. (2010). Still missing Marcuse: Hamnett's foggy analysis in London town. *City, 14*(1–2), 170–179. https://doi.org/10.1080/13604811003633719

Smith, N. (1996). The New Urban Frontier: Gentrification and the Revanchist City (1st ed.). Routledge. https://doi.org/10.4324/9780203975640

Thackway, W., Ng, M., Lee, C. L., & Pettit, C. (2023). Implementing a deep-learning model using Google Street view to combine social and physical indicators of gentrification. *Computers, Environment and Urban Systems, 102*, Article 101970. https://doi.org/10.1016/j.compenvurbsys.2023.101970

Tiwari, A. (2022). Supervised learning: From theory to applications. *Artificial Intelligence and Machine Learning for EDGE Computing, 23–32.* https://doi.org/10.1016/b978-0-12-824054-0.00026-5

Trebuňa, P., Halčinová, J., Fil'O, M., & Markovič, J. (2014). *The importance of normalization and standardization in the process of clustering.* SAMI 2014—*IEEE 12th International Symposium on Applied Machine Intelligence and Informatics*, Proceedings, 381–385. https://doi.org/10.1109/sami.2014.6822444

Tyagi, K., Rane, C., Sriram, R., & Manry, M. (2022). Unsupervised learning. *Artificial Intelligence and Machine Learning for EDGE Computing, 33–52.* https://doi.org/10.1016/b978-0-12-824054-0.00012-5

UPM. (2023). Unsupervised learning. Clustering. Retrieved from Introduction to Machine Learning: https://dcain.etsin.upm.es/~carlos/bookAA/03.1_Clustering-K-Means.html. Accessed April 30, 2023

Vlachos, M. (2017). Dimensionality Reduction. In C. Sammut, G. I. Webb (Eds.), *Encyclopedia of Machine Learning and Data Mining* (pp. 354–361). Springer. https://doi.org/10.1007/978-1-4899-7687-1_71

Williams, K. (2015). Toward a universal operationalization of gentrification. *Sociation Today, 13*(2), 1.

Wong, G., Greenhalgh, T., Westhorp, G., Buckingham, J., & Pawson, R. (2013). RAMESES publication standards: Realist syntheses. *BMC Medicine, 11*, 1–14. https://doi.org/10.1186/1741-7015-11-21

Wu, L., & G. Rowe, P. (2022). Green space progress or paradox: identifying green space associated gentrification in Beijing. *Landscape and Urban Planning, 219*, 104321. https://doi.org/10.1016/j.landurbplan.2021.104321

Wyly, E. K., & Hammel, D. J. (1998). Modeling the context and contingency of gentrification. *Journal of Urban Affairs, 20*(3), 303–326. https://doi.org/10.1111/j.1467-9906.1998.tb00424.x

Zeng, J., Yue, Y., Gao, Q., Gu, Y., & Ma, C. (2022). Identifying localized amenities for gentrification using a machine learning-based framework. *Applied Geography, 145*, Article 102748. https://doi.org/10.1016/j.apgeog.2022.102748

Zhang, Y., Bouadi, T., & Martin, A. (2018). An empirical study to determine the optimal k in Ek-NNclus method. In S. Destercke, T. Denoeux, F. Cuzzolin, & A. Martin (Eds.), *Belief Functions: Theory and Applications*. BELIEF 2018. *Lecture Notes in Computer Science* (pp. 260–268). Springer, Cham. https://doi.org/10.1007/978-3-319-99383-6_32

Zuk, M., Bierbaum, A. H., Chapple, K., Gorska, K., & Loukaitou-Sideris, A. (2018). Gentrification, displacement, and the role of public investment. *Journal of Planning Literature, 33*(1), 31–44. https://doi.org/10.1177/0885412217716439

Chapter 9
Clustering for Sustainable Development Goals and Organic Residuals Reduction

9.1 Introduction

A number of initiatives to foster a more environmentally conscious society have been underway for decades. Burton (1987) describes a prototype of development that meets present global requirements, although guaranteeing that the requirements of those in the future are not put at risk. Improving the quality of life and fostering environmental, economic, as well as social balance have been fueled by sustainability.

Implemented in 2015, the 2030 United Nations Agenda outlines a shared vision for sustainable global progress, calling on all countries to report key topics including poverty, social injustice, as well as food insecurity. The implementation of SDGs will also aid in the reduction of nonrenewable resources by safeguarding the environment. In essence, countries must accomplish 169 nested objectives and 17 SDGs as part of the 2030 Agenda in order to attain integrated development (United Nations, 2023).

Researchers throughout the world are being driven to work and study on such SDGs as renewable resources (Elavarasan et al., 2022), water (Singh et al., 2022), food-related sustainability (Ardra et al., 2022), responsible usage practices (Ortiz-de-Montellano et al., 2023), global health concerns (Addo-Atuah et al., 2020), and ecological impacts linked to climate change (Campbell et al., 2018). Furthermore, scholars have assessed the advancement of the SDGs in certain areas, including Latin America (Ibrahim et al., 2023), Europe (Huang, 2023), and Asia (UNDP, 2023). To keep the focus on enhancing the objectives' synergies and minimizing their trade-offs, additional study is necessary to comprehend the relationships between each SDG.

Given its close ties to the SDGs, the food supply system is essential in the achievement of a sustainable development (FAO, 2018). Land usage, climate change, nutrition, human well-being, energy and water use are all significantly impacted by the food sector (Garcia-Garcia et al., 2021). Furthermore, there are a number of steps within the food production chain that require improvements in order to remain sustainable, including organic residuals and losses. According to Chakrabarty and

A. Y. Raya-Tapia et al., *Machine Learning and Clustering for a Sustainable Future*, Studies in Computational Intelligence 1233,
https://doi.org/10.1007/978-3-032-03876-0_9

Das (2020), organic residuals in Europe and North America is expected to be around 115 kg per person annually. Thus, significant transformations in the food supply system are mandatory to enhance the sustainability of food manufacture, consumption, and progress toward the SDGs. Furthermore, learning more around in what way the food organization works is essential to guide the design of policies that enable the attainment of set goals and to take into account public, private, and social stakeholders (Petruzzelli et al., 2023). Nevertheless, techniques to assess the ranking, the advancement and the assistance that countries have produced in their sustainable development have to be established in order to assess the advancements of the SDGs and their connection to the food industry. Scherer et al. (2018) assert that quantitative analysis that makes it possible to identify the connections between the SDGs is essential and need to be given top priority in research.

In this sense, several methods may be used to quantify and assess the efforts made by nations. Clustering methods are practical algorithms that enable a collection of items to be grouped according to common properties. Numerous domains, including energy systems (Matenga, 2022), human social dynamics (Jiang et al., 2014), investment trend identification (Nemitz et al., 2016), and consumer behavior studies (Schösler et al., 2014), have applied clustering techniques as a methodological tool. Clustering algorithms have proven helpful in classifying significant elements that may need further focus to guarantee the attainment of desired values. As a result, we used cluster analysis to examine the SDGs' current state and the contributions of various countries. First, we assessed how the SDGs were doing in various nations at the time. We next assessed how the SDGs and organic residuals in each nation were related to one another. Lastly, the vulnerable sectors were identified and the countries' degrees of sustainable development were grouped using cluster analysis. The goal of this work is to build a framework which rewards stakeholders and encourages the collaboration between countries and international organizations with the aim to accomplish sustainable development. Furthermore, this study applies advanced clustering methods to inform decision-making and implementation strategies aimed at global sustainability goals, providing a clear assessment of national-level advancements toward them. With the use of this method, the study effectively pinpoints the discrepancy between each SDG's intended goals and existing state, enabling well-informed decision-making and focused actions.

9.2 Methodology Implemnted

The applied approach, which made use of two sets of data, is depicted in Fig. 9.1. The worldwide food damage index was added to the prior database in the second set, while the acceptance ratings for all 17 SDGs were utilized in the first set. After performing a correlation analysis, the data were normalized. The Elbow and Silhouette procedures were used to locate the best cluster total, k. Ultimately, clustering was used, and each database's groupings were described.

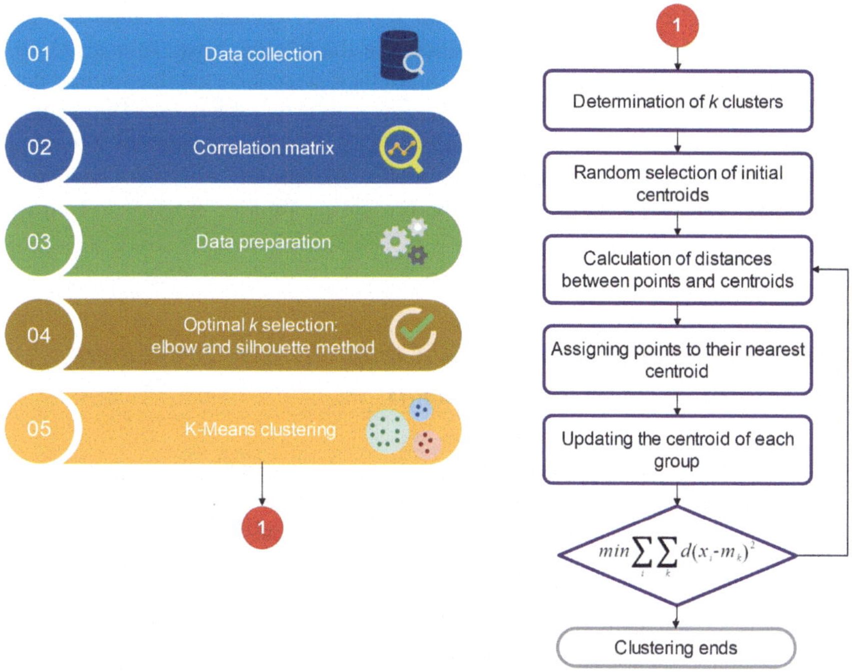

Fig. 9.1 Diagrammatic illustration of the entire process

9.3 Addressed Case Study

In anticipation of the 2023 Summit on the Sustainable Development Goals, as well as how the implementation period for the 2030 Agenda reaches its midpoint, global efforts to address critical challenges have intensified significantly. The United Nations 2030 Agenda in two thousand fifteen is intended to drive global transformation and improve the lives of all individuals through seventeen global objectives and one hundred sixty-nine concrete targets (UN, 2015). Each of these objectives encompasses a set of cross-cutting goals that reflect a universal commitment to human prosperity, environmental protection, and social well-being. Whether the plan succeeds or fails will entirely depend on how well the economic, social, and environmental domains are balanced.

Nevertheless, the population faces new and difficult obstacles in realizing the aspirational goal of human growth. Ending poverty, improving health, reducing hunger, and protecting the environment remain pressing global priorities. Addressing these challenges also requires efficient resource management and the capacity to deal with uncertainty associated with large-scale data processing. To meet these objectives, a coordinated administrative strategy is required. This approach should strengthen synergies and encourage investment to ensure that benefits are equitably shared across multiple sectors. Through this approach, the use of data can help with evidence-based decision-making by revealing information about the Agenda's accomplishments, highlighting areas of lack, and comprehending the relationships. To analyze 143 nations, the usage of machine learning techniques, were selected due to their data availability and because of their transparency within two scenarios. The analysis accounts for the setbacks resulting from the COVID-19 pandemic, sociopolitical conflicts, natural hazards, as well as economic disruptions, with the objective of identifying critical factors that define the present position of progress toward the Sustainable Development Goals.

9.3.1 Assessment Scenario One: Comprehensive Evaluation of the Seventeen Global Goals

The 2023 Sustainable Development Report's evaluation and reporting of the 17 SDGs' indicators is examined (Sachs et al., 2023). The chosen nations have adopted actions in an action plan aimed at creating a resilient and sustainable planet and function as a cooperative association. It is crucial to note that every nation embraced the Agenda while keeping in mind its own realities, levels of development, and adherence to policies and goals. There may be benefits and drawbacks to this. It is evident that low-income and less developed countries face limitations in allocating resources to research and technological innovation. Their immediate priorities often prevent them from placing environmental demands above more urgent social and economic needs. Global collaboration will, however, enable less developed economies to absorb knowledge and adapt practices from more developed nations, which will enhance sustainable development's fundamental dimensions: environmental protection, social inclusion, and economic growth.

9.3.2 Assessment Scenario Two: Focused Evaluation of the Seventeen Global Goals and Target Twelve Point Three

This analysis explores the seventeen global goals, taking as a central reference Target Twelve Point Three, which addresses the reduction of food waste and loss. The approach is grounded in the broader aim of promoting sustainability through responsible and efficient resource management. Considering the projected increase in global population, Sustainable Development Goal Twelve, which promotes accountable production and consumption, is regarded as a strategic priority. Within this framework, Target Twelve Point Three is especially relevant, as it focuses on decreasing food waste at the user and retail levels, as well as decreasing food losses across the supply chain. It is about addressing fundamental necessities, accelerating the supply chain at every level, doing more with less, and encouraging the decrease of waste and its undesirable properties on the ecosystem in urban areas. At the moment, each individual produces 0.28 kg of organic residuals, which represents 17% of the global food production (WHO, 2022). Even more concerning is the continued rise in food insecurity, with 23% of the global inhabitants affected by undernourishment (FAO, 2023b). This underscores the urgent need to advance food management and production strategies over the integration of technical innovation, didactic initiatives, and practical solutions.

9.3.3 Data Collection

The 2023 Sustainable Development Report presents compliance scores for the seventeen Sustainable Development Goals across one hundred forty-three countries (Sachs et al., 2023).

Using 122 indicators, the 17 SDGs are assessed. The majority of the information is taken from databanks maintained by globally recognized agencies such as the World Bank, the Organisation for Economic Co-operation and Development, Food and Agriculture Organization, International Labour Organization, United Nations Children's Fund, World Health Organization, and others that adhere to stringent and thorough procedures which will validate the data. The Food and Agriculture Organization provided data on global food damage (FAO, 2023a). Because they don't have enough data, some nations aren't included.

9.3.4 Matrix of Interdependencies Among the SDGs

The correlation matrix illustrates the direction and strength of the linear association among variables. In this context, it reflects the interactions among the seventeen Sustainable Development Goals. Correlation values vary between one to minus one: values near to minus one specify an inverse relationship, wherever a rise in one variable matches to a reduction in the other value; values close zero represent a weak or no linear relationship; and values near to one correspond a direct association, where both variables tend to increase or decrease together.

9.3.5 Data Preparation

This model might induce biases regarding variables with a wider variation range when the variables are on different scales. To ensure that each variable values are on comparable scales, it is crucial to normalize the data. The variables were scaled between the range [0, 1] using Eq. 9.1 in the MinMax approach, where $x_{\min}$ and $x_{\max}$ are the minimum and maximum values for each variable.

$$x_{norm} = \frac{x - x_{min}}{xmin_{max}} \quad (9.1)$$

9.3.6 Clustering Using the K-Means Approach

The K-means approach, a widespread clustering technique, allocates each data point to the adjacent cluster after identifying a certain number of centroids. The following is a description of the K-means clustering approach:

1. Given an amount of clusters k, the original centroids are arbitrarily designated $m_1^t, \ldots, m_{kd}^t$, here t = 0.

For point X of the data set in the d-dimensional space R_d, $X = \{x_i | x_i \in R_d, i = 1, 2, \ldots, n\}$, its distance is determined concerning each centroid m_{kd}^t. Equation 9.2 is used to determine the Euclidean distance, which is a degree of likeness.

$$d(x_{id}, m_{kd}) = \sqrt{\sum\nolimits_d (x_{id} - m_{kd})^2} \, \forall i, \forall k \quad (9.2)$$

2. Subsequently, each x_i is allocated to the cluster C_k with the minimum distance from its centroid m_k.
3. Eq. 9.3 illustrates how the arithmetic mean of the group's points is used to update the centroid of each group.

$$m_k^{t+1} = \frac{1}{|C_k^t|} \sum\nolimits_{x_i \in C_k^t} x_i \, \forall k \quad (9.3)$$

Until a stopping criterion is met the steps two through four are repeated. In this case, the stopping criterion corresponds to the sum of Squared Euclidean Distances, or "SED" (Eq. 9.4), which is chosen as the objective function to be reduced. Reaching a specific number of iterations might be an additional choice.

$$SED = \sum\nolimits_i \sum\nolimits_k d(x_i - m_k)^2 \quad (9.4)$$

9.3.7 Validation

The "optimal number of clusters" is defined as the cluster total where each group member is the most similar to each other and most different from those of other groups. There is no objective criterion for choosing the cluster total, but various approaches propose this number (Wu, 2012). The Elbow and Silhouette approach establishes the best cluster total, k.

9.3.7.1 Elbow Method

According to Ketchen and Shook (1996), the Elbow technique gives greater weight to intra-cluster distances; the narrower the gap, the more compact the clusters that are created. The "distortion" that the Elbow technique employs is the sum of squared distances between each centroid and its cluster point (Eq. 9.3). It plots the distortion versus the cluster total in a linear graph using the distortion values that are acquired subsequently implementing the K-means to varying amounts of clusters (from 1, 2, 3,..., j). This graph should show a dramatic shift in slope (the elbow) at the best cluster total. At this stage, the intra-cluster variance declines more slowly, suggesting that cluster compactness will not be enhanced by adding more clusters.

9.3.7.2 Silhouette Method

Clustering quality is measured by the silhouette method (Rousseeuw, 1987). This method encompasses two clustering factors (Handl & Knowles, 2005). The first factor being separation (the mean distance to the adjacent cluster), and the second factor being compactness (the mean distance inside the cluster).

The silhouette coefficient for x_i is given by $s(x_i)$ as shown in Eq. 9.5.

$$s(x_i) = \frac{b(x_i) - a(x_i)}{max\{a(x_i), b(x_i)\}} \tag{9.5}$$

where $a(x_i)$ is the mean distance from x_i to all other points in the same cluster, this way, it denotes the mean intra-cluster distance. $b(x_i)$ corresponds to the mean distance from x_i to all other points in the closest cluster. The score of $s(x_i)$ is between $[-1, 1]$.

A faulty clustering is shown by negative values because the points, nearing other clusters, are allocated to the incorrect cluster. If it is zero, it means that the point is on the border among two clusters or extremely near to them. Finally, since the points are far from the surrounding clusters and near to their own cluster, a value of 1 denotes a good clustering.

Consequently, Eq. 9.6 yields the Silhouette coefficient for the whole clustering.

$$SC = \frac{1}{n}\sum\nolimits_{i=1}^{n} s(x_i) \tag{9.6}$$

The mean of each point's silhouette coefficients over a range of k values is determined using the Silhouette technique. For a k value range, the cluster total k that maximizes the average of the coefficient silhouette is the best number. When the majority of the points acquire great silhouette scores, an appropriate clustering procedure is reached. If not, there can be too many or too few clusters in the clustering configuration.

9.4 Results

This section presents the correlation between the 17 SDGs, their relationship to the proposed home organic residuals index, and cluster analysis.

9.4.1 Correlation

A positive connection is shown by the color red, while a negative correlation is indicated by the color blue in Fig. 9.2, which displays the correlation matrix between the household organic residuals indicator and the 17 SDGs. The connection increases with the intensity of the color. SDG 15 (life on land), SDG 10 (reduced inequalities), and SDG 14 (life below water), all show extremely thin lines on the home organic residuals indicator, suggesting that these goals are not highly related to one another.

To conduct the analysis, the pairings among Sustainable Development Goals were organized in ascending order based on their correlation values, allowing for a clearer interpretation of their interrelationships. SDG 3 was shown to have a more positive correlation with other goals; that is, at what time SDG 3 has a high value, other objectives such as SDG 16 (peace, justice, and strong institutions), SDG 11 (sustainable cities and communities), SDG 7 (affordable and clean energy), SDG 6 (clean water and sanitation), SDG 9 (industry, innovation, and infrastructure), SDG 4 (quality education), as well as SDG 1 (no poverty) will also have high scores.

A peculiar trait of SDG 12 is that it has a negative correlation with other goals. When SDG 12 (responsible production and consumption) has a high score, other goals such as SDG 16 (peace, justice, and strong institutions), SDG 11 (sustainable cities and communities), SDG 9 (industry, innovation, and infrastructure), SDG 7 (affordable and clean energy), SDG 6 (clean water and sanitation), SDG 4 (quality education), and SDG 3 (good health and well-being) will also have low values.

Additionally, it was found that SDG 9 (industry, innovation, and infrastructure) and SDG 3 (good health and well-being) have the strongest positive link, suggesting that nations with a bigger SDG 9 score also have a higher SDG 3 score. The relationship among SDG 9 and SDG 3 may be explained by the circumstance that human well-being is surely supported by advancements in sustainable technology, resilient infrastructure, communication, and financial services. This was demonstrated during the COVID-19 epidemic. Countries that made progress in these areas were able to take a more operative hands-on-approach to stop the spread of the disease and to reduce the number of fatalities. Additionally, some countries were able to carry out

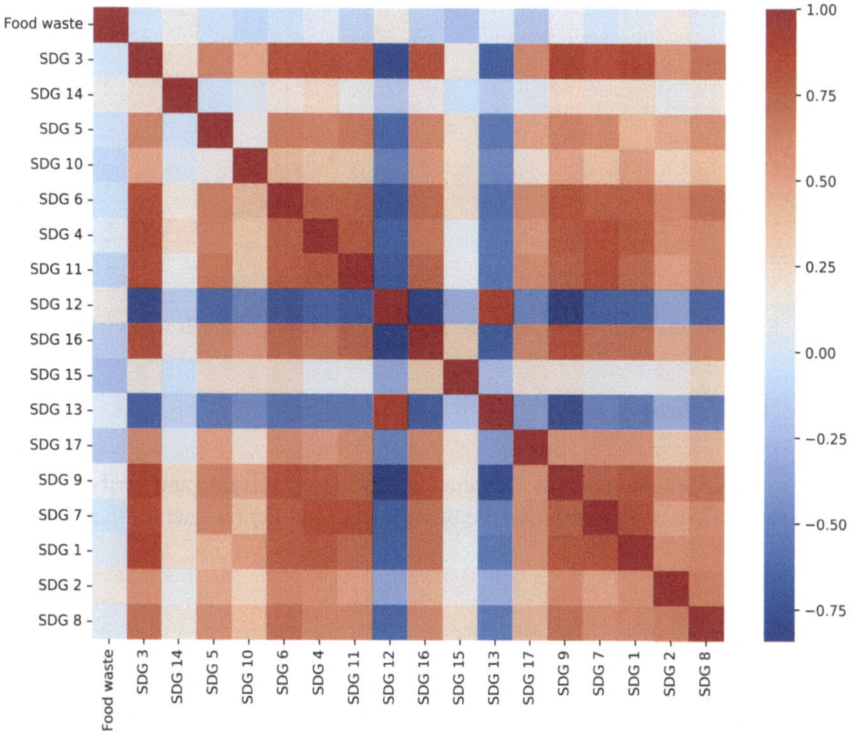

Fig. 9.2 The correlation matrix between the home organic residuals indicator and the 17 SDGs

research to develop vaccines and then effectively distribute those vaccines to the populace.

However, SDG 12 (responsible production and consumption) and SDG 9 (industry, innovation, and infrastructure) have the most negative link, suggesting that nations with high SDG 12 scores have a low industry, innovation, and infrastructure rating. The effective usage of resources and related objectives may be the cause of this. For instance, goal 12.2 on the material footprint has been bigger in more developed nations than in less developed ones; similarly, goals 12.3 on organic residuals, 12.4 on life cycle intensification, as well as 12.5 on recycling activities have been inferior in low-income nations where people must make do with less and find value in waste (UN, 2023).

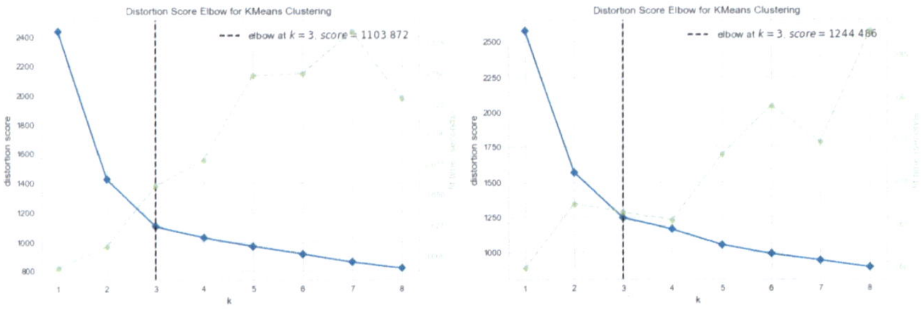

Fig. 9.3 Elbow results for 17 SDGs (left) and 17 SDGs and organic residuals scores (right)

9.4.2 *Validation*

The Elbow and Silhouette methods were employed to determine and verify the best cluster total k. We must determine the best cluster total (k) for each of the two sets of data.

9.4.2.1 Elbow Approach

The Elbow technique represented in Fig. 9.3 displays the amount of groups (k) on the x-axis, the distortion (SED) on the y-axis, and the computing time required to train the clustering approach on the green dashed line. The elbow approach's findings indicate that the best cluster total for the 17 SDGs database is three. Three is also the best cluster total for organic residuals indicator and the 17 SDGs database.

9.4.2.2 Silhouette Approach

The findings of the silhouette approach are illustrated employing a bar chart, where the y-axis corresponds to the average silhouette coefficient and the x-axis indicates the total quantity of clusters, k. The maximum mean silhouette coefficient for the database with the 17 goals is roughly 0.34 at $k = 2$, which is the best cluster total. However, it was determined that there were too few groups to segment, so $k = 3$, which has a coefficient of 0.26, was chosen.

The average silhouette coefficient for the database of the 17 goal and organic residuals scores was 0.34 when $k = 2$, but we wanted more segmentation for our purposes; thus, the next best value was $k = 4$.

The Silhouette Visualizer presents the silhouette coefficient for every example and for every group. This permits users to visually assess the density and group separation. Each graph uses a different shade to indicate each cluster and a dotted line is used to show the common silhouette coefficient. Figure 9.4 displays the groupings using the 17 SDGs database. The best cluster selections are $k = 3$ and $k = 2$, because

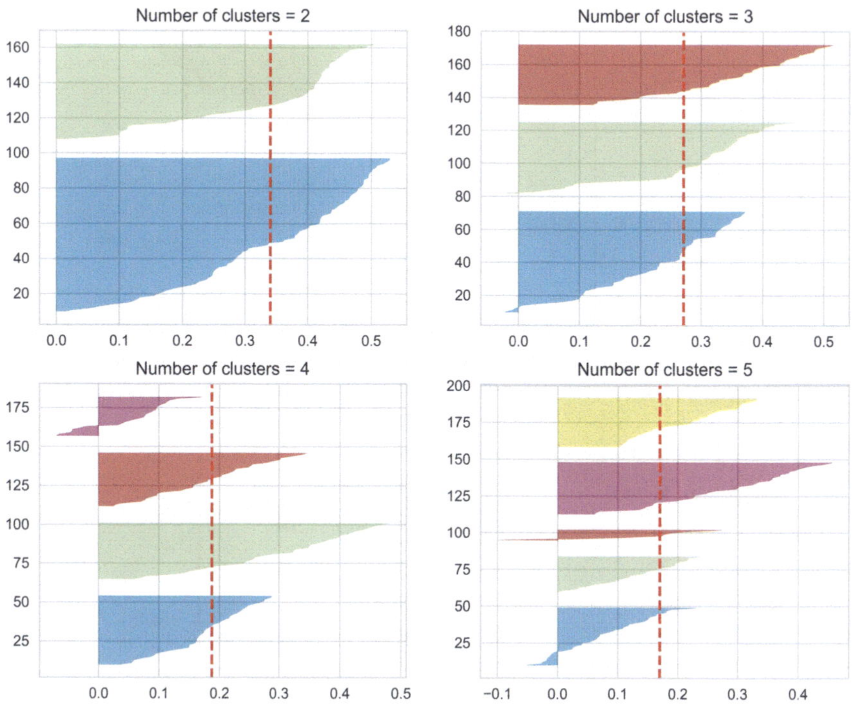

Fig. 9.4 K-means clustering silhouette chart for 17 SDGs in k groups, $k = 2, 3, 4$ and 5

each group is beyond the mean silhouette coefficient and presents a decent density of the groups and distribution. A group in cluster k = 4 has values below the average silhouette coefficient as well as numerous values below zero in cluster k = 5. This indicates that the data could have been incorrectly allocated to the incorrect group.

The 17 SDGs silhouette plots and the organic residual indicator for k = 2, k = 3, k = 4, and k = 5 are displayed in Fig. 9.5. Even though k = 2 and k = 3 contain the greatest average silhouette coefficients, k = 4 and k = 5 contain many values below zero. This is an indication that the data could have been incorrectly assigned to the wrong group. This situation worsens because k = 5 has negative values when a group fails to achieve the mean silhouette coefficient. It has been determined that grouping the database with its subsequent cluster (which has the greatest silhouette coefficient (k = 4)) was the best option since the database does not include numerous negative scores, as well as all of the databases' groups surpass the average silhouette coefficient.

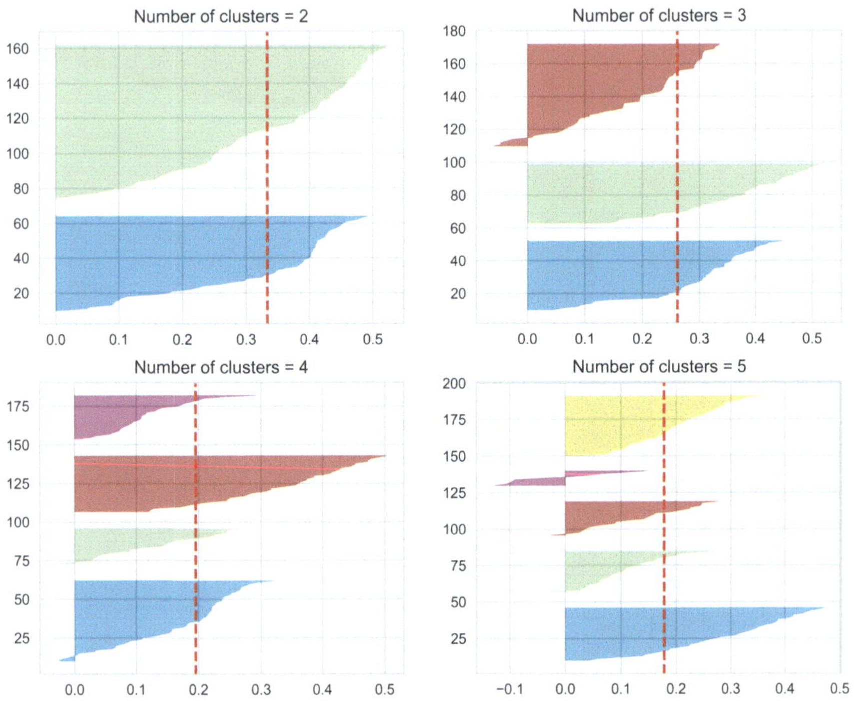

Fig. 9.5 Silhouette chart of K-means grouping of 17 SDGs and the organic residuals indicator in k groups, $k = 2, 3, 4$, and 5

9.4.3 Group Analysis

9.4.3.1 Database of the 17 SDGs

Groups 2, 1, and 0 were classified based on the amount of high, medium, and low mean scores of the nations that make up each group. This is so that they can be appropriately named in accordance with the SDGs.

Group 2, for example, is known as the "High Sustainability" group for it has the highest-scoring sustainable development objective. Group 1 is known as the "Medium-scoring Sustainability" group, and group 0 is known as the "Low Sustainability" group for it has the lowest-scoring sustainable development objective.

Countries categorized within the high sustainability group generally achieve strong results across multiple Sustainable Development Goals, including partnerships for the goals (Goal 17), peace, justice and strong institutions (Goal 16), life below water (Goal 14), life on land (Goal 15), sustainable cities and communities (Goal 11), decent work and economic growth (Goal 8), industry, innovation and infrastructure (Goal 9), affordable and clean energy (Goal 7), clean water and sanitation (Goal 6), reduced inequalities (Goal 10), gender equality (Goal 5), quality education (Goal 4), zero hunger (Goal 2), good health and well-being (Goal 3), and no poverty (Goal 1). However, their performance tends to be lower in areas related to responsible production and consumption (Goal 12) in addition to climate action (Goal 13).

With the exception of SDG 15 (life on land) where they have poor ratings, the nations in the group with intermediate sustainability have middle values for practically every objective.

With the exception of SDG 12 (responsible consumption and production) and SDG 13 (climate action), which have the highest ratings, countries in the low sustainability group perform poorly on the majority of the goals. Additionally, they get middling ratings in the SDG 15 (life on land).

9.4.3.2 Database of the 17 SDGs and the Organic Residuals Indicator

Nations in the high sustainability category scored highly on fifteen SDGs while receiving low ratings on two of them: SDG 13 (climate action) as well as SDG 12 (responsible consumption and production). Furthermore, these nations exhibit exceptionally low levels of household organic residuals, which is a crucial factor in their high sustainability. This finding demonstrates how well their initiatives are working to encourage sensible eating habits and lessen the negative environmental effects of organic residuals.

The intermediate-high sustainability group's nations have above-mean but inferior scores than the great sustainability group; they have the most household organic residuals and poor scores on SDG 15 (life on land).

Countries in the medium–low sustainability group had below-average ratings on nearly every objective, including SDG 10 (reduced inequalities) as well as SDG 14 (life below water).

With the exception of SDG 12 (responsible consumption and production) as well as SDG 13 (climate action), nations in the poor sustainability category score the lowest on the majority of the objectives. Their organic residuals are likewise higher than typical.

9.4.4 Country Clustering

With the use of the 17 SDGs database, we were able to display K = 3's clustering results. The groups were named High Sustainability Level, Medium Sustainability Level, and Low Sustainability Level based on their compliance with the SDG rating. Figure 9.6 shows the 37 nations that have a "High Sustainability Level" rank in green, the 63 nations that have a "Medium Sustainability Level" rank in yellow, and the 43 nations that have a "Low Sustainability Level" rank in red.

With k = 4, all the nations classification using the 17 SDGs database as well as the organic residuals indicator are displayed. The groups were named High Sustainability

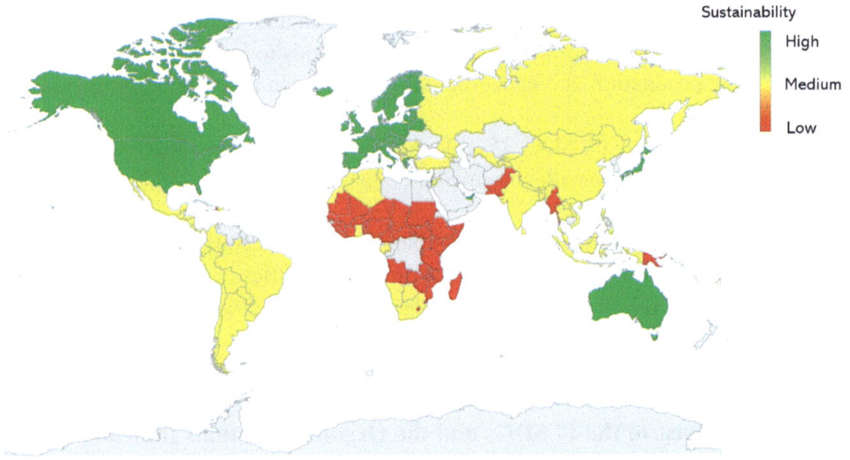

Fig. 9.6 Clustering of countries according to their score of sustainability (Database of 17 SDGs)

Level, Medium Sustainability Level, and Low Sustainability Level based on their compliance with the SDG rating. Figure 9.7 shows the 36 nations that have a "High Sustainability Level" rank in green, the 52 nations that have a "Medium Sustainability Level" rank in yellow, and the 26 nations that have a "Low Sustainability Level" rank in red.

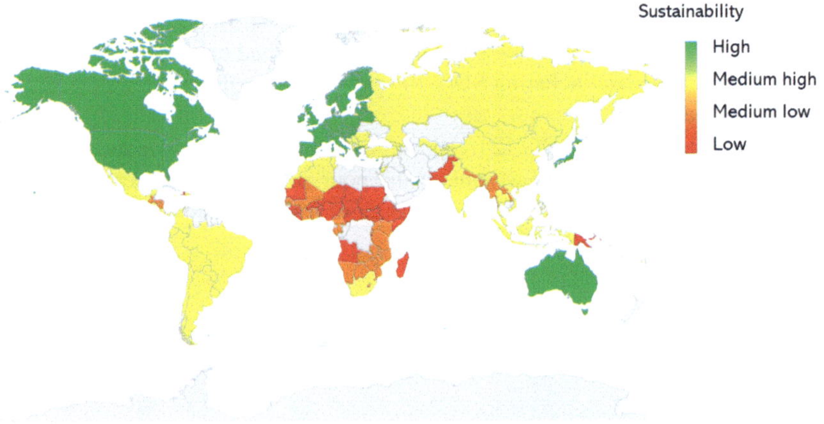

Fig. 9.7 Clustering of countries based on their level of sustainability (Database of 17 SDGs and organic residuals indicator)

9.5 Conclusions

Countries are required to advance on each SDG so as to achieve the Agenda's 2030 aims, but it may be very beneficial to have a framework that enables them to evaluate their progress and the ways in which the SDGs connect to one another in order to determine how best to move forward. A framework for classifying nations' tendencies toward their SDG status has been proposed in this paper. Using two databases—one containing the values of the 17 SDGs and the other including the Household Organic residuals Index—a clustering of nations has been created using the K-means approach. When the SDGs and the Household Organic residuals Index were correlated, it was discovered that SDG 3 (good health and well-being) and SDG 9 (industry, innovation, and infrastructure) had the strongest positive correlation, while SDG 12 (responsible consumption and production) as well as SDG 9 (industry, innovation, and infrastructure) had the strongest negative correlation.

The clustering created three groups using the database of the 17 SDGs' progress ratings. SDG 13 (climate action) as well as SDG 12 (responsible consumption and production) received the lowest ratings, while the group of nations with the highest sustainability earned the maximum compliance scores across the fifteen SDGs. The fact that they are developed nations is another noteworthy feature. In addition to being developing nations, the group of countries with intermediate sustainability has above-mean scores in the majority of the SDGs, as well as a poor score in SDG 15 (life on land). Lastly, in addition to being impoverished, low-sustainability nations have the lowest results for fourteen SDG compliance, an intermediate score for SDG 15, and good scores for SDGs 12 and 13. However, the clustering of the Household Organic residuals Indicator, which is correlated with population size, and the 17 SDGs database reveals the creation of four groups. Developed nations with high sustainability have the least amount of organic residuals in their households. Organic residuals are higher in emerging nations, which are in the medium–high sustainability group.

The results of this study give stakeholders and policymakers crucial direction by educating them about methods and best practices to improve the results of sustainable growth. The research specifically highlights areas for improvement in the food sector's efforts to reduce organic residuals, SDG 13 (climate action), as well as SDG 12 (responsible consumption and production). By addressing these issues, a course, for an improved and viable future for our world can be achieved.

References

Addo-Atuah, J., Senhaji-Tomza, B., Ray, D., Basu, P., Loh, F. H. E., & Owusu-Daaku, F. (2020). Global health research partnerships in the context of the Sustainable Development Goals (SDGs). *Research in Social and Administrative Pharmacy, 16*(11), 1614–1618. https://doi.org/10.1016/j.sapharm.2020.08.015

Ardra, S., & Barua, M. K. (2022). Halving organic residuals generation by 2030: The challenges and strategies of monitoring UN sustainable development goal target 12.3. *Journal of Cleaner Production, 380*, 135042. https://doi.org/10.1016/j.jclepro.2022.135042

Burton, I. (1987). Report on reports: Our common future: The world commission on environment and development. *Environment: Science and Policy for Sustainable Development, 29*(5), 25–29. https://doi.org/10.1080/00139157.1987.9928891

Campbell, B. M., Hansen, J., Rioux, J., Stirling, C. M., & Twomlow, S. (2018). Urgent action to combat climate change and its impacts (SDG 13): Transforming agriculture and food systems. *Current Opinion in Environmental Sustainability, 34*, 13–20. https://doi.org/10.1016/j.cosust.2018.06.005

Chakrabarty, A., & Das, U. S. (2020). Mission SDGs through organic residuals management: Nature and approaches. In P. Pattnaik, R. Kumar, & S. Pal (eds.), *Internet of Things and Analytics for Agriculture, Volume 2* (pp. 249–260). Springer, Singapore. https://doi.org/10.1007/978-981-15-0663-5_12

Elavarasan, R. M., Pugazhendhi, R., Irfan, M., Mihet-Popa, L., Campana, P. E., & Khan, I. A. (2022). A novel Sustainable Development Goal 7 composite index as the paradigm for energy sustainability assessment: A case study from Europe. *Applied Energy, 307*, Article 118173. https://doi.org/10.1016/j.apenergy.2021.118173

FAO. (2018). *Sustainable Food Systems*. Retrieved from https://www.fao.org/3/ca2079en/CA2079EN.pdf. Accessed August 4, 2023

FAO. (2023a). *Technical Platform on the Measurement and Reduction of Food Loss and Waste*. Retrieved from https://www.fao.org/platform-food-loss-waste/food-loss/food-loss-measurement/en. Accessed June 25, 2023

FAO. (2023b). *Global Report on Food Crises: Number of people facing acute food insecurity rose to 258 million in 58 countries in 2022*. Retrieved from https://www.fao.org/newsroom/detail/global-report-on-food-crises-GRFC-2023-GNAFC-fao-wfp-unicef-ifpri/es. Accessed July 26, 2023

Garcia-Garcia, G., Azanedo, L., & Rahimifard, S. (2021). Embedding sustainability analysis in new food product development. *Trends in Food Science & Technology, 108*, 236–244. https://doi.org/10.1016/j.tifs.2020.12.018

Handl, J., & Knowles, J. (2005). Improvements to the scalability of multiobjective clustering. *2005 IEEE Congress on Evolutionary Computation, 3*, 2372–2379. https://doi.org/10.1109/CEC.2005.1554990

Huang, R. (2023). SDG-oriented sustainability assessment for Central and Eastern European countries. *Environmental and Sustainability Indicators, 100268*,. https://doi.org/10.1016/j.indic.2023.100268

Ibrahim, M. D., Alola, A. A., & Ferreira, D. C. (2023). Assessing sustainable development goals attainment through energy-environmental efficiency: The case of Latin American and Caribbean countries. *Sustainable Energy Technologies and Assessments, 57*, Article 103219. https://doi.org/10.1016/j.seta.2023.103219

Jiang, Y., King, J. M., & Prinyawiwatkul, W. (2014). A review of measurement and relationships between food, eating behavior and emotion. *Trends in Food Science & Technology, 36*(1), 15–28. https://doi.org/10.1016/j.tifs.2013.12.005

Ketchen, D. J., & Shook, C. L. (1996). The application of cluster analysis in strategic management research: An analysis and critique. *Strategic Management Journal, 17*(6), 441–458. https://doi.org/10.1002/(SICI)1097-0266(199606)17:6%3c441::AID-SMJ819%3e3.0.CO;2-G

Matenga, Z. (2022). Assessment of energy market's progress towards achieving sustainable development goal 7: A clustering approach. *Sustainable Energy Technologies and Assessments, 52*, Article 102224. https://doi.org/10.1016/j.seta.2022.102224

Nemitz, M. C., Argenta, D. F., Koester, L. S., Bassani, V. L., von Poser, G. L., & Teixeira, H. F. (2016). The international scenario of patents concerning isoflavones. *Trends in Food Science & Technology, 49*, 85–95. https://doi.org/10.1016/j.tifs.2016.01.008

Ortiz-de-Montellano, C. G. S., Samani, P., & van der Meer, Y. (2023). How can the circular economy support the advancement of the sustainable development goals? A comprehensive analysis. *Sustainable Production and Consumption, 40*, 352–362. https://doi.org/10.1016/j.spc.2023.07.003

Petruzzelli, M., Ihle, R., Colitti, S., & Vittuari, M. (2023). The role of short food supply chains in advancing the global agenda for sustainable food systems transitions. *Cities, 141*, Article 104496. https://doi.org/10.1016/j.cities.2023.104496

Rousseeuw, P. J. (1987). Silhouettes: A graphical aid to the interpretation and validation of cluster analysis. *Journal of Computational and Applied Mathematics, 20*(C), 53–65. https://doi.org/10.1016/0377-0427(87)90125-7

Sachs, J. D., Lafortune, G., Fuller, G., & Drumm, E. (2023). Sustainable Development Report 2023: Implementing the SDG Stimulus, SDSN. Retrieved from https://policycommons.net/artifacts/4445283/2023-sustainable-development-report/5242513/. Accessed August 15, 2023.

Scherer, L., Behrens, P., de Koning, A., Heijungs, R., Sprecher, B., & Tukker, A. (2018). Trade-offs between social and environmental sustainable development goals. *Environmental Science & Policy, 90*, 65–72. https://doi.org/10.1016/j.envsci.2018.10.002

Schösler, H., de Boer, J., & Boersema, J. J. (2014). Fostering more sustainable food choices: Can Self-Determination Theory help? *Food Quality and Preference, 35*, 59–69. https://doi.org/10.1016/j.foodqual.2014.01.008

Singh, S., & Jayaram, R. (2022). Attainment of water and sanitation goals: A review and agenda for research. *Sustainable Water Resources Management, 8*(5), 146. https://doi.org/10.1007/s40899-022-00719-9

United Nations. (2023). *Transforming our world: the 2030 Agenda for Sustainable Development*. Retrieved from https://sdgs.un.org/2030agenda. Accessed July 15, 2023

UN. (2015). *Sustainable development goals*. Retrieved from https://www.un.org/sustainabledevelopment/es/2015/09/la-asamblea-general-adopta-la-agenda-2030-para-el-desarrollo-sostenible/. Accessed July 26, 2023

UN. (2023). *Ensure sustainable production and consumption*. Retrieved from https://unstats.un.org/sdgs/report/2019/goal-12/. Accessed August 4, 2023

UNDP. 2023. *Delivering on the Sustainable Development Goals through Solutions at the Energy, Food and Finance Nexus: 2023 Asia-Pacific SDG Partnership Report*. Retrieved from https://www.undp.org/asia-pacific/publications/delivering-sustainable-development-goals-through-solutions-energy-food-and-finance-nexus-2023-asia-pacific-sdg-partnership#:~:text=March%2026%2C%202023, and %20the%20worsening%20climate%20crisis. Accessed August 5, 2023

WHO. (2022). *UN Report: Global hunger numbers rose to as many as 828 million in 2021. 2022*. Retrieved from https://www.who.int/news/item/06-07-2022-un-report--global-hunger-numbers-rose-to-as-many-as-828-million-in-2021. Accessed July 26, 2023

Wu, J. (2012). Cluster analysis and K-means clustering: An introduction. In: *Advances in K-means Clustering* (pp.1–16). Springer. https://doi.org/10.1007/978-3-642-29807-3_1

Chapter 10
Spatiotemporal Clustering of Dam Filling Patterns

10.1 Introduction

Water is a vital resource for all forms of life, making its effective management essential to meet the growing needs of the world's inhabitants. According to estimates, the planet has approximately 42,700 km^3 of freshwater available each year. In theory, if this volume were evenly distributed across both time and geography, it would be sufficient to fulfill global needs (UN, 1998). However, in practice, such a balance does not exist. There is a significant mismatch between the locations of freshwater availability and population centers, compounded by substantial spatial and temporal variations in global water distribution (UNDP, 2006).

One of the main strategies used to cope with the geographical and temporal variability of water, which has historically resulted in severe droughts and floods, has been the controlled distribution and storage of water (Ho et al., 2017). Dams are essential infrastructures that hold social, economic, and environmental importance, and due to their complexity and strategic value, they require continuous real-time monitoring and maintenance (Pipitone et al., 2018; Zhu et al., 2021). The placement of a dam depends on several factors, including the topography, the hydrological characteristics of nearby streams, geological conditions, and the intended water usage (EAM, 2021). Dams support a variety of purposes, such as electricity production, flood regulation, fish farming, aquaculture, navigation, recreation, and supplying water for public consumption and agricultural irrigation (Schulz & Adams, 2021). In recent years, multipurpose dams have become increasingly important, as many are now designed to serve more than one function (Chen et al., 2016).

The rising demand for food, energy, and water, driven by rapid population growth, has led to the building of large-scale dams (Poff & Hard, 2002). In system planning, it is essential to consider the food-energy-water (WEF) nexus, as these resources are fundamentally interconnected. A disruption in one area can produce cascading effects on the others (Raya-Tapia et al., 2023). For example, both food and energy production rely heavily on water availability, while the processes involved in water

A. Y. Raya-Tapia et al., *Machine Learning and Clustering for a Sustainable Future*,
Studies in Computational Intelligence 1233,
https://doi.org/10.1007/978-3-032-03876-0_10

extraction, conveyance, treatment, and distribution require significant energy inputs. Water systems are directly linked to Sustainable Development Goal (SDG) 6, which emphasizes clean water and sanitation, while they also influence SDG 1 (No Poverty), SDG 2 (Zero Hunger), and SDG 7 (Affordable and Clean Energy) indirectly (UN, 2024).

Globally, agriculture accounts for over 70% of freshwater consumption (World Bank, 2017). With the worldwide population expected to reach 9.7 billion by the year 2050, it is predictable that water withdrawals will need to rise by 15%, while agricultural output must rise by approximately 50% to meet growing demand (UN, 2022). Furthermore, economic development and rising individual incomes are leading to shifts in dietary patterns, which are expected to further increase per capita food consumption (FAO, 2012). As a result, higher food demand will place additional pressure on water resources. Insufficient water availability could reduce agricultural productivity, posing serious risks to both nutrition and food security (UN, 2023).

Growing concerns about climate change within the energy sector have intensified interest in hydroelectric power generation, as it emits significantly fewer greenhouse gases compared to hydropower systems reliant on fossil fuel combustion (Erlewein, 2014; Kaygusuz, 2009). Unlike other renewable sources such as solar and wind, hydropower also offers greater flexibility in operation (Singh & Singal, 2017). However, hydropower plants are highly vulnerable to shifts in climatic conditions (Obahoundje et al., 2022). Their water availability can be adversely affected by extreme hydrological phenomena, including droughts, floods, changes in precipitation, and rising temperatures (Hamududu and Killingtveit, 2016). In addition, thermal power plants may experience reduced efficiency due to temperature increases, which can elevate the temperature of water used in their cooling processes (Mideksa & Kallbekken, 2010).

Climate change has caused notable alterations in precipitation patterns, significantly affecting water resources and local ecosystems in various regions across the globe (Di Nunno et al., 2024; Fan et al., 2020; Tarroja et al., 2019). Its inherent unpredictability complicates planning processes and often gives rise to conflicts among nations, regions, and different user groups (Obahoundje et al., 2022).

Reduced precipitation has led to declines in both water levels in reservoirs used for energy generation and food production (Olsson, 2013). Lower reservoir levels have often caused power outages at various times throughout the day, disrupting productivity and a wide range of socioeconomic activities (Handayani et al., 2019; Hunt et al., 2018; Scanlon et al., 2013). Drought conditions can emerge from several contributing factors, including insufficient water availability, excessive demand, inadequate infrastructure for water extraction, or ineffective management of operational systems and their corresponding regulations (Pedro-Monzonís et al., 2015). Drought has widespread consequences across environmental, social, and economic dimensions (Daneshmand et al., 2014). In recent decades, increased pressure on water resources driven by human activity has intensified the occurrence of droughts, leading to substantial economic losses (Alcamo et al., 2007; Cesano & Gustafsson, 2000).

For decision-makers and managers overseeing hydrological regions, it is essential to manage water resources effectively and efficiently in order to address the interests of all stakeholders and reconcile competing demands. This task becomes increasingly complex due to rising climate variability and the evolving expectations of stakeholders regarding system performance (Loucks & van Beek, 2017). Today, the objective extends beyond simply maximizing net economic benefits and ensuring equitable distribution; environmental and ecological considerations must also be integrated into water management strategies. As noted by Loucks and van Beek (2017), sustainable water systems are designed and operated with the well-being of both present and future generations in mind.

Regardless of the water planning strategy adopted to optimize allocation benefits, it is crucial to understand the interconnections between climate variables, sector-specific demands, and infrastructure management (Pedro-Monzonís et al., 2015). This understanding not only enables decision-makers to anticipate and respond to irregular precipitation patterns, but also enhances their ability to grasp the complexities of water dynamics and formulate appropriate policies (Tajbar et al., 2024).

Water system data are often large in scale and complex in nature, as they are collected at high temporal resolutions (Cominola et al., 2018; Wang et al., 2024). Investigating such high-dimensional datasets necessitates advanced modeling and data processing techniques capable of handling temporal dependencies and variability in system behavior (Bennett et al., 2013). Clustering, an unsupervised machine learning technique, is particularly useful for revealing hidden structures within large datasets. It has become an essential tool in climate analysis and regional classification, supporting hydrological planning, enabling comparisons across studies, and offering valuable insights into similar weather patterns (Hoghoughinia et al., 2024).

Various clustering methods have been applied to evaluate the spatiotemporal variability of monthly precipitation. For example, between 1998 and 2019, a hierarchical clustering approach was used to group 156 grids from the Tropical Rainfall Measuring Mission located in Pakistan's Khyber Pakhtunkhwa province (Tajbar et al., 2024). Similarly, Ward's clustering method was utilized to analyze precipitation records from 32 locations in the Brazilian state of Mato Grosso do Sul, covering the period from 1954 to 2013 (Teodoro et al., 2016). In Taiwan, Lin and Chen (2006) employed both Ward's method and K-means clustering to delineate homogeneous areas for frequency analysis by means of national precipitation data. Hoghoughinia et al. (2024) also applied clustering techniques to precipitation records collected from 1991 to 2020 at 65 synoptic stations across Iran, aiming to group sub-daily precipitation distribution patterns. In another study, Di Nunno et al. (2024) used both hierarchical and K-means clustering algorithms along with the Standardized Precipitation-Evapotranspiration Index (SPEI) at 6-, 12-, and 24-month timescales to identify drought-prone homogeneous zones in Ethiopia.

In the context of water infrastructure, clustering techniques have been applied to group dams based on multiple variables. For instance, Hermawan and Limantara (2024) classified 12 dams using the agglomerative hierarchical algorithm within a specific time. Their analysis incorporated a water conservation index that considered local climatic conditions such as temperature and precipitation, along with water quality indicators including pH, dissolved oxygen (DO), and total dissolved solids (TDS). In another study, Sharghi et al. (2018) utilized approximately 13,000 records of streamflow, precipitation, and temperature data collected monthly from 2001 to 2015 to cluster 27 watersheds located in Utah. Their research involved a comparison of several clustering techniques, including Ward's method, self-organizing maps (SOM), and K-means. Additionally, urban water consumption patterns have been identified using time series clustering, specifically applying the K-means algorithm in combination with dynamic time warping (DTW) distance (Wang et al., 2024).

Given the current absence of studies that integrate clustering techniques with seasonal analysis to examine variability in dam behavior, this research introduces a novel approach to exploring water level dynamics in reservoirs. The primary contributions of this study can be summarized as follows:

- Grouping time series for seasonal evaluation: Introducing a method that organizes dam water level time series to detect seasonal patterns both within individual groups and across different groupings.
- Identifying behavioral differences among dams: Highlighting distinct behavioral trends across dam clusters, offering insights into how water levels respond over time under varying conditions.
- Enhancing pattern understanding beyond traditional approaches: Unlike conventional analyses that focus on single dams or static evaluations, this study provides a broader understanding of water level behavior through comprehensive pattern recognition.
- Tools to support sustainable decision-making: Developing actionable tools that can assist policymakers in formulating improved water management strategies centered on conservation and efficient resource use.
- Bridging a research gap with applied data science: This approach integrates advanced analytical techniques with practical applications in water resource planning, addressing a notable gap in existing research on sustainability.

The approach adopted in this chapter offers a valued contribution to the field of sustainable water resource management and establishes a foundation for future investigations.

10.2 Problem Statement

Accelerated urban growth and rising population levels have led to increased water consumption, driving greater demand not only for public water supply but also for industrial use, agriculture, power generation, and other sectors. This has placed additional pressure on global freshwater resources. At the same time, climate change is contributing to more frequent and intense droughts, complicating efforts to manage water sustainably and efficiently. The construction of dams has strengthened society's ability to plan and regulate water availability, indirectly supporting food and energy security. Given its critical role across multiple sectors, the water stored in dams holds strategic importance, particularly for agricultural and energy-related activities.

Enhancing water storage capacity contributes directly to system resilience and economic development. Due to its inherent variability and complexity, water remains one of the most valuable and challenging resources to manage. Climate change, increasing demand across energy-related sectors, and the pursuit of emerging environmental and social objectives require planning processes to remain dynamic and adaptive. This study proposes organizing a dataset of 87,024 records by clustering all dams in Mexico based on their historical filling percentages, the variability of those percentages over time, and seasonal characteristics defined by monthly patterns. By applying time series clustering, it becomes possible to identify seasonal trends in reservoir filling behavior and to analyze how patterns shift between clusters over time. This approach also allows for the assessment of the influence of temperature fluctuations and precipitation on dam storage levels. The results can serve as a practical tool for policymakers, enabling more effective and sustainable management of water systems and potentially fostering economic development at both regional and national scales.

10.3 Case Study Addressed

Mexico is bordered by Belize and Guatemala to the south, and by the United States to the north. To the east lies the Gulf of Mexico, while the Pacific Ocean stretches along its western coast (Fig. 10.1). At the beginning of the twentieth century, agriculture was not only the country's primary source of sustenance but also a driving force for national development. The earliest dams in Mexico were built to store water for irrigation in regions where geographical conditions limited natural water availability. The generation of hydroelectric power through dams began to emerge during the 1930s (Domínguez, 2019).

By 2024, Mexico has registered a total of 222 dams, most of which are located in the northern part of the country and primarily serve agricultural and energy production purposes. Of the national water storage capacity, estimated at 150 million cubic meters, 80% is concentrated in 181 of these dams. The operation of dams is governed by two reference levels: the Ordinary Maximum Water Level (NAMO) and the Minimum Operating Water Level (NAMINO). As defined by CONAGUA (2016), NAMO represents the highest level at which a dam can be operated to meet water demand. The selection of dam sites is influenced by numerous issues, including the region's topography, the hydrological behavior of nearby rivers, geological conditions, and the intended use of the water. The largest dam in Mexico is Belizario Domínguez, located in the state of Chiapas (INEGI, 2018).

In Mexico, CONAGUA (the National Water Commission) is the authority responsible for overseeing the administration of water resources and the public benefits associated with them. The agency is also tasked with promoting dam safety by implementing regulatory frameworks aimed at minimizing risks and potential damage to

Fig. 10.1 Administrative division of Mexico by federal states

third parties. These regulations address threats related to water releases, as well as the partial or total failure of dam structures, which could lead to loss of life, destruction of infrastructure, and environmental degradation (Molina-Aguilar et al., 2019).

Drought is a frequent natural event in Mexico, with significant impacts on livestock production, hydropower generation, water availability, and irrigation for seasonal crops. In years of severe drought, hydroelectric output has been observed to decline by as much as 20%. Dam storage levels can also be influenced by other meteorological factors, including the La Niña phenomenon, cyclones, hurricanes, rainfall patterns, and ambient temperature. Climate change is predictable to rise the occurrence of droughts in the future (Breña-Naranjo, 2021). Low dam storage levels may contribute to broader issues such as political instability, migration, and threats to food security. Nevertheless, the implementation of effective drought management strategies can help mitigate these consequences. Based on data provided by CONAGUA, the present study applies K-means clustering (an unsupervised machine learning method) to examine various characteristics of dam filling levels in Mexico over the period from 1991 to 2024.

Each characteristic was classified into distinct categories based on the behavior of filling percentages over time, the variability observed at different periods, and the temporal evolution of dam storage levels. This categorization allows for the identification of relationships between dam filling patterns in Mexico and meteorological variables such as average temperature and accumulated precipitation.

10.4 Methodology Implemented

The different stages of the methodology applied in this study are illustrated in Fig. 10.2. The first stage involves the collection of data obtained from the official website of CONAGUA. This dataset contains the monthly filling percentages for 222 dams across Mexico, covering the period from January 1991 to August 2024, and resulting in a total of 87,024 data entries (a complete list of the dams is provided in the supplemental material). In the second stage, data preprocessing is carried out. This includes evaluating the amount of missing data for each dam. Dams with more than 35% missing values are excluded from the analysis, while missing values in the remaining records are filled in using the average filling percentage for each respective dam. This process defines the structure of data set A. For data set C, the inverse matrix of data set A is used. Data set B is constructed by calculating the percentage variation in dam filling, as defined in Eq. 10.1.

$$\%C = \frac{V_1 - V_0}{V_0} \cdot 100 \tag{10.1}$$

where V_1 is the present filling rate at period t and V_0 is the filling rate at period $t-1$.

In the third stage, the data is normalized using the Min–Max scaling method provided by the preprocessing module of the scikit-learn library (ScikitLearn, 2024). This technique, as defined in Eq. 10.2, adjusts the feature values to fall within the range of zero to one.

$$x_i^N = \frac{x_i - x_{min}}{xmin_{max}} \tag{10.2}$$

where x_i is the actual score of the attribute, whereas x_{min} and x_{max} are the minimum and maximum values, and x_i^N is the normalized score.

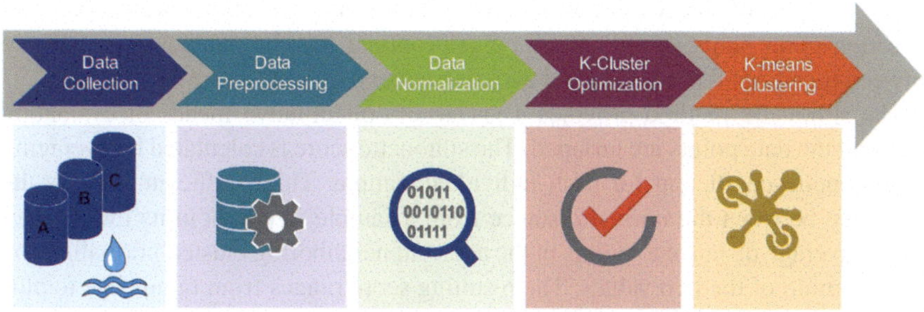

Fig. 10.2 Diagram illustrating the methodology applied in this study

After standardizing the data, the best number of clusters for each dataset is identified by means of the elbow approach and the silhouette coefficient. Once the appropriate number of clusters has been established, the K-means approach is applied to group the data points accordingly. Following the clustering process, each resulting group is characterized and interpreted.

10.4.1 Clustering

K-means clustering is an unsupervised machine learning approach that organizes items based on shared characteristics, as illustrated in Fig. 10.3. The steps followed by the K-means algorithm are shown in the same figure. Initially, K centroids are randomly selected. Next, the Euclidean distance between each data point and the centroids is determined. Each data point is then allocated to the group with the nearest centroid. After all data points have been assigned, the centroid of each group is updated by computing the mean position of the points within that group. This process is repeated until a stopping criterion is met. The stopping condition may consist of reaching a predefined number of iterations or minimizing the sum of squared Euclidean distances. By reducing the squared distances between data points and their respective centroids, the K-means algorithm effectively solves an optimization problem.

10.4.2 Optimal Number of K Clusters

The best amount of clusters for each dataset was determined using both the elbow approach and the silhouette method. The elbow technique involves applying the K-means approach across a range of K values to evaluate the point at which further decreases in the within-cluster sum of squares (WCSS) become negligible. In other words, the plot of WCSS begins to level off, forming an elbow-like shape. This inflection point helps identify the most appropriate K value, achieving a balance between meaningful clustering and avoiding excessive subdivision of the data.

The silhouette method provides a visual and quantitative means of evaluating how well the data points are grouped. The silhouette score is calculated by averaging the silhouette coefficient for each individual sample. This coefficient reflects the difference between the normal distance from a sample to points in its own cluster and the average distance to points in the adjacent neighboring cluster, normalized by the maximum of the two values. The resulting score ranges from minus one to plus one. A value close to plus one indicates that samples are well matched to their own cluster and clearly separated from others, while values closer to minus one suggest overlap or misclassification.

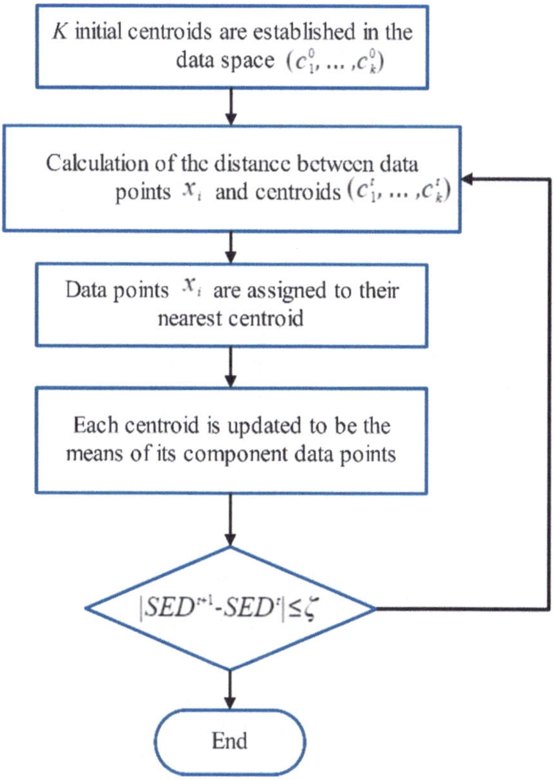

Fig. 10.3 Overview of the K-means clustering method

10.5 Results Obtained

10.5.1 Dataset A, Which Groups Dams According to Percent Fill at Different Times

Following the exclusion of dams with more than 35% missing data, a total of 187 dams remained for analysis. For these dams, the dataset includes 404 monthly filling rate records covering the period from January 1991 to August 2024. In the resulting data matrix, each row corresponds to a dam, while each column represents a specific point in time. This dataset was then used to calculate the best number of clusters and to perform the grouping of the dams accordingly.

10.5.1.1 Optimal Cluster Number K

The elbow method identifies the best amount of clusters for this dataset as $k = 3$. This is illustrated by the black dotted line in Fig. 10.4, where the y-axis represents the

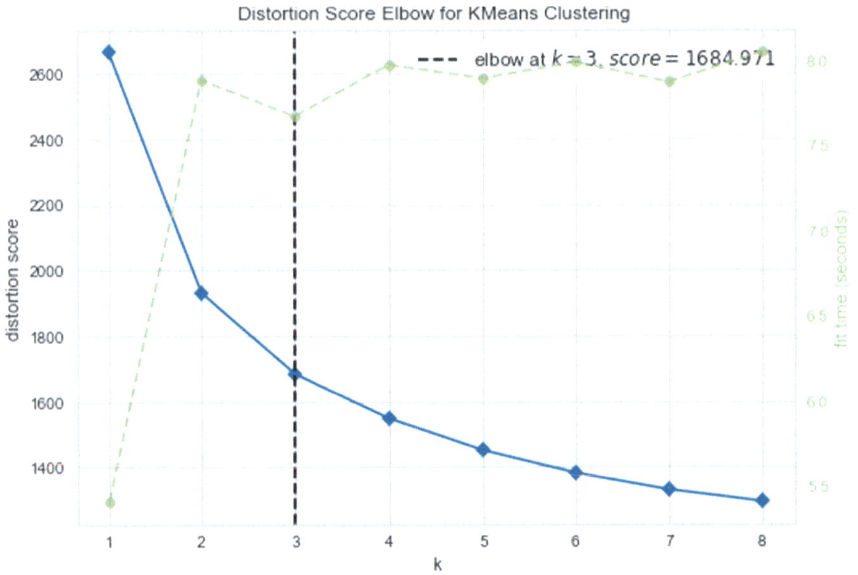

Fig. 10.4 Elbow method results for data set A

WCSS values and the x-axis corresponds to the number of clusters (k). Figure 10.5 shows the average silhouette coefficient for each cluster count, represented by red dotted lines. These values also serve as a guide to calculate the appropriate number of clusters. The highest average silhouette coefficient of 0.22 occurs at k = 2, followed by k = 3 with a value of 0.16. For k = 4 and k = 5, the clusters display significant imbalance in size, and in some cases, the presence of a sparse group is only indicated by a narrow distribution, making them unsuitable. Based on this analysis, the most appropriate amount of clusters for this dataset is calculated to be k = 3.

10.5.1.2 Cluster Analysis

Figure 10.6 presents the centroids, which represent the average behavior of each group, to analyze the clusters formed using the K-means method. The x-axis indicates time, while the yellow, blue, and red lines correspond to the different dam groups. The graph illustrates how each group behaves over the study period. The dams with the highest average filling rates belong to cluster 2, shown in yellow. Those with moderate filling levels are part of cluster 0, represented in blue. Cluster 1, displayed in red, includes the dams with the lowest filling rates throughout the analyzed period. In terms of distribution, 51% of the dams are assigned to cluster 0, 27% to cluster 1, and 22% to cluster 2. The relative sizes of these clusters are illustrated in Fig. 10.5.

Figure 10.7 displays the classification of each dam, with colors corresponding to their assigned clusters. Yellow dots represent dams located in the central-southern

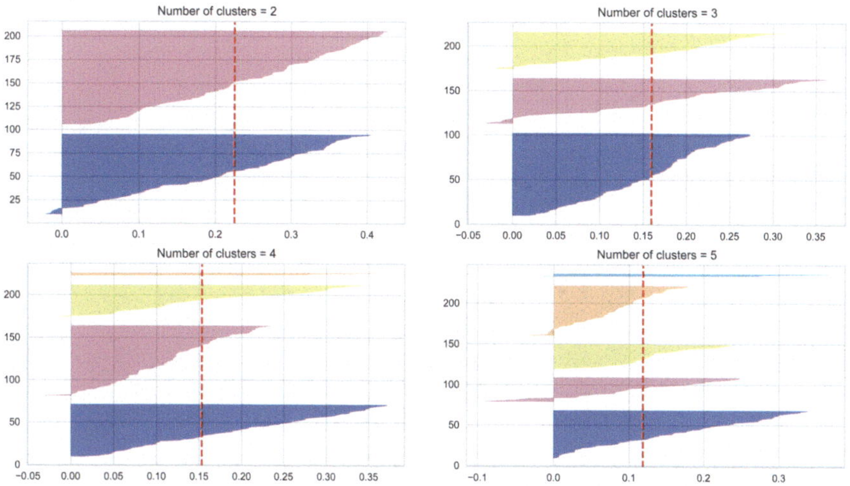

Fig. 10.5 Silhouette method results for data set A

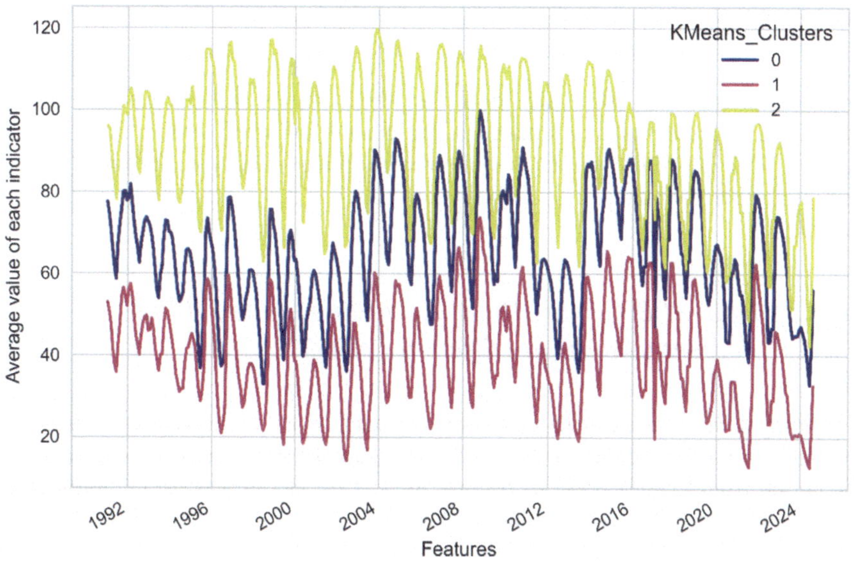

Fig. 10.6 Cluster-wise averages for data set A

area of the country, where the highest filling rates are observed. One possible explanation for this pattern is the frequent occurrence of cyclones along Mexico's Pacific coastline. Between 1991 and 2023, 114 weather events affected the Atlantic Ocean, while 147 impacted the Pacific Ocean (SMN, 2024b). Tropical storms are responsible for transporting large amounts of moisture from the oceans to the country's semi-arid zones, and in some regions, these cyclonic systems account for the majority of

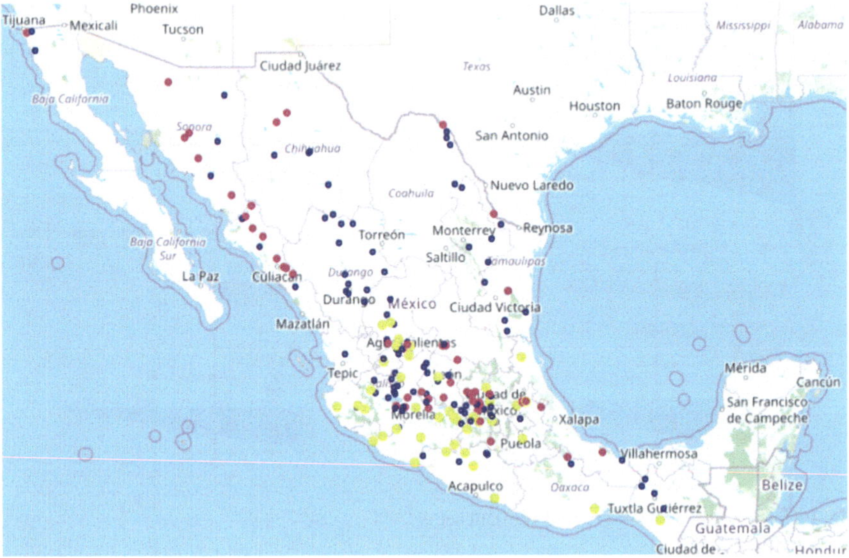

Fig. 10.7 Spatial distribution of dams by cluster within the country (Data set A)

annual rainfall (EAM, 2021). Red dots mark the locations of dams with the lowest filling levels, predominantly situated in the central and northwestern parts of the country. Blue dots represent dams with average filling rates, typically found in the central-northern region.

10.5.2 *Dataset B Groups Dams According to the Filling Percent Change at Different Periods*

Excluding dams with more than 35% missing data, a total of 184 dams remained in the analysis. For these dams, 404 monthly filling rate records were collected, covering the period from January 1991 to August 2024. This dataset was then employed to calculate and apply the best amount of clusters.

10.5.2.1 Optimal Cluster Number K

Using the elbow method, the best amount of clusters is identified as k = 3, as shown in Fig. 10.8. The silhouette method indicates that k = 2 yields the highest average silhouette coefficient, with a value of 0.2, followed by k = 3, k = 4, and k = 5, each with values around 0.07 (Fig. 10.9). However, k = 5 is excluded from consideration because it fails to clearly represent all five clusters. Likewise, k = 2 is also disregarded, as one of its groups exhibits a higher proportion of negative silhouette coefficients, reaching as low as −0.29. This suggests that the data points in that cluster are likely to overlap with another group and may in fact belong to it.

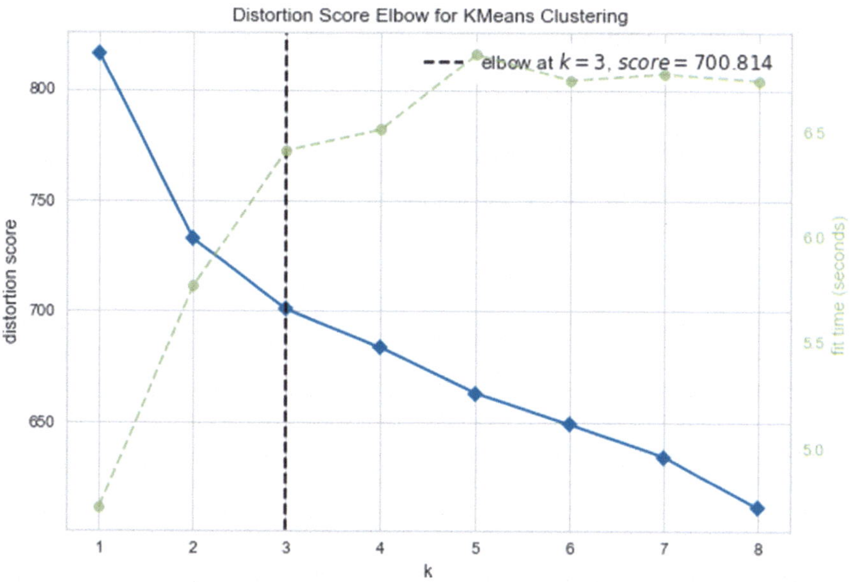

Fig. 10.8 Elbow method results for data set B

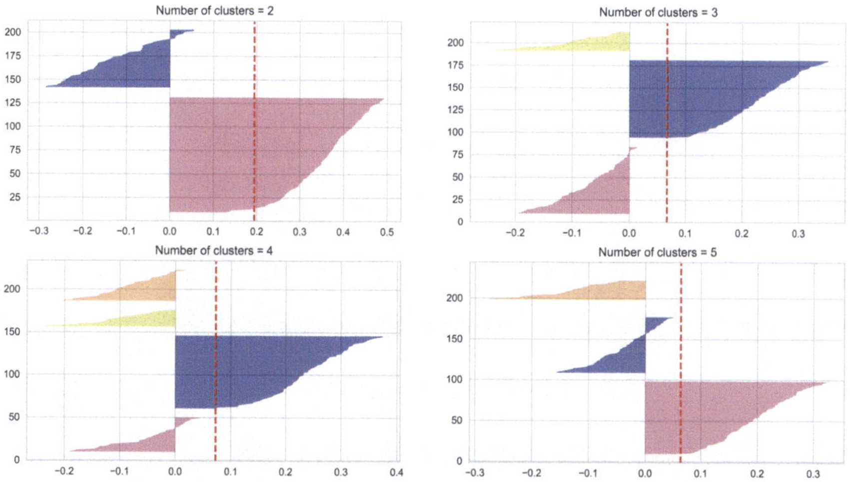

Fig. 10.9 Silhouette method results for data set B

The final selection was between k = 4 and k = 3. However, k = 3 was chosen because it was also supported by the results of the elbow method for this dataset. Cluster 0 includes 49% of the dams, Cluster 1 comprises 40%, and Cluster 2 accounts for the remaining 11%. The relative sizes of the clusters are illustrated in Fig. 10.9.

10.5.2.2 Cluster Analysis

The groups were examined based on their centroids, as shown in Fig. 10.10, where the x-axis represents the time period under analysis. Group 2, shown in yellow, includes the dams with the highest variability in percentage change of filling levels. Group 0, represented in blue, displays moderate variation. Group 1, shown in red, contains the dams with the lowest degree of fluctuation. Compared to the other two groups, the dams in group 1 demonstrate greater stability, as they exhibit minimal changes in their filling percentages over time.

The volume of water stored in dams across the country can vary depending on operational regulations, runoff levels, and precipitation. These management policies are designed to ensure water availability for different uses (EAM, 2021). Figure 10.11 shows the location of each dam and associates each color with its corresponding group.

Cluster 2, shown in yellow, contains the smallest number of dams and is primarily located in the central region of the country. These dams display significant variation in their filling percentages. The main purpose of the dams in Cluster 2, which include four located in the state of Queretaro, two in the State of Mexico, and one

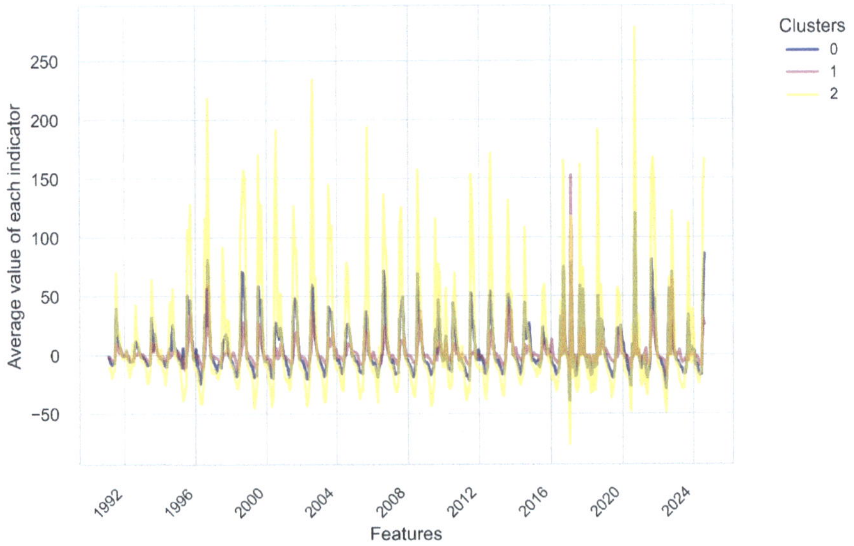

Fig. 10.10 Cluster-wise average filling variation in data set B

Fig. 10.11 Spatial distribution of dams across the country by cluster (Data set B)

in Morelos, is agricultural irrigation (CONAGUA, 2024). According to the results from data set A, the dams in Cluster 0 are also part of the group with consistently low filling levels. In regions where rainfall is scarce or irregular and temperatures are high, these hydrological imbalances have a negative impact on agricultural production, making farming increasingly difficult. The livestock industry has also faced rising production costs due to higher prices of essential inputs such as feed, driven by lower agricultural yields. Water stored in other Cluster 2 dams, including three located in Sinaloa and two in Sonora, is allocated for flood control, agricultural irrigation, public consumption, energy generation, and recreational use (CONAGUA, 2024). According to findings from data set A, these dams also showed low filling rates throughout the study period. The largest number of dams belongs to Cluster 0, represented in blue, which is mostly located in the northeastern region of the country. This group is characterized by less variability when compared to Cluster 2. Cluster 1, shown in red and situated mainly in the central and northwestern regions, contains the most stable dams.

10.5.3 Dataset C, Which Groups Time as a Function of the Filling Rate of the Dams

After excluding dams with more than 35% missing data, 187 dams remained in the analysis. For these dams, 404 monthly filling rate records were available, covering the period from January 1991 to August 2024. In the resulting data matrix, time is represented by the rows, and the dams are represented by the columns. This dataset was used to calculate the best number of clusters and to carry out the grouping accordingly.

10.5.3.1 Optimal Cluster Number K

For this dataset, the elbow method identified k = 3 as the best number of clusters (Fig. 10.12). According to the silhouette method, the highest average silhouette coefficient is obtained at k = 2, with a value of 0.205, followed by k = 3, which yields a value of 0.155. The silhouette method is used to evaluate clustering quality by measuring both the density within clusters and the separation between them. Across the clustering solutions from k = 2 to k = 5, this dataset generally shows positive silhouette coefficients and a reasonable distribution of cluster sizes. Since k = 3 was also supported by the elbow method, it was selected as the most appropriate value for this analysis. Approximately 75% of the time-related data points are assigned to Cluster 2, 33% to Cluster 0, and 36% to Cluster 1. The relative size of each cluster is illustrated in Fig. 10.13.

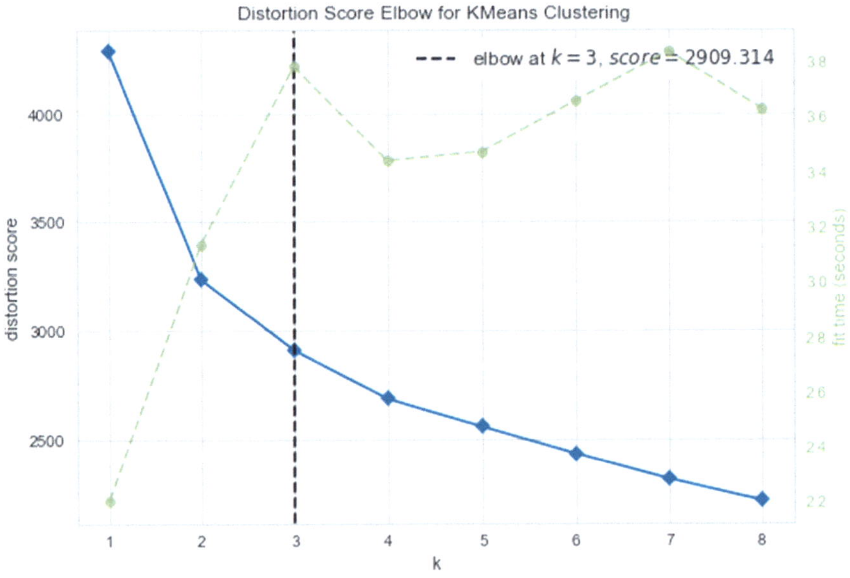

Fig. 10.12 Elbow method results for data set C

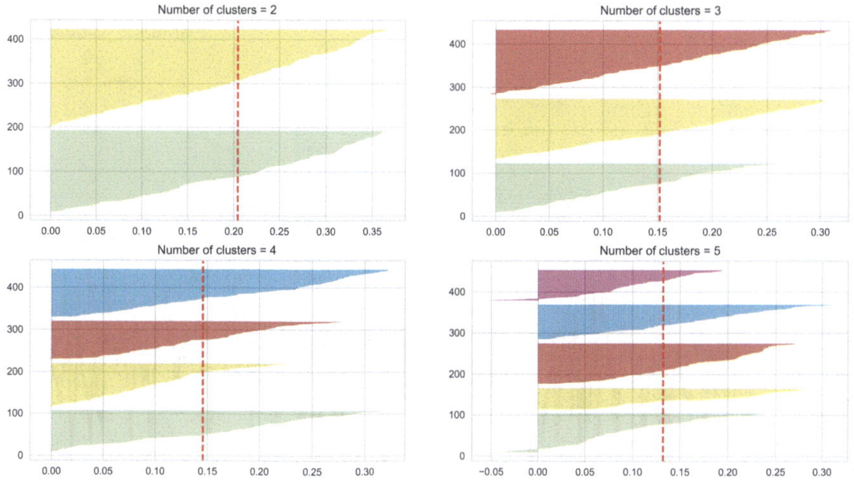

Fig. 10.13 Silhouette method results for data set C

10.5.3.2 Cluster Analysis

The cluster centroids are used to analyze the behavior of the groups. Figure 10.14 illustrates the centroid patterns for each cluster, where time intervals with different filling levels are shown using yellow, red, and green lines, and the x-axis represents individual dams. In most dams, Cluster 1, shown in red, corresponds to the time period with the lowest filling rates. Cluster 2, displayed in green, represents the time intervals with the highest filling levels in the majority of dams. Cluster 0, indicated in yellow, generally reflects periods with intermediate filling rates, falling between the other two groups.

Figure 10.15 displays the three temporal groups identified based on dam filling rates. The red group, associated with the lowest filling levels, is most frequently observed during the months of April, May, June, and July. However, since 2016, this pattern has expanded to include additional months. One possible explanation is the rainy season, which typically spans from May to October across much of Mexico and accounts for approximately 80% of the country's total annual precipitation. Still, it is common to observe a temporary decrease in rainfall during July and August, a phenomenon known as the midsummer drought or "dogdays." Notably, the years 2020, 2021, 2023, and 2024 were marked by severe drought conditions.

The green group, which corresponds to periods with the highest filling rates, occurs most frequently in January, February, and March, and also appears in the post-summer months of August through December. The yellow group, which reflects moderate filling levels, was relatively prevalent between 1991 and 2002. Periods with no available data are indicated by the blue areas.

The integration of clustering analysis with heat map visualization offers a powerful device for understanding the complex dynamics of dam filling across the country. By

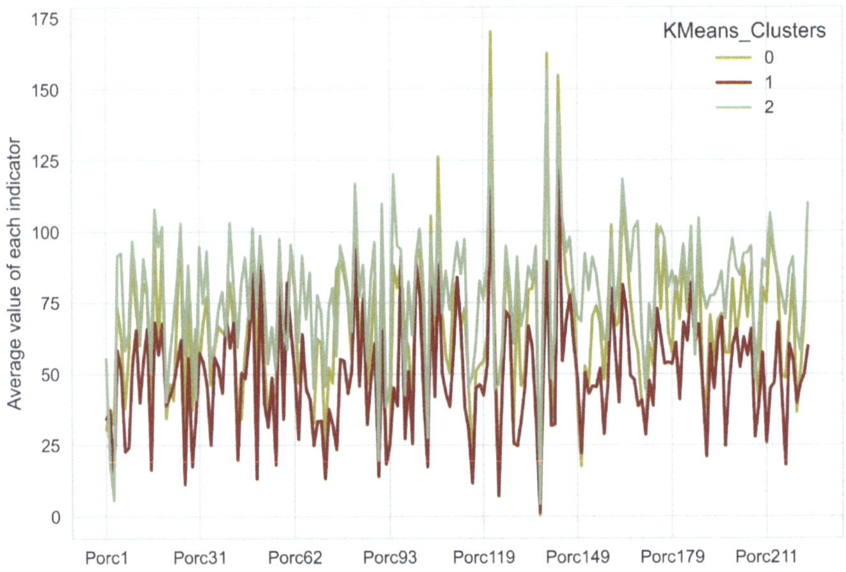

Fig. 10.14 Cluster-wise average values for data set C

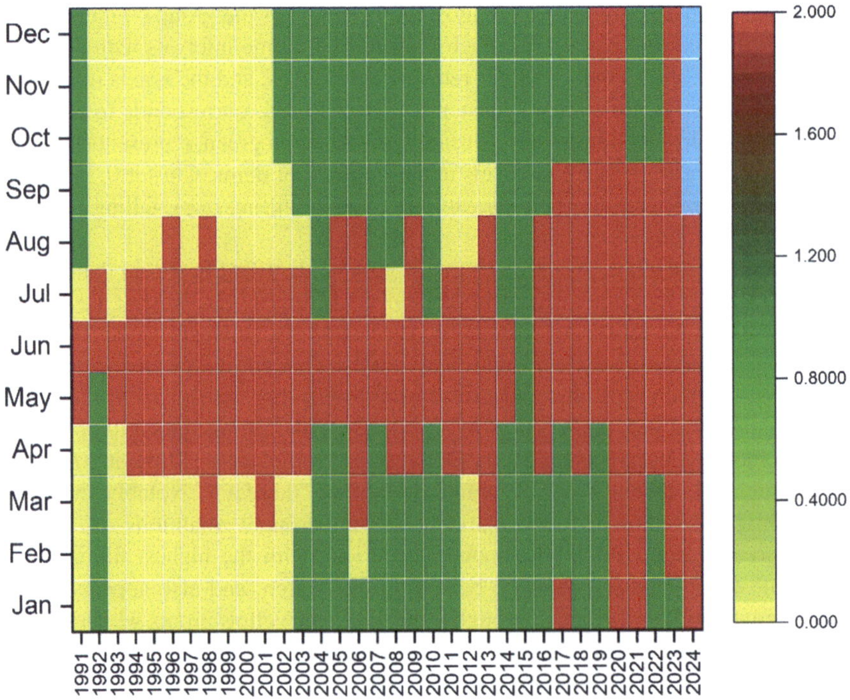

Fig. 10.15 Temporal distribution of filling groups across the years

identifying and visualizing the three distinct filling profiles, it becomes possible to gain insights into both the frequency and seasonal timing of reservoir storage patterns. This information can be used to design targeted water conservation strategies that are tailored to specific seasonal trends and hydrological conditions.

In 2012, most dams fell within the medium and low filling rate categories, corresponding to the yellow and red groups, respectively. This pattern is attributed to the influence of the La Niña phenomenon over the Pacific Ocean. When this event occurs, it typically brings drier conditions to central and northern regions of Mexico. According to SMN (2012), the year was also classified as a warm La Niña year. The rainfall distribution exhibited significant deviations compared to the previous year. Nationwide precipitation in 2012 totaled 742.34 mm, representing a 4.5% decrease compared to the historical average from 1941 to 2011. Approximately 49% of the country received less than the average annual rainfall, with the central-northern regions being the most affected. Only 15% of the national territory experienced a rainfall surplus, while 36% recorded rainfall near the historical average.

The national average temperature in 2012 was 21.9 degrees Celsius, which was 1.2 degrees above the average for the period 1971 to 2000. With the exception of February, all months in 2012 recorded above-average temperatures. Geographically, the northeastern and other northern regions of the country experienced the most pronounced temperature anomalies (SMN, 2012).

In 2015, most of Mexico's dams recorded high filling rates, corresponding to the green group. Total annual precipitation reached 872.0 mm, which was 92.7 mm above the historical average for the period from 1941 to 2014. Rainfall across the country ranged from normal to above-average levels. During that year, eight months registered precipitation above the long-term average, while only four months fell below it. March stood out as the wettest March on record since 1941, contributing 7.9% of the total annual rainfall for 2015, which is more than four times the typical rainfall for that month (SMN, 2015).

Throughout the 2015 tropical cyclone season, four cyclones impacted Mexico's Pacific coastline, and Tropical Storm Bill affected the Yucatán Peninsula. The most intense storm of the year was Hurricane Patricia, a category 4 hurricane on the Saffir-Simpson scale, which brought heavy rainfall to the central-northern and northwestern regions of the country. As a result of Hurricane Patricia, October 23 was recorded as the wettest day of the year (SMN, 2015).

With an anomaly of 2.3 °C above the normal average, the national average temperature in 2015 was 22.1 °C, 1.1 °C higher than the 1981–2010 normal average. This continued the trend of positive anomalies during the preceding twelve years. In addition, that summer was the warmest on record, with temperatures above the mean of the two preceding summers, 2014 and 2013 (SMN, 2015).

In 2020, dam filling rates across the country were notably low. The total annual precipitation reached 722.5 mm, marking the second consecutive year with rainfall levels below the national average. From northern to central Mexico, precipitation was generally below average, with the most severe deficits occurring in the northwestern region and along the Pacific coastal states. In contrast, the southern and southeastern parts of the country experienced higher than normal rainfall (SMN, 2020).

In several regions, rainfall was either delayed or entirely absent, disrupting the typical rainy season cycle. On average, the country experienced a rainy season that was 30 to 60 days shorter than the climatological norm. From June to December 2020, large portions of the country faced persistent rainfall shortages. During this seven-month period, five months recorded precipitation below the historical monthly average (SMN, 2020).

In 2020, several tropical systems impacted the coasts of the Gulf of Mexico and the Caribbean Sea, including Hurricanes Hanna, Delta, Gamma, and Zeta, as well as Tropical Storm Cristobal. However, no major natural events directly affected the Pacific coast during that year. With a national average temperature of 22.4 degrees Celsius, 2020 was classified as a warm and dry year. From January through November, monthly average temperatures consistently exceeded the historical monthly averages for the period 1981 to 2010. During the summer of 2020, the national average temperature reached 26.5 degrees Celsius, which was 2.6 degrees above the 1981–2010 summer average, although slightly lower than that recorded in the summer of 2019. Additionally, in 27 of the 31 states in the country, at least one day in 2020 registered a maximum temperature of 40.0 degrees Celsius or higher (SMN, 2020).

In 2023, eleven months are marked in red in Fig. 10.15, indicating that most dams experienced low filling rates throughout the year. The average annual precipitation for the period from 1991 to 2020 was 747.6 mm, while the total recorded rainfall in 2023 was 589.9 mm. This represents a deficit of 157.7 mm compared to the long-term average. Historical records identify 2023 as the driest year to date, reflecting the lowest overall precipitation levels. Most regions of the country experienced below-average rainfall when compared to the 1991 to 2020 climatological baseline.

During the tropical cyclone season of 2023, two cyclones, Harold and Idalia, made landfall along the Atlantic coast of Mexico in August. In the Pacific Ocean, six systems developed: Hurricane Beatriz in June and July, Hurricane Hilary in August, Hurricane Lidia, Tropical Storm Max, Hurricane Norma, and Hurricane Otis, all in October. Five of these tropical cyclones brought increased humidity to the country during October, making it the hottest October since 1953 and the tenth wettest October on record (SMN, 2023).

According to historical records covering the period from 1941 to 2024, September 2023 registered the greatest national rainfall deficit and was the hottest month on record between 1953 and 2023. This extreme condition was largely due to the presence of a persistent anticyclonic system that dominated much of the country throughout the month, preventing the inflow of moisture from surrounding seas. Throughout 2023, total precipitation remained consistently below the climatological average, with the shortfall becoming particularly evident starting in June. By the end of the year, the annual precipitation deficit reached 157.7 mm, marking a significant departure from the long-term average (SMN, 2023).

In 2023, the rainy season ended earlier than usual in northern, northeastern, Bajío, central, and Gulf coast regions of Mexico. This was primarily due to limited cyclonic activity in the Atlantic Ocean and during the summer months. In several areas of the country, rainfall was insufficient to signal a clear end to the rainy season. Overall, the rainy period was shortened by approximately 30 to 90 days across much of the national territory.

The national average temperature for 2023 was 22.7 degrees Celsius, which was 1.3 degrees higher than the climatological average for the period 1991 to 2020. Based on both precipitation and temperature data, 2023 was the driest year since 1941 and the hottest year since 1953, making it an exceptionally dry and warm year (SMN, 2023).

Data on annual cumulative precipitation, average yearly temperature, and annual storage volume for 187 dams were collected across the country for the period from 1991 to 2023 (CONAGUA, 2024; SMN, 2024a). The objective was to identify potential relationships among these variables. In Fig. 10.16, each of the three y-axes corresponds to one of the variables, each with its own scale. The trends in dam storage volume and accumulated precipitation show similar patterns in their behavior.

Figure 10.17 presents a heat map that visualizes the relationships between the variables. It highlights a substantial positive correlation of 0.72 between the total annual precipitation and the volume stored in dams. Conversely, there is a modest negative correlation between the mean annual temperature and the dams' storage capacity.

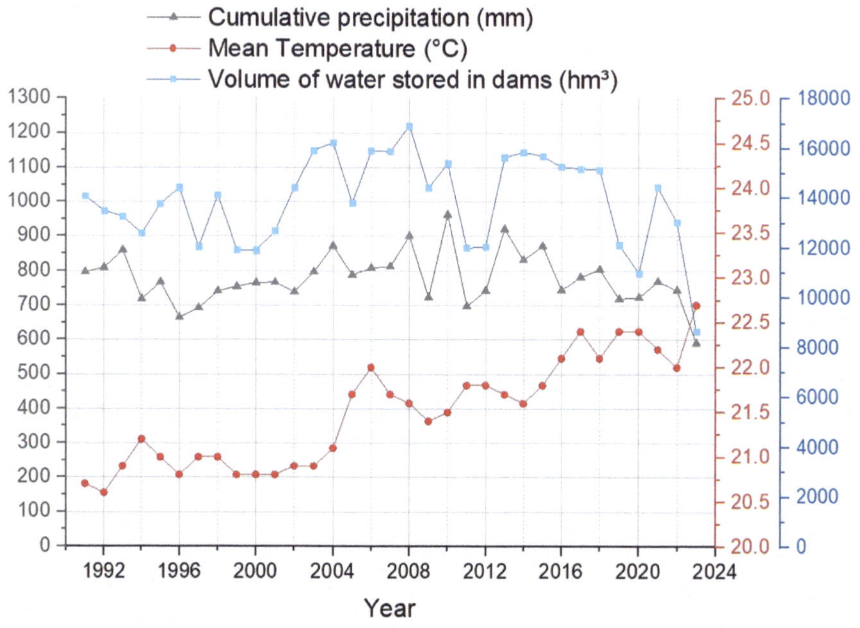

Fig. 10.16 Graph illustrating annual average temperature, cumulative precipitation, and dam storage volumen

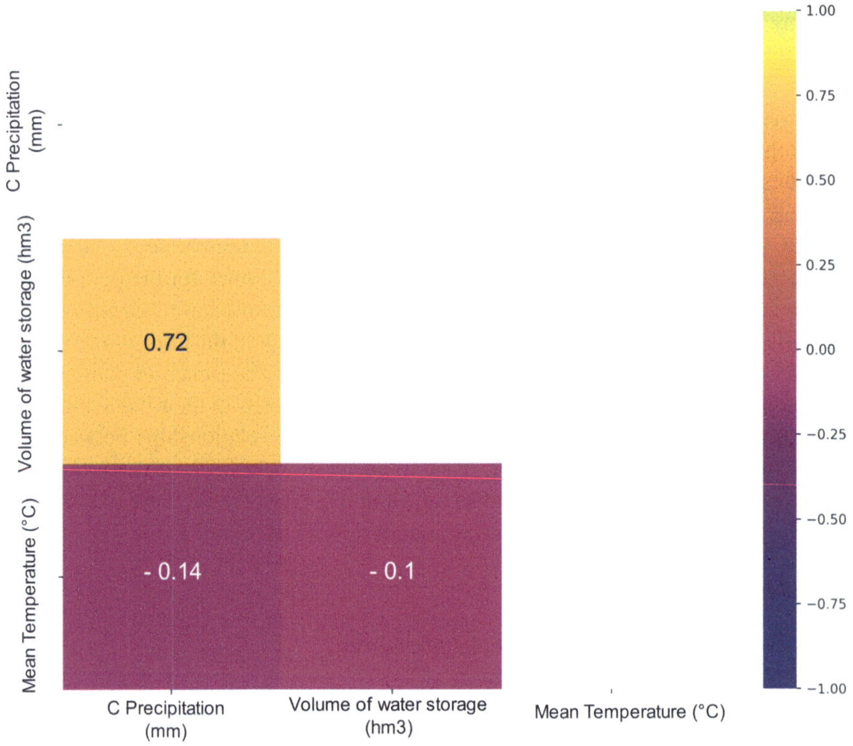

Fig. 10.17 Correlation heat map among the variables

10.6 Conclusion

Population growth intensifies the challenges related to the sustainable administration of interdependent energy, food, and water resources. Addressing these challenges requires integrated planning that acknowledges their complex interactions, predominantly in the face of climate change and increasing demand. This chapter highlights several key strategies and considerations essential for improving water resource management and ensuring long-term sustainability.

- Population increase can cause shortages in food, energy, and water, necessitating integrated resource planning that considers their interdependence.
- Enhancing water productivity in food production reduces the overall consumption of food, energy, and water, achievable through adoption of climate-adaptive and sustainable production methods.
- Expanding water availability via dam construction, reservoir capacity enhancement, inter-basin transfers, treated wastewater reuse, and desalination presents viable solutions, but must be balanced against economic, social, and environmental factors.

- Climate change intensifies the occurrence and harshness of dangerous weather events such as floods and droughts, complicating sustainable water management and requiring adaptive strategies.
- Decision-makers must prioritize equitable water distribution among diverse users while accounting for evolving climatic and demand patterns.
- Seasonal clustering analyses of dam filling percentages reveal dynamic hydrological variability, enabling identification of regions at risk for severe water stress.
- Targeted water conservation policies, including seasonal pricing and education campaigns, should be implemented to promote efficient water use across industrial, agricultural, and residential sectors.
- Cross-sector cooperation is critical to establish water use restrictions during periods of scarcity and to maximize resource sustainability.
- Water governance frameworks must be continuously updated to reflect changing goals and challenges at the local and national levels.
- The lack of homogeneous data collection and reporting methods across utilities and regions hampers effective management and integration of water resource information.
- Advanced data management techniques are needed to handle complex, high-dimensional datasets from multiple sources, ensuring results are reliable, interpretable, and actionable.

10.7 Appendix

See Tables 10.1, 10.2 and 10.3.

Table 10.1 List of dams in Mexico (Part 1)

1	Abelardo L. Rodríguez., Ags.	47	El Gallo, Gro.
2	Abelardo L. Rodríguez, B.C.	48	El Granero, Chih.
3	Abelardo Rodríguez Lujan, Son.	49	El Molino, Méx.
4	Abraham González, Chih.	50	El Moralillo, Ver.
5	Achimec, Jal.	51	El Niagara, Ags.
6	Achimec, Zac.	52	El Palote, Gto.
7	Adolfo López Mateos, Sin.	53	El Potosino, S.L.P.
8	Adolfo Ruiz Cortines, Son.	54	El Rejón, Chih.
9	Agostitlán, Mich.	55	El Rodeo, Mor.
10	Agua Caliente, Son.	56	El Salto, Jal.
11	Aguamilpa Solidaridad, Nay.	57	El Tigre, Dgo.

(continued)

Table 10.1 (continued)

12	Álvaro Obregón, Son.	58	El Tintero, Chih.
13	Andrés Figueroa, Gro.	59	El Tule, Jal.
14	Ángel Albino Corzo, Chis.	60	Emilio López Zamora, B.C.
15	Aristeo Mercado, Mich.	61	Endhó, Hgo
16	Basilio Vadillo, Jal.	62	Estudiante Ramiro Caballero Dorantes, Tamps.
17	Caboraca, Dgo.	63	Francisco I. Madero, Chih.
18	Cajón de Peña, Jal.	64	Francisco José Trinidad Fabela, Méx.
19	Canseco, Ver.	65	Francisco Villa, Dgo.
20	Cañada del Lobo, S.L.P.	66	Francisco Zarco, Dgo.
21	Chihuahua, Chih.	67	Gobernador Leobardo Reynoso, Zac.
22	Chila, Jal.	68	Gonzalo N. Santos, S.L.P.
23	Cincuenta Aniversario, Ags.	69	Guadalupe, Méx.
24	Cointzio, Mich.	70	Guaracha, Mich.
25	Constitución de 1917, Qro.	71	Gustavo Díaz Ordaz, Sin.
26	Constitución de Apatzingán, Jal	72	Huapango, Méx.
28	Copándaro, Mich.	73	Hurtado, Jal.
29	Corrinchis, Jal.	74	Ignacio Allende, Gto.
30	Cuauhtémoc, Son.	75	Ignacio R. Alatorre, Son.
31	Cuchillo Solidaridad, N.L.	76	Ignacio Ramírez, Méx.
32	Cuquío, Jal.	77	Independencia Nacional, Zac.
33	Danxho, Méx.	78	Infiernillo, Mich.
34	De Gonzalo, Mich.	79	Ing. Alfredo Elías Ayub, Jal.
35	Derivadora Las Blancas, Tamps.	80	Ing. Andrew Weiss, Chih.
36	Derivadora Pabellón, Ags.	81	Ing. Aurelio Benassini Vizcaíno, Sin.
37	Doctor Belisario Domínguez, Chis.	82	Ing. Benjamín Ortega Cantero, Dgo.
38	El Batán, Qro.	83	Ing. Carlos Ramírez Ulloa, Gro.
39	El Bosque, Mich.	84	Ing. Elías González Chávez, Jal.
40	El Carrizo, B.C.	85	Ing. Fernando Hiriart Balderrama, Hgo.
41	El Cazadero, Zac.	86	Ing. Guillermo Blake Águilar, Sin.
42	El Centenario, Coah.	87	Ing. Guillermo Lugo Sanabria, Jal.
43	El Centenario, Qro.	88	Ing. Julián Adame Alatorre, Zac.
44	El Chique, Zac.	89	Ing. Luis L. León, Chih.
45	El Cuarenta, Jal.	90	Ing. Rodolfo Félix Valdés, Son.
46	El Estribón, Jal.	91	Ing. Santiago Camarena, Jal.

Table 10.2 List of dams in Mexico (Part 2)

92	Internacional Falcón, Tamps.	137	Lic. Eustaquio Buelna, Sin.
93	Internacional La Amistad, Coah.	138	Los Angeles, Mich.
94	Jalpan, Qro.	139	Los Moraleños, Zac.
95	Jaripo, Mich.	140	Los Naranjos, Dgo.
96	Javier Rojo Gómez, Hgo.	141	Los Olivos, Mich.
97	Jocoque, Ags.	142	Los Reyes, Hgo.
98	José Antonio Alzate, Méx.	143	Luis Donaldo Colosio, Sin.
99	José López Portillo, N.L.	144	Madín, Méx.
100	José López Portillo, Sin.	145	Malpaís, Mich.
101	José María Morelos y Pavón, Mich.	146	Manuel Ávila Camacho, Pue.
102	José María Morelos, Zac.	147	Manuel Felgueres, Zac.
103	Josefa Ortiz de Domínguez, Sin.	148	Manuel M. Diéguez, Jal.
104	Juan Guerrero Alcocer, Sin.	149	Manuel Moreno Torres, Chis.
105	Juan Sabines, Chis.	150	Mariano Abasolo, Gto.
106	La Boquilla, Chih.	151	Marte Rodolfo Gómez, Tamps.
107	La Calera, Gro.	152	Media Luna, Ags.
108	La Cangrejera, Ver.	153	Melchor Ocampo, Mich.
109	La Codorniz, Ags.	154	Miguel Alemán, Zac.
110	La Colonia, Jal.	155	Miguel de la Madrid Hurtado, Oax.
111	La Concepción, Méx.	156	Miguel Hidalgo, Sin.
112	La Esperanza, Hgo.	157	Necaxa, Pue.
113	La Fragua, Coah.	158	Nexapa, Pue.
114	La Golondrina, Gto.	159	Nezahualcóyotl, Chis.
115	La Laguna, Hgo.	160	Ñadó, Méx.
116	La Llave, Qro.	161	Palomas, Zac.
117	La Muñeca, S.L.P.	162	Paso de Piedras, Ver.
118	La Purísima, Gto.	163	Pátzcuaro, Mich.
119	La Red, Jal.	164	Pedro José Méndez, Tamps.
120	La Sauceda, Jal.	165	Peña del Águila, Dgo.
121	La Soledad, Gto.	166	Peñuelitas, Gto.
122	La Soledad, Pue.	167	Pico del Águila, Chih.
123	La Venta, Qro.	168	Piedras Azules, Chih.
124	Lago de Chapala, Jal.	169	Plutarco Elías Calles, Ags.
125	Laguna de Amela, Col.	170	Plutarco Elías Calles, Son.
126	Laguna de Tuxpan, Gro.	171	Presa derivadora Pabellón, Ags.
127	Laguna de Yuriria, Gto.	172	Presa Falcón parte Mexicana.
128	Laguna del Fresno, Mich.	173	Presa la Amistad parte Mexicana.
129	Las Auras, B.C.	174	Presidente Alemán, Oax

(continued)

Table 10.2 (continued)

130	Las Blancas, Tamps.	175	Presidente Benito Juárez, Oax.
131	Las Lajas, Chih.	176	Presidente Guadalupe Victoria, Dgo.
132	Las Vírgenes, Chih.	177	Pucuato, Mich
133	Lázaro Cárdenas, Dgo.	178	Ramón Corona Madrigal General, Jal.
134	Lázaro Cárdenas, Son.	179	Ramón López Velarde, Zac
135	Leonardo Rodríguez Alcaine, Nay.	180	República Española, Tamps.
136	Lic. Emilio Portes Gil, Tamps.	181	Requena, Hgo.

Table 10.3 List of dams in Mexico (Part 3)

182	Revolución Mexicana, Gro.	204	Tacotán, Jal.
183	Rodrigo Gómez, N.L.	205	Tampico, Tamps.
184	Sabaneta, Mich.	206	Taxhimay, Hgo.
185	Salinillas, N.L.	207	Taxhimay, Méx.
186	San Andrés Tepetitlán, Méx.	208	Tenango, Pue.
187	San Gabriel, Dgo.	209	Tenasco, Jal.
188	San Ildefonso, Qro.	210	Tepuxtepec, Mich.
189	San José Atlanga, Tlax.	211	Tercer Mundo, Mich.
190	San José, S.L.P.	212	Urepetiro, Mich.
191	San Juanico, Mich.	213	Valentín Gama, S.L.P.
192	San Miguel, Coah.	214	Valerio Trujano, Gro.
193	San Rafael, Nay.	215	Valle de Bravo, Méx.
194	Sanalona, Sin.	216	Venustiano Carranza, Coah.
195	Santa Elena, Dgo.	217	Vicente Aguirre, Hgo.
196	Santa Rosa, Jal.	218	Vicente C. Villaseñor, Jal.
197	Santa Rosa, Zac..	219	Vicente Guerrero, Gro.
198	Santiago Bayacora, Dgo.	220	Vicente Guerrero, Tamps.
199	Santiago, Zac.	221	Villa Hidalgo, Dgo.
200	Sierra de Guadalupe, Méx.	222	Villa Victoria, Méx.
202	Solidaridad, Jal.	223	Yosocuta, Oax.
203	Solís, Gto.	224	Zicuirán, Mich.

References

Breña-Naranjo, A. (2021). Water security in times of drought. IMTA perspectives (Mexican Institute of Water Technology). https://www.imta.gob.mx/gobmx/DOI/perspectivas/2021/b-imta-perspectivas-2021-14.pdf

Alcamo, J., Flörke, M., & Märker, M. (2007). Future long-term changes in global water resources driven by socio-economic and climatic changes. *Hydrological Science Journal, 52*, 247–275. https://doi.org/10.1623/hysj.52.2.247

Bennett, C., Stewart, R. A., & Beal, C. D. (2013). ANN-based residential water end-use demand forecasting model. *Expert Systems with Applications, 40*, 1014–1023. https://doi.org/10.1016/j.eswa.2012.08.012

Cesano, D., & Gustafsson, J. E. (2000). Impact of economic globalisation on water resources: A source of technical, social and environmental challenges for the next decade. *Water Policy, 2*, 213–227. https://doi.org/10.1016/s1366-7017(00)00003-9

Chen, J., Shi, H., Sivakumar, B., & Peart, M. R. (2016). Population, water, food, energy and dams. *Renewable and Sustainable Energy Reviews, 56*, 18–28. https://doi.org/10.1016/j.rser.2015.11.043

Cominola, A., Giuliani, M., Castelletti, A., Rosenberg, D. E., & Abdallah, A. M. (2018). Implications of data sampling resolution on water use simulation, end-use disaggregation, and demand management. *Environmental Modelling and Software, 102*, 199–212. https://doi.org/10.1016/j.envsoft.2017.11.022

CONAGUA. (2016). Dams: water containers. https://www.gob.mx/conagua/articulos/presas-contenedores-de-agua?idiom=es. Accessed 12 October 2024.

CONAGUA. (2024). National water information system: Monitoring of Mexico's main dams. https://sinav30.conagua.gob.mx:8080/Presas/. Accessed 1 Jun 2024.

Daneshmand, F., Karimi, A., Nikoo, M. R., Bazargan-Lari, M. R., & Adamowski, J. (2014). Mitigating socio-economic-environmental impacts during drought periods by optimizing the conjunctive management of water resources. *Water Resource Management, 28*, 1517–1529. https://doi.org/10.1007/s11269-014-0549-7

Di Nunno, F., Yıldız, M. B., Afework, Y. G., de Marinis, G., & Granata, F. (2024). Mapping drought evolution in Ethiopia: Trends, clustering, and Bayesian estimation of abrupt changes. *Natural Hazards, 1–29*,. https://doi.org/10.1007/s11069-024-06935-0

Domínguez, J. (2019). Dam construction in Mexico. Evolution, current situation and new approaches to make water infrastructure viable. *Management and Public Policy, 28*(1), 3–37. https://doi.org/10.29265/gypp.v28i1.551

EAM. (2021). Water statistics in Mexico.

Erlewein, A. (2014). The promotion of dams through the clean development mechanism: between sustainable climate protection and carbon colonialism. In M. Nüsser (eds.), *Large Dams in Asia. Advances in Asian Human-Environmental Research.* Springer, pp. 149–168. https://doi.org/10.1007/978-94-007-2798-4_8

Fan, J. L., Hu, J. W., Zhang, X., Kong, L. S., Li, F., & Mi, Z. (2020). Impacts of climate change on hydropower generation in China. *Mathematics and Computers in Simulation, 167*, 4–18. https://doi.org/10.1016/j.matcom.2018.01.002

FAO. (2012). Economic growth, hunger and malnutrition. https://www.fao.org/4/i3027s/i3027s03.pdf. Accessed 22 October 2024.

Hamududu, B. H., & Killingtveit, Å. (2016). Hydropower production in future climate scenarios; the case for the Zambezi River. *Energies, 9*, 502. https://doi.org/10.3390/en9070502

Handayani, K., Filatova, T., & Krozer, Y. (2019). The vulnerability of the power sector to climate variability and change: Evidence from Indonesia. *Energies, 12*, 3640. https://doi.org/10.3390/en12193640

Hermawan, D., & Limantara, L. M. (2024). Dam water conservation equation index using AHP and cluster analysis methods: A study on selected dams in Java. *IOP Conference Series Earth Environment Science, 1311*, Article 012050. https://doi.org/10.1088/1755-1315/1311/1/012050

Ho, M., Lall, U., Allaire, M., Devineni, N., Kwon, H. H., Pal, I., Raff, D., & Wegner, D. (2017). The future role of dams in the United States of America. *Water Resources Research, 53*, 982–998. https://doi.org/10.1002/2016wr019905

Hoghoughinia, K., Saghafian, B., & Aminyavari, S. (2024). Clustering the temporal distribution pattern of sub-daily precipitations over Iran. *Earth Science Information, 17*, 1949–1956. https://doi.org/10.1007/s12145-024-01261-2

Hunt., J. D., Stilpen, D., & de Freitas, M. A. V. (2018). A review of the causes, impacts and solutions for electricity supply crises in Brazil. *Renew. Sustainability Energy Review, 88*, 208–222. https://doi.org/10.1016/j.rser.2018.02.030

INEGI. (2018). Water. Tell me about Mexico, National Institute of Statistics and Geography. https://cuentame.inegi.org.mx/territorio/agua/presas.aspx?tema=T. Accessed 10 October 2024.

Kaygusuz, K. (2009). The role of hydropower for sustainable energy development. Energy sources. *Part B: Economics Planning Policy, 4*, 365–376. https://doi.org/10.1080/15567240701756889

Lin, G. F., & Chen, L. H. (2006). Identification of homogeneous regions for regional frequency analysis using the self-organizing map. *Journal of Hydrology, 324*, 1–9. https://doi.org/10.1016/j.jhydrol.2005.09.009

Loucks, D. P., & van Beek, E. (2017). Water resources planning and management: An overview. In *Water Resource Systems Planning and Management*. Springer, pp. 1–49. https://doi.org/10.1007/978-3-319-44234-1_1

Obahoundje, S., Diedhiou, A., Kouassi, K. L., Ta, M. Y., Mortey, E. M., Roudier, P., & Kouame, D. G. M. (2022). Analysis of hydroclimatic trends and variability and their impacts on hydropower generation in two river basins in Côte d'Ivoire (West Africa) during 1981–2017. *Environment Research Communication, 4*, Article 065001. https://doi.org/10.1088/2515-7620/ac71fa

Olsson, G. (2013). Water, energy and food interactions-Challenges and opportunities. *Frontiers of Environmental Science & Engineering, 7*, 787–793. https://doi.org/10.1007/s11783-013-0526-z

Pedro-Monzonís, M., Solera, A., Ferrer, J., Estrela, T., & Paredes-Arquiola, J. (2015). A review of water scarcity and drought indexes in water resources planning and management. *Journal of Hydrology, 527*, 482–493. https://doi.org/10.1016/j.jhydrol.2015.05.003

Pipitone, C., Maltese, A., Dardanelli, G., Brutto, M. L., & Loggia, G. L. (2018). Monitoring water surface and level of a reservoir using different remote sensing approaches and comparison with dam displacements evaluated via GNSS. *Remote Sensing, 10*, 71. https://doi.org/10.3390/rs10010071

Poff, N. L., & Hart, D. D. (2002). How dams vary and why it matters for the emerging science of dam removal: An ecological classification of dams is needed to characterize how the tremendous variation in the size, operational mode, age, and number of dams in a river basin influences the potential for restoring regulated rivers via dam removal. *BioScience, 52*, 659–668. https://doi.org/10.1641/0006-3568(2002)052[0659:hdvawi]2.0.co;2

Raya-Tapia, A. Y., Cansino-Loeza, B., Sánchez-Zarco, X. G., Ramírez-Márquez, C., & Ponce-Ortega, J. M. (2023). A systematic approach for assessing water, energy and food security during the COVID-19 pandemic: Case study of Mexico. *Environment, Development and Sustainability, 25*, 14381–14409. https://doi.org/10.1007/s10668-022-02671-2

Mideksa, T. K., & Kallbekken, S. (2010). The impact of climate change on the electricity market: A review. *Energy Policy, 38*, 3579–3585. https://doi.org/10.1016/j.enpol.2010.02.035

Molina-Aguilar, J. P., Gutiérrez-López, A., & Cruz Paz, I. M. (2019). Canonical correlation between storage volumes in dams and precipitation intensities during hurricanes. *Mexican Institute of Water Technology*. https://doi.org/10.24850/j-tyca-2019-06-02

Scanlon, B. R., Duncan, I., & Reedy, R. C. (2013). Drought and the water–energy nexus in Texas. *Environmental Research Letters, 8*, Article 045033. https://doi.org/10.1088/1748-9326/8/4/045033

Schulz, C., & Adams, W. M. (2021). In search of the good dam: Contemporary views on dam planning in Latin America. *Sustainability Science, 16*, 255–269. https://doi.org/10.1007/s11625-020-00870-2

ScikitLearn. (2024). MinMaxScaler, class sklearn.preprocessing. https://scikit-learn.org/1.5/modules/generated/sklearn.preprocessing.MinMaxScaler.html. Accessed 12 October 2024.

Sharghi, E., Nourani, V., Soleimani, S., & Sadikoglu, F. (2018). Application of different clustering approaches to hydroclimatological catchment regionalization in mountainous regions, a case study in Utah State. *Journal of Mountain Science, 15*, 461–484. https://doi.org/10.1007/s11629-017-4454-4

Singh, V. K., & Singal, S. K. (2017). Operation of hydro power plants-a review. *Renewable and Sustainable Energy Reviews, 69*, 610–619. https://doi.org/10.1016/j.rser.2016.11.169

SMN. (2012). *Weather report for Mexico.* National Meteorological Service.

SMN. (2015). *Weather report for Mexico.* National Meteorological Service.

SMN. (2020). *Weather report for Mexico.* National Meteorological Service.

SMN. (2023). *Weather report for Mexico.* National Meteorological Service.

SMN. (2024a). National Meteorological Service, from Monthly Summaries of Temperatures and Rainfall. https://smn.conagua.gob.mx/es/climatologia/temperaturas-y-lluvias/resumenes-mensuales-de-temperaturas-y-lluvias. Accessed 1 October 2024.

SMN. (2024b). National Meteorological System, from Rains associated with tropical cyclones. https://smn.conagua.gob.mx/es/ciclones-tropicales/lluvias-asociadas-a-ciclones-tropicales. Accessed 12 September 2024.

Tajbar, S., Tajbar, A., Pashaie, Z., Khorshiddoust, A. M., & Rafiq, L. (2024). Application of hierarchical cluster analysis to spatiotemporal variability of monthly precipitation over Khyber Pakhtunkhwa. *Pakistan. Acta Geophysics, 72*, 1159–1174. https://doi.org/10.1007/s11600-023-01161-x

Tarroja, B., Forrest, K., Chiang, F., AghaKouchak, A., & Samuelsen, S. (2019). Implications of hydropower variability from climate change for a future, highly-renewable electric grid in California. *Applied Energy, 237*, 353–366. https://doi.org/10.1016/j.apenergy.2018.12.079

Teodoro, P. E., de Oliveira-Júnior, J. F., da Cunha, E. R., Correa, C. C. G., Torres, F. E., Bacani, V. M., Gois, G., & Ribeiro, L. P. (2016). Cluster analysis applied to the spatial and temporal variability of monthly rainfall in Mato Grosso do Sul State. *Brazil. Meteorology Atmospheric Physics, 128*, 197–209. https://doi.org/10.1007/s00703-015-0408-y

UN. (2022). A growing population (United Nations). https://www.un.org/es/global-issues/population#:~:text=Est%C3%A1%20previsto%20que%20la%20poblaci%C3%B3n,y%2010.400%20millones%20en%202100. Accessed 20 October 2024.

UN. (1998). World Water Resources: A New Appraisal and Assessment for the 21st Cen. Educational, Scientific and Cultural Organization (UNESCO).

UN. (2023). Less water… less food (United Nations). https://news.un.org/es/interview/2023/03/1519582. Accessed 22 October 2024.

UN. (2024). Sustainable development goals (United Nations). https://www.un.org/sustainabledevelopment/es/objetivos-de-desarrollo-sostenible/. Accessed 14 October 2024.

UNDP. (2006). Human Development Report 2006–Beyond Scarcity: Power, Poverty and the Global Water Crisis. United Nations Development Programme.

Wang, R., Zhao, X., Qiu, H., Cheng, X., & Liu, X. (2024). Uncovering urban water consumption patterns through time series clustering and entropy analysis. *Water Research, 262*, Article 122085. https://doi.org/10.1016/j.watres.2024.122085

World Bank. (2017). Chart: Globally, 70% of freshwater is used for agriculture. https://blogs.worldbank.org/es/opendata/el-70-del-agua-dulce-es-utilizada-para-la-agricultura#:~:text=En%20la%20mayor%C3%ADa%20de%20las,en%20las%20extracciones%20de%20agua. Accessed 24 October 2024.

Zhu, Y., Niu, X., Gu, C., Dai, B., & Huang, L. (2021). A fuzzy clustering logic life loss risk evaluation model for dam-break floods. *Complexity, 2021*, 7093256. https://doi.org/10.1155/2021/7093256

The manufacturer's authorised representative in the EU is Springer Nature Customer Service Centre GmbH, Europaplatz 3, 69115 Heidelberg, Germany. If you have any concerns regarding our products, please contact ProductSafety@springernature.com

Printed and bound by CPI Group (UK) Ltd, Croydon, CR0 4YY
30/09/2026
02225591-0013